The Origins of Modern Science

The Origins of Modern Science is the first synthetic account of the history of science from Antiquity through the Scientific Revolution in many decades. Providing readers of all backgrounds and students of all disciplines with the tools to study science like a historian, Ofer Gal covers everything from Pythagorean mathematics to Newton's *Principia*, through Islamic medicine, medieval architecture, global commerce and magic. Richly illustrated throughout, scientific reasoning and practices are introduced in accessible and engaging ways with an emphasis on the complex relationships between institutions, beliefs and political structures and practices. Readers gain valuable new insights into the role played by science both in history and in the world today, placing the crucial challenges to science and technology of our time within their historical and cultural context.

Ofer Gal is Professor of History and Philosophy of Science at the University of Sydney and has been teaching the history of science for over a quarter century. He has won numerous prizes and has published monographs, edited volumes and articles, especially about early modern physical sciences, but also on global knowledge, eighteenth-century chemistry and various philosophical issues.

The Origins of Modern Science

From Antiquity to the Scientific Revolution

Ofer Gal   University of Sydney

CAMBRIDGE
UNIVERSITY PRESS

University Printing House, Cambridge CB2 8BS, United Kingdom

One Liberty Plaza, 20th Floor, New York, NY 10006, USA

477 Williamstown Road, Port Melbourne, VIC 3207, Australia

314–321, 3rd Floor, Plot 3, Splendor Forum, Jasola District Centre, New Delhi – 110025, India

79 Anson Road, #06–04/06, Singapore 079906

Cambridge University Press is part of the University of Cambridge.

It furthers the University's mission by disseminating knowledge in the pursuit of
education, learning, and research at the highest international levels of excellence.

www.cambridge.org
Information on this title: www.cambridge.org/9781316510308
DOI: 10.1017/9781108225205

First published 2021

Printed in the United Kingdom by TJ Books Limited, Padstow Cornwall

A catalogue record for this publication is available from the British Library.

ISBN 978-1-316-51030-8 Hardback
ISBN 978-1-316-64970-1 Paperback

For Yi

Contents

Figures

Note from the Publisher

This book attempts to introduce to its readers major chapters in the history of science. It tries to present science as a human endeavor – a great achievement, and all the more human for it. In place of the story of progress and its obstacles or a parade of truths revealed, this book stresses the contingent and historical nature of scientific knowledge. Knowledge, science included, is always developed by real people, within communities, answering immediate needs and challenges shaped by place, culture and historical events with resources drawn from their present and past.

Chronologically, this book spans from Pythagorean mathematics to Newton's *Principia*. The book starts in the High Middle Ages and proceeds to introduce the readers to the historian's way of inquiry. At the center of this introduction is the Gothic Cathedral – a grand achievement of human knowledge, rooted in a complex cultural context and a powerful metaphor for science. The book alternates thematic chapters with chapters concentrating on an era. Yet it attempts to integrate discussion of all different aspects of the making of knowledge: social and cultural settings, challenges and opportunities; intellectual motivations and worries; epistemological assumptions and technical ideas; instruments and procedures. The cathedral metaphor is evoked intermittently throughout, to tie the many themes discussed to the main lesson: that the complex set of beliefs, practices and institutions we call science is a particular, contingent human phenomenon.

The wide scope and varied audience of this book required sacrificing footnotes for the sake of fluency – not without some professional anxiety – and I provide exact references only for direct quotations. The place of referencing within the text is taken by a list of Suggested Readings at the end of each chapter, and the book's main resources are in the "Secondary Sources" part of these. For any factual error I bear full responsibility. The "Primary Texts" listed in the Suggested Readings are easily accessible, English translations of sources from the period or theme discussed. For the instructor, they should serve as suggestions for tutorial readings; for the student, they present an exercise in the interpretation of texts remote in place and time. The discussion questions are offered to help the instructor in preparing for tutorials, and the reader may find in them clues to the main insights that the story attempts to convey.

Acknowledgements

This book is a tribute to the intellectual value of university teaching, and in that, indebted to almost everyone whose classroom I attended over the years – formally or informally, literally or figuratively. This assembly of scholars far too large to reconvene here, so beyond the authors of the works populating the book's reading lists, I will have to directly thank only those from whose scholarship I have benefited directly, as my immediate teachers or colleagues: Rivka Feldhay, Sabbetay Unguru, J. E. McGuire, Peter Machamer, Bernard Goldstein and the late Marcelo Dascal belong to the former category; Alan Chalmers, John Schuster, Hanan Yoran, Ohad Parnes, Daniela Helbig, Dominic Murphy, Victor Boantza, Snait Gissis and especially Raz Chen-Morris – to the latter. Victor volunteered to serve as a scientific editor at the very last stage of writing, saved me some serious factual embarrassments and forced me to shorten sentences and sharpen arguments.

With some scholars I have not had an opportunity to study in an official setting, but still consider my teachers: Hal Cook, Ben Elman, Dan Garber, Tony Grafton, Simon Schaffer and the late Sam Schweber. For the very idea of a bold yet careful account of *science*, wide ranging but rich with details, I am the venerating disciple of the people who first taught me such courses: the late Amos Funkenstein and Yehuda Elkana. I am still in awe at their erudition, depth and intellectual courage.

Blessed as I have been with the intellectual support for this book, for which Hagar and Yi provided an essential (if sometimes unwitting) foundation, it would still not have come into being without the professional and dedicated help of the people at Cambridge University Press: Lucy Rhymer, Lisa Pinto, Maggie Jeffers, Sophie Rosinke and Charlie Howell, to whom I am deeply grateful. Many librarians went out of their way to help assemble the book's images, and among those I owe special thanks to Tom Goodfellow of Sydney University Library and Urte Brauckmann and her team at the Max Planck Institute for the History of Science in Berlin.

But first and foremost, this book is indebted to the graduate students who have served as my tutors throughout the years in Sydney. The weekly meeting with them has always been the most intellectually exciting hour of the week. Alan Salter taught me history of medicine, and Ian Wills the history of technology. James Ley is my teacher of classical thought and many matters pedagogical, and Kiran Krishna of all things medieval. Jennifer Tomlinson, who helped me tremendously by editing early versions and gathering exciting images, also forced me to find the voice of women – especially that of Jane Sharp. Sahar Tavakoli joined her with instruction about early modern midwives and Megan Baumhammer about the power and mysteries of the visual. To Ian Lawson I owe my understanding of the delicate balance of instruments and their environs and to Claire Kennedy – the fascination with maps. Arin Harman, a visitor from better rooted disciplines, was my master of pedagogical smooth sailing, and Nick Bozic contributed a powerful philosophical mind. Paddy Holt taught me all I know about the Royal Society; to Laura Sumrall I owe whatever insight I have into magic and Cindy Hodoba-Eric retaught me early modern natural philosophy. Cindy also contributed skills, intelligence and passion without which the book would never have been completed: editing for style and content; gathering images and copyrights; designing and drawing complicated diagrams.

It is the joy of working with my tutors that I try to capture in this book, and it is dedicated to them.

1 Cathedrals

> They climbed on sketchy ladders towards God,
> With winch and pulley hoisted hewn rock into heaven,
> Inhabited the sky with hammers, defied gravity,
> Deified stone, took up God's house to meet him,
> And came down to their suppers and small beer ...
>
> John Ormond, *The Cathedral Builders*

The Cathedral

This is the Gothic cathedral, the marvel that inspired Ormond's poem. The one in the picture (Figure 1.1) is perhaps the grandest of them all: the Notre Dame Cathedral at Chartres, south-west of Paris. It is a breathtaking accomplishment: 130 meters in length, it would cover a Manhattan block and a half; its vaults are 37 meters in height, higher than a modern ten-story building; its southern, Romanesque tower is 107.5 meters tall, and the northern, Jehan de Beauce tower, 114 meters – a 30-storey skyscraper of "hewn rock ... hoisted into heaven."

We don't know much about the people who "climbed on sketchy ladders" to build the Chartres Cathedral. We know that the Cathedral was founded on the site of an ancient temple and an early medieval church. We think that its construction in its current form commenced in 1145, stopped and then resumed after a fire in 1194. We know that it was completed in the early part of the thirteenth century – very rapidly in medieval terms – but very little beyond that.

This is not just an unfortunate gap in our historical knowledge; it is an integral part of the story of the Chartres Cathedral in particular and the Gothic cathedral in general. The cathedral has no architect. Its construction has no exact beginning or end, and no blueprint for it to follow. It is an achievement of many hands – a grand achievement, and all the more human for it. This is also true of science, and is the main reason we have started the book with the cathedral: it will serve us as an ongoing metaphor with which to think about the coming-into-being of science.

Figure 1.1 Chartres Cathedral, southern façade. The bridge-like structures between the Romanesque tower and the transept (marked by the rose window) and around the rear on the eastern side are the 'flying buttresses' supporting the walls.

We need such a guiding metaphor as it is difficult to see, at first sight, what a history of science might be. We commonly use 'science' to mean the correct and proper way in which we know how the world is and works; so in what sense does science have a history? We may think of such a history as a list of all the ways our predecessors used to get it wrong until we got it right, but little insight into our predecessors' ways of attaining knowledge – and our own indebtedness to these ways – can be gained by such an approach. And even if we took this approach as an idle but innocent curiosity, it is completely unclear where such a list should start and where it should end. Why should we bother about any mistake and superstition more than any other?

The English word 'science' comes from the Latin *scientia*: 'true and well-supported knowledge.' But *scientia* is science's ideal. In reality, science is a cathedral: it is an achievement – grand, yet *human*. Like all human knowledge, like all beliefs and ways of developing and supporting them, science is particular, local and historical.

Philosophy: The Cathedral that is Science

It *is* difficult to think of science as having a history. It is obvious that people in earlier epochs knew the world differently than we do. But since we are rightly impressed by our own knowledge of the world – by our science – it is very tempting to assume that they were simply wrong, and that science's past consists of a march towards its present. In other words, it seems as if the reason for *our* beliefs lies in their truth. It seems that if we gather and produce knowledge the way we do, it is simply because these are the proper methods to do so. It seems that if we support our knowledge by the evidence and arguments we use, it is because these are reliable evidence and valid arguments. And it seems that if people of yore thought otherwise, it is because they didn't yet know what we already do.

This is a tempting thought, but an entirely misleading one. It puts the cart before the horse: we think the way we do because we followed a route laid down by our predecessors. We think that bodies have 'mass' because Isaac Newton needed to explain to himself and to his rivals how bodies attract one another. We no longer think of matter or mass the way Newton thought – he didn't 'get it right.' But we still use his concept and his mathematics – had Newton lost some of his debates, which might very well have happened, we would have had a different physics. We have the 'unconscious' because Sigmund Freud was intrigued and fascinated by women who were blind despite healthy eyes or paralyzed despite healthy limbs. But we no longer have hysteria, the malady that Freud explained psychologically; Freud didn't discover for us something we now know to be true. Yet 'unconscious' is still as integral a part of psychology as it is of general culture. Had there not been Freud, we would have had a different psychology.

This is what it means for science to have a history. It means that we are not the aim and final cause of the work of people of the past. Rather, we are the product of their work. Our thoughts and beliefs are not the correction of their mistakes but the outcome of their struggles with the challenges that *they* faced, in *their* time, with the resources available to *them*. Had they come up with different ways to accommodate those challenges, we would have different beliefs. We rightly hold our beliefs as true, but they are true because of the effort put into making them so. Their truth is not the cause of this effort but its outcome.

This is a difficult insight – that our knowledge is truly contingent, determined by its history, like every other human affair. Thinking about cathedrals makes this idea easier to come to terms with. Here is an example.

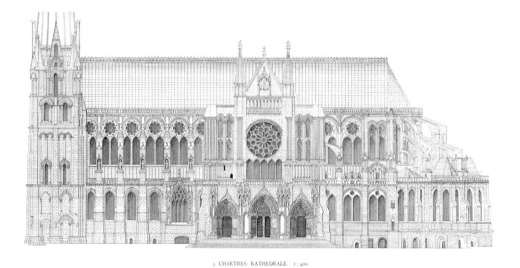

j. CHARTRES: KATHEDRALE. 1 : 400.

Figure 1.2 The Dream of Harmony: a modern, scaled diagram of the south elevation of the Chartres Cathedral, from G. Dehio and G. von Bezold, *Die Kirchliche Baukunst des abendlandes* (Stuttgart: Cotta, 1887–1902).

Observed with an admiring eye, the Gothic cathedral looks like the marvel of order and harmony represented in Figure 1.2. The cathedral does reflect a very particular, strict idea of order: it aspires to be a cross whose parts relate to each other in musical proportions (more on this below). But even a casual examination of the actual building, rather than its idealized drawings, reveals that it falls far short of this ideal. Look at Figure 1.3 and you'll note the uneven spires, the differently sized windows and, in general, the asymmetry and inconsistency. Yet these imperfections should not be looked down upon. They are marks of a living, evolving human undertaking: fire or earthquake damage repaired; a new spire erected; balustrades and cornices added or remodeled; a pipe organ installed or removed. Moreover: the cathedral doesn't have to be, and never is, complete. Perfect order may be the proper representation of the worshipped God, but it is not required for actual worship: ancient chapels can be used while the grandiose nave is under construction; a wooden roof can be installed if a proper dome isn't affordable. Like any human artifact, especially one created over centuries, the cathedral carries all the marks of the changing opportunities, resources and aspirations of the many people building it, as well as the difficulties which faced them and their imperfect solutions.

This is not to say that the cathedral does not embody ideals of order, perfection and harmony. Quite the opposite: these were exactly the ideals of the people who "climbed on sketchy ladders towards God." But it is this very

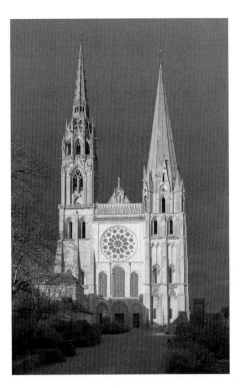

Figure 1.3 The Asymmetrical Reality: the Chartres Cathedral's façade. Note that while the center
– built more or less continuously – is properly symmetrical, the two towers, and especially the
spires – built some four centuries apart – are growingly asymmetrical.

point that needs to be stressed: harmonious order was an ideal to which the
people building the cathedral subscribed, not a template they could follow.
This is the way we need to think about science. We know that the claims of
our science are 'true,' but this truth is an ideal which guides science. It is not
some *thing*, existing prior to and independently of the inquiry, waiting to be
revealed. When evidence and argument convince scientists that some claim
is true, it does not mean that they have reached something that was always
there, obscured by error and misjudgment. It definitely does not mean that
this 'something' will always be there. What it does mean is that they have
succeeded in using their resources to solve a current challenge in a way
that they find satisfactory, even though they are most likely well aware that
this solution is temporary, and so is the challenge. Like the building of the
cathedral, scientific work has moments of satisfaction, even glory, but is
never completed. And like harmony for the building of the cathedral, truth
is an ideal that guides scientific research; like the cathedral, science is the
product of people striving towards this ideal, not its accomplishment.

Yet another reason to turn to the cathedral as a metaphor for science is that like the cathedral, and like most human accomplishments, science is a work of many hands (Figure 1.4). It has no single architect, no single design, no single vision it follows. It is forgivable that we concentrate, in this book and in the historiography of science in general, on the exciting contributions of great thinkers such as Aristotle, Galileo and Newton. But great contributions should not be construed as great leaps forward. In science, as in the building of the cathedral, all hands are necessary and every "winch and pulley" is indispensable. Of course, some craftspeople are more skilled than others, and some crafts are more difficult to acquire and replicate. But at no point can the master mason completely transcend the work

Figure 1.4 Work of many hands: Jean Fouquet's *The Construction of the Temple of Jerusalem by King Solomon* (c. 1475) – a miniature illustration from a manuscript edition of Josephus Flavius' *The Antiquities of the Jews* (man. fr. 247, fol. 163 v. BN, Paris. Josephus' text is from c. 94). Fouquet depicts the king visiting what he imagines as a construction site of a cathedral (modeled on the Notre Dame in Paris), populated by a swarm of workers engaged in different facets of the construction, requiring different tools and expertise.

of the youngest and least experienced stone-chipper; it is always the "hewn rock" produced by one that the other has to use. A Copernicus or a Kepler may construct a new way to look at the astronomical relations between Heaven and Earth, but he can only do so using the available intellectual resources developed by the astronomers, mathematicians and natural philosophers of his and earlier generations.

The way in which knowledge is rooted in a particular time and place is yet another philosophical insight about science and its history that the cathedral metaphor illustrates. Science is the most global of all modern-day endeavors: laboratories and computer programs, theories and empirical procedures, are fundamentally the same in the United States and China, in Australia and Sweden. We tend to assume that this globality means that scientific knowledge is in its very essence *universal*. We are commonly told that by developing a 'scientific method,' allegedly independent of time and place, we are now discovering 'scientific truths' that are independent of time and place. Looking at the cathedral, we can once again see that this assumption of universality confuses historical cause and effect.

Like science, one finds Gothic cathedrals all over the globe: they dot Central and Western Europe, there are many in Asia, and especially South America. There is even one, built using traditional materials and techniques, on West 110 St., New York. Yet no one would suggest that there is something inherently universal about cathedrals. We can easily see that they originated in a certain place and at a certain time, taking their shape for the religious, aesthetic and practical reasons in play there and then (we will consider these briefly below). The reasons why they are to be found in this distinctive shape in so many places and so far away from their origins are different. The global presence of Gothic cathedrals no longer relates to the preferences of the people who originally shaped them in the independent communities of small and enclosed Europe, but to what these cathedrals came to symbolize in the political and religious circumstances of empire and mission, two, three and four centuries later. The globality of the Gothic cathedral has little to do with what actually makes it a Gothic cathedral.

Similarly, the globality of scientific theories and procedures is *not* a sign of an inherent universality; not any more than the globality of the Gothic cathedral is a sign of an inherent universality of the pointed arch or the flying buttress (see Figure 1.1). The universal aspirations of scientists are important for understanding the global reach of their work, and the religious aspirations of those who "took up God's house to meet him" are important for understanding why they built similar cathedrals around the globe. But the point is that science and cathedrals are everywhere because they were *exported*. The universality of science is the *outcome* of its history, not its cause. This is what it means for science to have a history.

Science has a history because it's a unique human cultural phenomenon. It is the unique – indeed diverse and incoherent – cluster of beliefs and practices being taught, exercised and sanctioned by the relevant social institutions of today: universities, governmental research institutes, scientific journals. But just as the global presence of science doesn't imply that scientific knowledge is inherently universal – that its claims and procedures are independent of the places and times where they are produced and implemented – the unique character of science doesn't imply that scientific knowledge has a unique access to truth or a unique hold on rationality.

This claim is in no sense a retreat from the point stressed above: that science is a marvelous achievement. It is just a reminder that it is a human achievement, and as such its accomplishments can only be measured against the challenges it was set to meet by the people who set them. There were and there are many other such clusters of beliefs, techniques, tools and institutions; such 'systems of knowledge,' complex and rich in their own right, fulfilling a crucial role in their own cultures, satisfying the curiosity and practical needs of their practitioners. As on-lookers from our own culture of science, we may find some of them particularly admirable and recognize that they comprise capacities that science doesn't provide: Polynesian seafaring and navigation techniques; ancient Chinese medicine; Australian Aboriginal fire technology; Inca astronomy; and certainly many more.

But none of these types of knowledge is 'science.' To say that science has a history means that we don't use this term as an honorific title. We don't bestow it to label types of knowledge that impress us or that we believe in, nor deny it from those we don't approve of. 'Science' is a proper name – the name by which we call those things taught and exercised in contemporary science faculties. It is the history of these beliefs, practices and institutions that we will follow below, and we will touch upon other beliefs and practices only if they have crossed paths with that particular history, as resources, competition, context or alternative for science-to-be.

History: The Cathedral as a Turning Point

The cathedral helps us understand what it means for science to have a history, and it's also a good place to start telling this history. This is because science's history, like all histories, has no clear beginning but many interesting turning points, and the great Gothic cathedral represents one of them. It was in the time of the cathedral – the era in which Chartres was built (see above), 'the High Middle Ages'[1] – and often inside and around the cathedral, that many of the modes, practices and institutions of science began to emerge.

The time of the grand cathedral is pivotal in the emergence of science, first, because it is a moment in European history when building a cathedral has become materially and socially feasible. The commitment towards such an undertaking – as science would also become – is extreme. It takes resources, which in agrarian communities are always scarce, and divests them into a very expensive project that takes not only years but generations to complete, and which would not contribute to the material welfare of the community even in the long run. The communities in urban centers like Paris, Cologne, Florence and Barcelona, but also in more rural areas like Noyon, Soisson and of course Chartres were strong, affluent and independent enough to undertake such a venture. The reasons are many and complex, but some of them are directly relevant to us: they have to do with knowledge. New technologies – wind and water mills, looms and deep ploughs (which we will discuss below) – significantly improved the economic conditions of European peasants and burghers, and brought with them dramatic social changes. The ambitious project that science would develop into could not have been imagined before these changes.

The age of the Gothic cathedral is also the culmination of an era in which the central and unifying force in Europe, culturally and politically, was an institution whose core was *intellectual*: the Catholic Church. The medieval Catholic Church drew its legitimacy and claim to power from the erudition of its leaders and the knowledge – mainly divine but also profane – which its emissaries garnered and imparted as priests, scholars and educators. The Church of this era represents a unique bridge between the political and the learned and between the mundane and the abstract that was essential to the coming to being of science and the intellectual and institutional form it would take. One of the expressions of this bridge is the establishment of the institution of research and teaching most identified with science from its inception until today: the university. Fundamentally religious, many of the early universities found their first homes in cathedrals.

---

[1] 'Middle Ages' and 'Dark Ages,' as will be discussed in Chapter 5, are titles coined in hindsight, by scholars of the fourteenth and fifteenth centuries. They used them to refer to the previous thousand years, stretching from the final collapse of the Roman Empire in the fifth century to their own era, which they called 'Renaissance' – rebirth. These terms demonstrate the problems with historians' categories: people living through this millennium obviously didn't experience their life and times as being just 'in between.' For them, these were by far the most important times.

Finally, the age of the cathedral also witnessed the appropriation into Christian learning of the great achievements of Greek thought (originating from the realm of Hellenistic[2] culture and originally written in Greek): metaphysics, astronomy, logic, cosmology, philosophy of nature, medicine, mathematics. If one has to choose a starting point for the history of modern science from the simple perspective of content, Greek knowledge is definitely the most obvious choice. But beyond content, the importation of the Greek corpus into Christian Europe is also a mark of the uniqueness of the cultural-historical moment represented by the Gothic cathedral. The European search for this knowledge was driven by the new universities' urge to acquire teaching material. The knowledge – in the form of manuscripts in Greek and Arabic – was available because of the gradual collapse of the Byzantine Empire and because of the shoulder-rubbing with the thriving Muslim culture in the west, south and east. And because of these cross-cultural sources, the quest for Greek knowledge set in motion a translation project unlike any in history.

Historiography: Culture and Knowledge

Finally, the cathedral is a synecdoche[3] of high medieval knowledge, and looking at it we can ask: what kind of knowledge does it take to build a cathedral? What knowledge of the world does the cathedral reflect? And more generally: what does it take to tell a history of knowledge? What kind of history is it?

One can start by asking the most straightforward type of question: how did the builders of the cathedral move such a grand amount of stone – some 80,000 tons for a large cathedral – from its place of quarry to the building site, which could be many miles away?

This is a simple question, and the short answer also seems simple: with horses. In fact, this answer turns out to be rich and complex.

The decline of horseback warfare in Europe in the second half of the Middle Ages[4] made draft horses cheaper and available to agriculture. Originally bred to carry heavily armored knights as well as their own armor, they were big, strong and agile, so they could operate a new kind of agricultural technology: the heavy plough (Figure 1.5, right). Replacing the shallow scratching of the traditional plough (Figure 1.5, left), the heavy plough dug deep into the ground and turned the soil, eliminating the need to let the soil

---

[2] Throughout the book I'll use 'Hellenic' to refer to the ethnically Greek and their indigenous realm and 'Hellenistic' to their culture as it spread beyond that.

[3] 'Synecdoche' means 'in a nutshell'; a part or a detail that represents the whole.

[4] Wars were obviously still abundant, but the knights seem to have noticed that they were being most un-chivalrously massacred by longbow-holding peasants on foot.

Figure 1.5 Ploughing. On the left: an engraving by M. van der Gucht (1660–1725) of a scratch plough drawn by oxen. Note the shallow furrows that the plough raises. On the right: *September* – an illumination by Simon Bening for Da Costa's 1515 *Book of Hours* (*Liber Chronicarum* – see Figure 1.12 for another example of this kind of text. This particular manuscript is now in Morgan Library, MS M.399, fol. 10v) of a heavy plough drawn by two horses. Note the wheels and the heavy flat frame weighing on the share.

'rest' and revitalize often. It thus enabled shifting from two to three planting seasons a year and growing grains like wheat and oats. Conversely, these crops enabled raising horses. Being finicky animals, horses can't subsist on grass, which is sufficient for oxen – the traditional, emblematic beasts of burden. The better mobility of the horse, compared to the ox, also allowed peasants to live further away from their fields and freed arable land for cultivation (look where the village is in Figure 1.5, right). The effects of all these changes were far-reaching: more and better food enabled population growth, and the concentration of residences away from the fields led to the growth of large villages. The need to invest in more complex equipment and expensive animals and crops created new economic and social networks in these communities, which were already relatively prosperous, and, because of their size and relative density, almost urban. These are exactly the essential conditions for investment in the cathedral: the workhorse enabled hauling stones for the cathedral, but it was also an integral part of the social, cultural and economic fabric that made the cathedral possible and desirable.

Hauling stones is a kind of *know-how*: the type of knowledge by which we do things such as make a cake or ride a bicycle. Know-how usually cannot be, and does not have to be, fully captured in words, because it is encoded in and operated by muscles, tools and materials. One may think of it as a simple kind of knowledge – manual, even menial. But as we saw with the horse, even an innocent question concerning one aspect of the know-how put into the cathedral cannot be answered without reference to an endlessly expandable web which relates economic and technological factors to social, cultural and even religious significations. The point becomes even clearer when one considers the meaning of the horse to medieval culture: not just in war and at work, but in defining social and gender roles in art and poetry. 'Chivalry' is the heart of medieval culture.

The ability to carry heavy loads is a very general form of knowledge, but the answer to 'how did they do it' does not become less historically intricate when we inquire about more specific challenges. For example: how did the builders of the cathedral "hoist hewn rock into heaven"? How did they build to such a height? The short answer here is: 'by arch and vault' (Figure 1.6), which is of course hardly a short answer. The history of the knowledge embedded in the arch is as intricate as that related to the use of the horse.

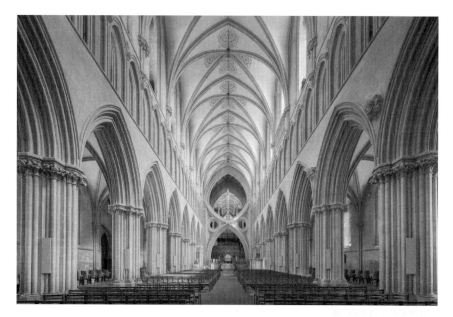

Figure 1.6 Arches and vaults in Wells Cathedral (south-west England) from the twelfth century. Note the pointed – 'Gothic' – arches, holding vaults of different rises and spans.

Ways of Knowing

Know-How: The Arch

It is not difficult to explain what makes the arch an efficient tool in the hands of the cathedral builders: it distributes the weight it carries obliquely, through its haunches, to the piers and onto the ground (Figure 1.7). The weight is thus supported not only by the strength of the building material – stone or brick in our case – but also by the structure. This enables much larger spans than could be carried by straight slabs on columns, such as those comprising the famous Greek temples. The medieval masons *knew how* to make use of this structural efficiency in constructing the arch: they'd fill the rise with scaffolding, arrange the stones and, once the keystone was put in place, the arch was stable and the scaffolding could be removed.

But as it was with the horse, this apparent simplicity of *know-how* is misleading. Like the horse, the arch was as intricately woven into the culture that adopted it.

Histories of the arch commonly begin with ancient Rome, but the Romans did not invent the arch. Arches can be found in various corners of the ancient world, and it is thought that the Romans came across it when they conquered the Middle East and adopted it from there. Yet the

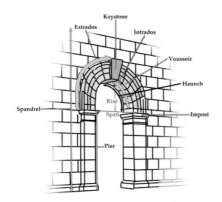

Figure 1.7 Elements of the Roman arch. The weight is distributed through the haunches and the piers to the ground, enabling much larger spans than horizontal beams could hold. However, because it is a segment of a circle (usually a semi-circle), the rise and span determine each other, and because weight in fact does not naturally distribute itself circularly, the piers need to be massive to resist the horizontal pressure.

Romans turned the arch, literally and metaphorically, into the foundation of their great political and material culture. Arches made the grandiose constructions – temples and theaters – by which their cities were known (Figure 1.8, left), as well as the walls that separated them from their immediate environs. Arches also made the grid of paths and routes which connected the city to its rural surroundings and remote frontier – the roads, bridges and aqueducts that brought provisions and taxes into the city and dispatched orders and soldiers to the provinces (Figure 1.8, right). Arches defined center and periphery and the relations between them; they were the infrastructure of an empire.

The Romans didn't only use the arch – they thought about it and considered it theoretically, and this allows us to think about the history of knowledge from a different perspective. Vitruvius, a Roman master of architecture and technology of the first century BCE, provides an excellent example. In the eighth book of his *De Architectura libri Decem* (*Ten Books on Architecture*), he explains how to "discharge the load of the walls by means of archings composed of voussoirs (see Figure 1.7) with joints radiating to the centre." Vitruvius engages with the same structure that is so important for the masons, but the knowledge he offers is of a different kind to theirs: we may call it *knowing that*.

The distinction between *knowing-how* and *knowing-that* is crucial for the history of science. Horse raising is a clear example of the former: medieval Europeans *knew how* to raise and use horses. It was a matter of practice, tradition, trial and error: it is hard to imagine the medieval farmer, almost certainly illiterate, having abstract principles or general equine theories. Vitruvius' is an example of *knowing-that*: knowledge carried and

Figure 1.8 Stacking arches vertically enabled the Romans to build the tall, storeyed buildings, which epitomized their cities, like the Colosseum in Rome on the left. Linking them horizontally enabled the construction of long structures connecting these cities to their rural periphery and to the provinces through roads, bridges and aqueducts like the Pont du Guard in Southern France on the right.

communicated in written words, developed and supported by arguments and evidence, aspiring for the general and often the abstract.

The ancient and medieval builders obviously *knew how* to build arches. But did they know, or bother with, the type of knowledge offered by Vitruvius? Did they *know that* the arch, as an idealized geometrical curve, is an efficient carrier of weight? Did they ask themselves *why* this curve was so efficient? Did they have the vocabulary to ask such questions? It is unlikely that medieval masons could read at all, let alone ancient Latin, or that they had access to such literature as a Roman technical treatise. Their knowledge was in their bodily skills, and it passed from generation to generation through tutoring and apprenticeship rather than bookish learning. This is in the very nature of *know-how*, distinguishing it from the scripted, transcribed, argued and taught *knowing that*, and it is well grounded in social and institutional structures.

But if farmers could raise horses and masons build cathedrals strictly with *know-how*, why should the history of science even bother with *knowing-that*? Or conversely, if science is knowledge like Vitruvius', why should we concern ourselves with the masons? We can better understand why both are necessary by returning to the question 'how did they build so high?'

The answer to that highlights another important feature of the historiography of science: the surprising contingency of many developments – namely, the fact that there was nothing necessary in them. Yet the ability to build the Gothic cathedral to such heights was not only a contingent development: it came about as a solution to a different challenge. The Roman arch is very simple to conceive, design and measure because it is a segment of a circle. But for the cathedral builder this harbored a problem. Unlike the ancient church, which, following Roman architecture, was a simple barrel-shaped basilica, the cathedral aspired to a cross shape and contained many separate spaces of different status and function. Naves, aisles, chapels, etc. called for different ceilings of different heights and openings of different spans. The Roman arch, being circular, had the same height and width – the same rise and span – so connecting the different hallways and enclosures required raising the smaller vaults on columns and supporting them with heavy piers (Figures 1.7 and 1.9, left) – dangerous, unseemly and expensive.

The solution contrived by the medieval masons was simple and ingenious. Like the other issues considered in this chapter, it represents a phenomenon that will capture our attention throughout the book: a local solution to one particular problem, which carries much wider and more general implications. It turned out that the arch does not have to be circular. One can pull the arch by its top, as it were, making it pointy, thereby

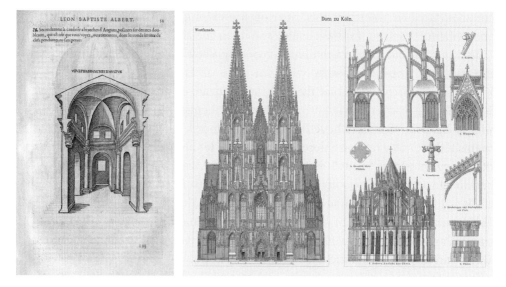

Figure 1.9 The advantages of the pointed Gothic arch over the circular Roman one. The diagram on the left, from Alberti's *De re aedificatoria* (originally 1485, the image is from the 1552 French edition: Leon Battista Alberti, *L'architecture et art de bien bastir* (Paris: Jacques Kerver, 1552)) shows clearly how all dimensions of the vaults are mutually determined. Gothic arches, as in the diagram on the right of Cologne Cathedral (whose construction started in 1248), enabled much higher vaults, as well as connecting vaults of different spans and heights. This image is from an 1897 wood engraving.

separating the rise and the span (Figure 1.9, right). That way, arches and vaults of different spans could be connected at one height, allowing for the intricate grid of paths and halls that is the Gothic cathedral. And it turned out that the pointed arch (named 'Gothic' actually as a derogative, by those who thought it was vulgar and graceless in comparison to the elegant Romanesque style it supplanted) had another, much more significant advantage: it was more stable than the circular arch and allowed building to the tremendous heights we mentioned at the very beginning.

But this account of *know-how* cannot tell us the 'why.' Why build so high? Why build arches of different heights? Why replace supporting walls with flying buttresses? In general: what did the builders of the cathedrals know *about* their world – and their place in it – that prompted them to build this way? The Gothic arch was a piece of know-how which enabled the medieval masons to construct very high buildings; it didn't dictate building them so.

We can't tell the history of science without paying close attention to both *knowing-how, knowing-that* and the relations between them, because both are necessary and neither one determines the other. Here is a more specific illustration of this idea: the *Mezquita* in Cordoba, Spain. It was built through the ninth and tenth centuries as the grand mosque of *Al-Andalus* – the southern part of the Iberian Peninsula, under Muslim rule from the eighth through the fifteenth centuries – and then turned into a cathedral in the thirteenth and fourteenth centuries, after the city was conquered by the Christians. The Muslim masons, the first to introduce the pointed arch, used it extensively in the original mosque, but they apparently were not completely happy with its aesthetics. They camouflaged it under circular ornaments (Figure 1.10, left) and were never tempted to use the added structural efficiency to "defy gravity" and stretch their mosques into the sky. They remained loyal to the sprawling, repetitive forest of circular arches which gives big mosques their almost hypnotic atmosphere. When the Christians took over, however, they not only adopted the engineering feat of the pointed arch; they turned it into their main architectural tool, piling soaring Gothic arches and domes on top of the circular, modestly elevated original ones (note how the bright white and gold Christian dome towers over the Muslim red and brick arches in Figure 1.10, right). "Hoist[ing] hewn rock into heaven" was a choice, determined not by *knowing how* to do so but by what the builders knew and thought *about* their world.

But what *did* the people who "inhabited the sky with hammers" know and think? Did they reflect on the fact that the most perfect curve – the circle – turned out not to be the most physically efficient one? Did they ask themselves why the pointed arch is more stable and how its stability relates to its shape? It is very hard to tell. They did not write down what they knew.

Figure 1.10 Know-how does not determine its application, as exemplified in the *Mezquita* in Cordoba. The structure on the left (known as *cinqfoil*) is a part of the original mosque. The Muslim masons who built it camouflaged the real weight-bearing elements, which are the five pointed arches, with four circular arches. Similarly, they obscured the pointed shape of the two arches on both sides (of which we see only half) with three circular adornments on each *voussoir* (see Figure 1.7). On the right is the high dome of the cathedral erected on top of the mosque. The Christian rulers who had it built made no attempt to hide the pointed arch. They celebrated the enormous height allowed by this piece of *Muslim* know-how and the superiority this height implied.

Knowing-That: The World of the Cathedral Builders

Yet some clues of what the masons, their employers and the members of their communities *knew about* the world they lived in are borne by the cathedral itself. Height is of course one of them: the cathedral builders were making a dangerous and costly effort to "inhabit the sky"; the reasons for this effort are that other kind of knowledge. Another clue is light. The Gothic cathedral is full of light. It's not simply lit, or situated so as to allow as much light as possible into its imposing halls. Rather, the cathedral draws light in through large, elevated windows, and dyes it with spectacular stained-glass panes. It was a new know-how – new pigments and new technologies of glass-making – that enabled this spectacle. But it is the *use* of this know-how – the drive for such new capabilities and this keen interest in light – that tells us what the cathedral builders knew *about* their world.

A third clue is the surprising cobweb appearance of the cathedral's masonry. It is as if the builders purposefully avoided matter, replacing it wherever possible with massless structures: instead of massive supporting walls they introduced slim flying buttresses; in the walls they could not avoid building, they tore huge windows; and in the remaining wall space they dug ornate apertures. The cathedral builders had a clear preference for structure over matter, and for certain structures in particular, which is the final clue: the high-medieval, Gothic cathedrals were built according to the fundamental musical consonances. Between the nave, the transept and the aisles, the builders used the ratios of 1:1, 1:2, 2:3 and 3:4; the ratios which constitute the harmonious intervals of tones – unison, eighth, fifth and fourth correspondingly – from which music can be produced (Figure 1.11; we will discuss later the Greek discovery of these ratios and the significance they assigned to them).

These are the clues about the world the cathedral builders knew *as they knew it*. It is this knowledge that interests us, so let's begin by summarizing what we know about it.

Politically, it was the world of the so-called Holy Roman Empire; famous for being neither holy, nor Roman, nor really an empire. It called itself Holy for its Christianity, Roman as a self-aggrandizing reference to the great empire that last ruled much of Europe and Empire for its own aspirations. In effect, it was a conglomerate of mostly Germanic kingdoms, princedoms and fiefdoms. They were dominated by the Franks, and their emperor was elected by the most powerful kings after the previous emperor died. The original Roman Empire was much more Mediterranean than 'European,' but the appropriation of the name was not innocent. It meant that the German

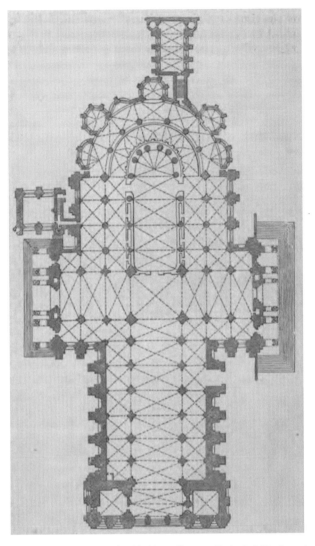

Figure 1.11 The musical harmonies embedded into Chartres Cathedral. The 'crossing' – the intersection of the nave and the transept – is a square, so the ratio between its length and its width is 1:1, which stands for the musical 'fundamental,' or unison. The ratio between the width of the nave and the width of the side aisles and between the length of the transept and its width is 2:1, corresponding to the octave, or eighth. Between the total length to the length of transept the ratio is 3:2 (fifth), and between choir's length and width: 4:3 (fourth).

rulers of medieval Northern Europe were claiming authority over the territories taken by the Romans at their height of power, around the turn of the Christian era. More crucial for us: they were claiming allegiance to the Greek culture that the Romans adopted when they conquered the Hellenistic realm at that time.

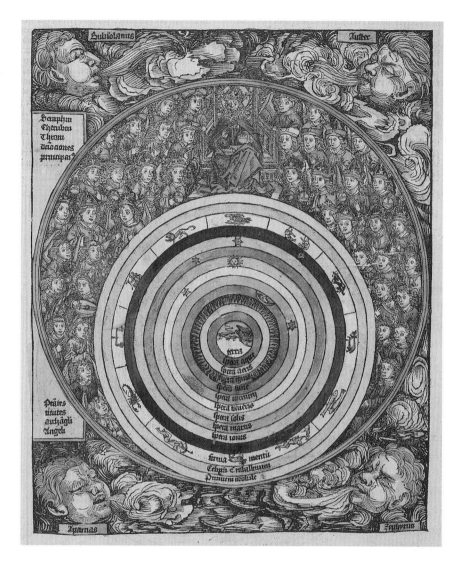

Figure 1.12 The Cosmos according to Hartmann Schedel's *Liber Chronicarum*, known as the *Nuremberg Chronicle* (folio 5v, woodcut by Michael Wolgemut's workshop; see also Figures 3.10 and 5.8, right). The *Chronicle* was printed in Latin and German in 1493, so it belongs to the first generation of printed books (*incunabula* – see Chapter 7), and is one of the very first illustrated ones. Comprising a Biblical paraphrase and a history of some European cities, books like that synthesized Pagan and Christian knowledge in general, and this image illustrates this synthesis in cosmology. It does not pretend to be to scale, but the compact orderliness it reflects is intended. At the very center rests the Earth, surrounded by the other three elements – Water, Air, then Fire. The realm of the elements is circumscribed by the lunar sphere (the Moon drawn at the upper right above the sphere of fire) and above it the planets (with the Sun directly above the Earth). Around the planets is the Biblical Firmament, identified with the sphere of the fixed stars and represented by the signs of the Zodiac; and around it the crystalline heaven and the unmoved mover – the *Primum Mobile*. The saints, angels and archangels populate the upper realm, presided over by God on his throne.

If politically it was a world with little structure and no clear center, it was, in the eyes of its dwellers, a very orderly whole. This order had the religious and aesthetic meanings that the builders encoded into the cathedral, because it was the benevolent creation of God. God made the world physically compact, perfectly round, with Man in its center and the Heavens encompassing it in concentric spheres. God, His angels and His saints were in the outer spheres, keeping a loving and caring eye from above (see Figure 1.12). (As is often the case, by the time this beautiful image was produced – a year after Columbus' first journey – the ideas it conveyed were quickly becoming obsolete.) The same order which God bestowed on the physical world permeated all aspects of life and thought: all that existed was organized in the 'Great Chain of Being,' with God at the top and matter at the bottom. Humans inhabited the middle region, below angels and above beasts, their reason partaking in the pure and active form which was God, their body made of profane, passive matter. Among humans, the political order reflected the metaphysical-ethical one: the king and the bishop were up and the peasant was down, each where they belonged, and the pope, God's emissary, was also the highest political authority.

The marvelous simplicity and perfection of the musical ratios were perhaps the most spectacular reflection of this harmonious order, and reason enough to embed them in the structure of the cathedral. These ratios were, indeed, embedded in the relations between other parts of the cosmos – the heavenly bodies in particular – and this meant the world was not only orderly and just, but also beautiful: the spheres literally sang the praises of the Lord. Light was another such marvel. It was the closest to pure form which people could find here, in the realm of matter; the best, most beautiful and harmonious entity in this world. Light was both the carrier and the symbol of God's grace: flowing freely and abundantly, without being diminished or exhausted, bringing life and happiness.

Tensions and Compromises

Belief and Authority: The Church

Science is like a cathedral: a grand human achievement; magnificent yet imperfect; purposeful yet contingent. It is an artificial creation, carrying signs of the particular historical processes by which it was made. The cathedral is like science: it embodies much *know-how* and represents much *knowing-that*. It was built by masonry, arches, horse breeding; each was an evolving tradition of knowledge, with complex relations to the culture, society, economy and religion in which it was embedded. Beyond the metaphor, we noted that the Gothic cathedral and the era of its construction – the High Middle Ages – are a good starting point for the very contingent story of science's coming into being. So we've now moved into inquiring about the world that the builders represented in their cathedral, their place within it and in relation to their God.

The religious aspect is crucial – and crucial for understanding science – because it was a world studied and taught by the Christian Church. The Church provided the rules of worship, the venues for congregation, an authoritative interpretation of the scriptures and much more. With its hierarchical-bureaucratic structure, its center in Rome and its loyalty to an ancient yet evolving code of law, the Church was the most centralizing political-cultural force of medieval Europe, and the kings and emperors were crucially dependent on the legitimation it granted them (see Chapter 4). The Church not only withheld and disseminated the scriptures: it provided an elaborate and extensive system of education. By the time of the cathedrals it had developed a sophisticated philosophy of nature, alongside mathematical sciences such as astronomy, music and mechanics. The synthesis of Aristotle's legacy with the fundamental principles of Christianity provided the intellectual framework for these studies, from the main topics and the basic assumptions, through modes of argumentation, to the criteria for evidence. This synthesis is an important part of the coming into being of modern science, and we will return to it in detail later.

Almost all scholarly labor during the age of the cathedral was done under the auspices of the Church, and the Church scholars venerated ancient knowledge. The most thorough formulation of this complex of ideas they found in a corpus that was, by their time, almost a millennium old. It was the works of St. Augustine, perhaps the most influential Christian thinker ever, whose life spanned the last decades of the Roman Empire (354–430).

Augustine was not European: apart from four years of schooling in Rome and Milan, he spent his life in the North African provinces of the declining Empire, where he became the bishop of Hippo, near Carthage (today's Annaba in Algeria). His life and thought are an emblem of this era of change, when Late Antiquity – the era of the rise and decline of the Roman Empire – was transforming into the Middle Ages. Augustine himself moved from the Paganism of Antiquity, through the Manichean Gnosticism of his time, to the up-and-coming Christianity; a painful metamorphosis of both reason and belief, which he recounted in his personal and moving *Confessions*. But he was a rhetorician by education, and the large part of his writings were anything but personal. Most powerful among them is his sprawling treatise on the superiority of Christianity: *The City of God*. He also wrote hundreds of letters and homilies elucidating the Christian doctrine in an era when many more religious options were still open and competing around the Mediterranean: Jewish monotheism and Canaanite paganism from Greater Syria; Hellenistic paganism from the Aegean; the worship of Isis and Osiris from Egypt; Zoroastrianism from Persia.

Augustine and the Problem of Evil

Augustine's intellectual and religious program was deeply rooted in this unique cultural setting, but it was shaped by a worry with which we can still identify and which was definitely relevant for the cathedral builders a millennium later. The cathedral aspires to capture and praise God's omnipresence and grace; but if, as Christianity tells, God is with us and He is all-good and all-powerful, then why is the world so evil? Why is there pain, sickness, famine, war? Why are God's creatures so cruel to one another? The medieval Europeans, with little medicine to protect them from malady or force to protect them from violence, experienced pain and suffering as a constant and overbearing presence. To the thinking Christian, in an era when other religions were still immediately present alternatives, these were pressing questions, especially for those who, like Augustine, had taken on Christianity only after conscious deliberation and choice.

The Greek and Roman pagans were of course as aware of the hardships of human life as their Christian successors. But for them these hardships did not present a philosophical or religious mystery: their gods were not particularly benevolent – in fact they were in many respects caricatures of human failings – and never avowed responsibility for human well-being. The Manicheans, the Gnostic sect to which Augustine belonged in his youth, did consider evil as a religious dilemma, but their religious-metaphysical solution was remote from anything a Christian would consider. At the core of Gnosticism in its various forms was the belief that *The Evil* is just as real and present in the world as *The Good*. In the Gnostic worldview, matter itself was evil and spirit had to be released from it to unite with The Good.

Christian thinkers could adopt neither the Gnostic solution nor the Greek indifference. Monotheism, despite its title, is distinguished from paganism *not* in having only one god, but by the role this god assumes. One can easily claim that Christianity, with its trinity, angels and saints, has as many demi-gods as any pagan religion. What makes it monotheistic is that its Godhead is all-powerful and all-good. Monotheism is the belief that God *created* this world and is *responsible* for it. For Christianity, in particular, this responsibility entails compassionate care – Grace. So it cannot accept that matter, the infrastructure of a world created by the benevolent, caring and omnipotent God, is bad. Moreover, central to Christian mythology is God's material embodiment as a man – thus matter cannot be evil. In terms of the people for and by whom cathedrals were built, if matter *were* essentially evil, there would be no cathedrals, because it would mean that human effort in this world is futile, and no material artifact can give glory to the immaterial God. From its very early days through the time of the cathedral builders over a millennium later, it has been essential for Christianity that God the Creator is morally responsible and involved, and the world, insofar as it is created in His image, must be perfect and perfectly good.

In his *Confessions*, Augustine undertakes this difficult intellectual task: to explain why, despite all the evil – both the evil afflicting people and the evil done by people – his readers should still believe that God is both benevolent and intimately concerned with every human person, attentive to every move that he or she makes. Augustine's strategy is to emphasize the transcendence of God: He is radically different from the world. God is everything the world is not – abstract, immaterial, eternal, unchanging – and thus cannot be held responsible for the failings of this world.

Yet this solution creates a problem which is crucially important for us here – a problem concerning our capacity to know. If God is absolutely transcendent, completely outside our grasp as temporal, material beings, how can He be known at all? Does this not mean that Christianity is irrational – demanding that we believe in the unknown and unknowable? To avoid this quandary Augustine introduces into Christian thought a very pagan metaphor: that of light. The knowledge of God, he explains, comes to us through direct 'enlightenment.' We know God through an unmediated absorption and recognition of His presence and goodness, just as we experience light. Unlike knowledge of this world, which demands careful and active inquiry, we receive divine illumination passively, the way we see light. We cannot but be enlightened by God if we open our mind's eye and allow the knowledge of His presence and goodness to flow into our souls. Unlike knowledge of this world, which is always tentative, our enlightenment by divine light is clear, certain and unequivocal.

These are epistemological considerations – considerations about knowledge – and are therefore crucial to what science would become. And with this the relation between the cathedral and science transcends the metaphorical: similar dilemmas were vested in both. Europe of the cathedrals was deeply Christian, and Christianity, from its earliest days, has always been deeply worried about what we can know and how. The

Church Fathers, before and including Augustine, shaped Christianity with the ambition to depose Greco-Roman paganism and take over its cultural place (the competition with Judaism ran on different grounds and is outside our scope here). It is therefore not surprising that they conceived the challenge in terms of contest about knowledge: knowledge – philosophy – was the great pride of Hellenistic culture. For early Christian thinkers, this contest allowed no compromise. Jesus was a relatively recent history, almost a living memory, and being only a century or two removed from God having been among them in the flesh they had no doubt that He would return very soon, and with him, the end of this world. Studying *this* world, as Greek philosophy did, was thus a diversion of attention from the eternal to the transient and insignificant. One of the most influential among these thinkers, Tertullian (160–230), who lived in Carthage, not far from where Augustine would reside two centuries later, put it succinctly: "philosophy is the material of the world's wisdom, the rash interpreter of the nature and dispensation of God." The philosophical diversion was not just a waste of time, but morally and religiously wrong: the mind should concentrate on the world beyond, "for heresies are themselves instigated by philosophy." It should be free and humble to accept the learning of the true religion, unhampered by the possible doubts philosophy might plant: "Our instruction comes from the porch of Solomon," says Tertullian, "who had himself taught that the Lord should be sought in simplicity of heart." His conclusions were fierce:

> What indeed has Athens to do with Jerusalem? What has the Academy to do with the Church? What have heretics to do with Christians? ... Away with all attempts to produce a Stoic, Platonic, and dialectic Christianity! We want no curious disputation after possessing Christ Jesus, no inquisition after receiving the gospel! When we believe, we desire no further belief. For this is our first article of faith, that there is nothing which we ought to believe besides.
>
> Tertullian, *Against the Heretics*, ch. 7 (www.newadvent.org/fathers/0311.htm)

Tertullian, a self-appointed sage of an oppressed sect, could treat worldly knowledge as abomination. But by Augustine's time Christianity was the Roman Empire's official religion, and Augustine was a bishop: he was not only a religious mentor, but also a mainstream political leader. Augustine had worldly responsibilities to his flock and had to seek and respect the knowledge necessary to discharge those responsibilities. The compromise he devised in order to fulfill this double role became the Catholic Church's official attitude towards knowledge for centuries to come and would determine its role in the shaping of science:

Pagan learning is not entirely made up of false teachings and superstitions. It contains also some excellent teachings, well suited to be used by truth. These are, so to speak, their gold and their silver, which they did not invent themselves, but which they dug out of the mines of the providence of God, which are scattered throughout the world.

De Doctrina Christiana, II, LX (https://faculty
.georgetown.edu/jod/augustine/ddc.html)

In the epistemology that Augustine formulated and that the Church has adhered to ever since, secular knowledge – knowledge of this world (*seculum* in Latin) – is always inferior to the knowledge that God bestowed on us directly, through his Book and his apostles. It is always hypothetical, discursive and subject to doubt, unlike the certainty of the sacred, revealed knowledge, which we receive by illumination. But *this* material world was also created by the benevolent God, and thus knowing it, even in this imperfect way, cannot be bad. So "pagan learning … contains also some excellent teachings" and can be useful, because it is also "dug out of the mines of the providence of God." The good Christian is not 'simple'; she's a person who knows. "A Christian," says Augustine, "is a person who thinks in believing and believes in thinking" (*De Praedestinatione*, 1:5, www.newadvent.org/fathers/15121.htm).

Augustine's concept of knowledge is crucial for two reasons. The first is historical: his compromise between sacred and secular knowledge, as we said, served the Church for almost two millennia. For most of this period it allowed the Church to take the acquisition of knowledge – not just of the divine but also of this world – as an integral part of its mission. The second reason is more philosophical: the ideas that allowed this compromise are still deeply entrenched in our concept of 'scientific knowledge.' With his notion of the supremacy of the abstract over the material and the metaphor of illumination, Augustine forms the link between his pagan predecessors and his scientific successors – us.

Augustine's Resources: Plotinus

We have been telling our story backwards: from the know-how of the medieval masons we proceeded to the arches of Roman antiquity; from the cultural hints they encoded in the cathedral to the teachings of Augustine, a millennium earlier. We'll discuss soon the reasons for this chronological reversal, but for the time being let's continue in this order and ask: what was there *before* Augustine? Or more specifically: what were the resources from which Augustine drew these solutions to his cluster of religious, moral and ultimately epistemological worries?

Augustine picked up these ideas from the writings of a pagan philosopher who lived a century before him: Plotinus (204–270). Like Augustine and Tertullian, Plotinus was a product of the North African periphery of the Roman Empire: born in Cairo, he was drawn to that great cultural center of antiquity, Alexandria. There he was not only mentored in the thought of Plato and his school, the *Academia* (which Tertullian mentioned above and to which we'll return in Chapter 2), but also studied Persian and Indian philosophy. He was almost 40 when he tried his luck as a soldier. When that failed, he moved to Rome in 245, where he spent the rest of his life and gained, already in his lifetime, a powerful reputation for the novelty and boldness of his interpretation of Plato. But unlike Augustine, as we said, Plotinus was a pagan, and his intellectual challenge was quite different. The problem of evil, which is such a theological dilemma for the Christian Bishop of Hippo, is a secondary (though interesting) question for the pagan philosopher. What mainly intrigues Plotinus is the metaphysical conundrums that troubled the Greek philosophical tradition he chose: how can our world be so varied and changing, yet still maintain unity and order? Or in the language favored by this tradition: how can *being* be reconciled with *becoming*?

Plotinus' answers to these questions will sound very esoteric to the modern ear, but it is worth paying attention to them. As our discussion of Augustine began to reveal, these worries and these deliberations played a crucial role in shaping what we consider as our own scientific thought. The fact that the roots of this thought spread so very far afield is part of what it means for science to have a genuine, human history. We'll return to discuss it below, so it's important to note exactly how particular and unique Plotinus' solution to the mystery of order vs. change is, how much *it is* an integral part of a specific tradition and a specific intellectual context.

The core of this solution is the conviction that existence is a consequence of thought. At the origin of everything, writes Plotinus, is "The One," which is pure thought – it thinks itself into existence; it exists in thinking. The One is also The Good and The Beautiful, because in this state of complete unity, all positive properties are one and the same. In thinking, The One creates an object of its thought and this separation – the separation of the Thinking One into the subject and the object of its own thought – is Plotinus' answer to the question 'how can *unity* breed *variety*?' In this state, of a subject-thinking-an-object, The One becomes what Plotinus calls the Nous – the intellect of the cosmos. The Nous comprises the ideal forms of all the many things in the world – variety in unity.

The Soul is the next phase of this descent of The One into the many; of the continuous differentiation by which being gives rise to becoming. Every living being partakes in this world soul; though only in humans does the soul strive towards the intellect. But whereas the intellect contemplates the forms purely, as they are in themselves, the soul also acts upon its thoughts. In its action the soul embodies the pure forms; it actualizes them into the realm of passive matter, differentiates them into particular material beings and relates these beings to one another. Creation is thus spontaneous activity of the soul, unceasing and undiminishing – an *emanation* like the flow of light. Materiality is nothing but the privation of this activity, the privation of form.

One can see what kind of intellectual tools a Christian thinker would find in this complex philosophy. To begin with, Plotinus' concept of "The One" provided a means to make sense of the difficult idea of a completely abstract God. With this concept at hand Augustine could explain how God could be both all-powerful and all-good while having no material existence. No less important, Plotinus provided an account of the place of the human soul within God's creation and in relation to Him. The soul occupies the middle of the Great Chain of Being, of which The One is at the top and matter at the bottom (see Figure 1.13). The soul should aspire to ascend towards its maker, but often loses itself and descends into the realm of matter, forgetting that it is illusory, a figment of its own creation. In this, the pagan Plotinus offered the Christian Augustine the wherewithal to understand evil – it is but an error, caused by human imperfection.

It was in this difficult, *Pagan*, and seemingly esoteric philosophy that Augustine found the intellectual resources to shape Christianity's approach to knowledge. Here was the strict dichotomy and hierarchy between the realm of matter and the realm of spirit and pure form: spirit stands for divinity, beauty and grace; matter is plurality and transience. And here also was the conviction that true knowledge is to be found only in the soul searching for the unity and eternity of the form, not in the body's dwellings in profane matter. Yet one important element of Plotinus' philosophy could not be accepted by Augustine: the idea that creation emanated; that it flows from the deity *spontaneously* – like light. This was exactly where Christianity distinguished itself from Pagan belief and philosophy: in the affirmation that God created the world by an act of *will* – that he *decided* to do so and could have decided otherwise. So Augustine modified the metaphor of light: it was no longer the way in which The One bestowed existence on everything; it was the way in which God bestowed knowledge to humans. The idea that the goodness of God flows marvelous and undiminished like light remains; but Pagan emanation is rejected for Christian creation.

Figure 1.13 The Christian version of Plotinus' Great Chain of Being, presenting the descent from pure divine form, through the mixtures of form with matter – humans, animals, plants, minerals – to pure profane matter. Christian and didactic, this image also presents Hell at the very bottom and falling angels on the right. It's taken from *Rhetorica Christiana* – a Franciscan textbook written and illustrated in Mexico by Didacus (Diego) Valadés (the scroll at the bottom right, just above hell, says *Valades fecit*), and published in Perugia, Italy in 1579. Valadés was the first Mestizo to join the Franciscan order and his book was the first of an American author to be printed.

Conclusion: Reflections on the History of Knowledge

Here, then, is the reason for the chronological reversal we noted above; for telling this history backwards. It is to turn our attention towards these types of compromises, to this attitude of thinkers towards their knowledge as tools and towards ideas from the past as resources – towards the deep relations between knowing-how and knowing-that. Resorting to the metaphor of the cathedral illuminates these considerations: in constructing his theology Augustine is not unlike the mason building a flying buttress or a pointed arch.

The mason constructs a stone structure and Augustine a religious-philosophical one, but both are using existing tools and materials to erect some new edifice. The mason's tools are material; Augustine's are intellectual – but they are tools nonetheless. Both builder and scholar face challenges: the mason recognizes a material-structural difficulty, like connecting arches of different spans to create a cross-shaped building; Augustine is troubled by an intellectual-theological problem – the problem of evil. They find their respective problems too complex to solve with the tools at their immediate disposal, so they reach for other tools. It can be an ancient tool, originally devised to meet other challenges: the mason may resort to the pointed arch, invented by his Muslim predecessors to solve a completely different stability problem (the Muslim masons obviously had no interest in crosses). Augustine, similarly, may find that Plotinus' solution to the problem of existence can be used to resolve the mystery of evil. Tools never fulfill their task perfectly, especially if adopted from one use to another, so they have to be modified and adapted to their new use. The mason would discover that the pointed arch allows building to great height, but also creates stability problems demanding flying buttresses for support. Augustine would find that Plotinus' philosophy carries unacceptable remnants of its pagan origins, forcing him to alter the way the metaphor of light is being used, from that of creation to that of knowledge. In both the material and the intellectual case, the modifications don't completely erase the marks of the original use: the Gothic cathedral will always carry the Muslim twist of its pointed arches; Augustine's theology will always be, like Plotinus' philosophy, Neo-Platonist (we will discuss the meaning of this term in Chapter 6).

Discussion Questions

. . . .

1. Does the cathedral make sense as a metaphor for science? Where does the analogy break? Can it be pushed even further?
2. How convincing is the distinction between *knowing-how* and *knowing-that*? Where is it useful? Where does it collapse?
3. Is there an important difference between thinking of the history of science in terms of 'resources' rather than 'sources' and 'influence'?
4. What kind of relations between knowledge and religion are suggested in this chapter? Is there a difference when the knowledge is science? When the religion is Pagan or Monotheistic?
5. Figure 1.4 is supposed to describe the building of the Temple of Solomon, but in fact depicts the construction of a cathedral (apparently *Notre Dame de Paris*). What can one learn from this about the way a culture conceives its real and imaginary past? How it relates it to its future? Does this entail specific lessons for the historiography of science?

Suggested Readings

. . . .

Primary Texts

Augustine, *On Christian Doctrine*, Book II, chs. 27–30; 39–42, www.newadvent.org/fathers/12022.htm.

Grosseteste, Robert, *On Light*, Clare C. Riedl (trans.) (Milwaukee, WI: Marquette University Press, 1942).

Secondary Sources

John Ormond's "Cathedral Builders":
Ormond, John, *Collected Poems*, Rian Evans (ed.) (Bridgend, Wales: Poetry Wales Press, 2015).

On cathedrals and their place in Medieval Science:
Heilbron, John L., *The Sun in the Church: Cathedrals as Solar Observatories* (Cambridge, MA: Harvard University Press, 1999).

On science as a human achievement:
Golinski, Jan, *Making Natural Knowledge* (Cambridge University Press, 1998).

On the impact of horses on medieval culture and on the relations between technology and culture in general:

White, Lynn Jr., *Medieval Technology and Social Change* (Oxford University Press, 1966).

On the uniqueness of *Know How*:
Polanyi, Michael, *Personal Knowledge: Towards a Post-Critical Philosophy*, corr. edn. (University of Chicago Press, 1974).

On Augustine:
Evans, Gillian Rosemary, *Augustine on Evil* (Cambridge University Press, 1982).
Hollingworth, Miles, *Saint Augustine of Hippo: An Intellectual Biography* (Oxford University Press, 2013).

On Plotinus:
Uždavinys, Algis, *The Heart of Plotinus: The Essential Enneads* (Bloomington, IN: World Wisdom, 2009).

On light and knowledge in Christianity:
Edgerton, Samuel, *The Mirror, the Window, and the Telescope* (Ithaca, NY: Cornell University Press, 2009).

2 | Greek Thought

Knowing-About as Know-How

The cathedral metaphor suggested an important distinction: between what its builders knew *how* to do – in order to erect the cathedral – and what they knew *about* the world – according to which they attempted to shape the cathedral. Going back in time, searching for the resources they drew on, we have also revealed the *limits* of the distinction: knowledge about the world, it turns out, is itself a kind of know-how. Both types of knowledge require skills, tools, materials; they involve recognizing problems and seeking solutions; searching for resources, adopting and adapting them to new use.

With this insight – that thinkers handle ideas like masons handle stones and chisels – comes another insight, crucial for the historiography of science: that ideas don't have life or history of their own. They don't evolve or 'influence' one another. People create them, use them, adopt or eschew them, change them for changing needs, sometimes simply neglect or forget them, and sometimes search for and rediscover them. Ideas have no agency: history of science and history of knowledge, like any history, is told of people.

Augustine, a Christian, in Hippo of the fifth century, needed to solve a problem about *knowledge*: how can we know a God completely detached from the world, and how can we reconcile our knowledge of Him with our knowledge of His creation? To that end, Augustine found, appropriated, modified and used ideas that Plotinus, a pagan in Rome of the third century, used to solve the mysteries of existence and order: how can the many come from the one? How can variety maintain uniformity? Plotinus, for his part, was working on interpreting and developing ideas that were first formulated some 700 years earlier, in fourth-century BCE Athens, by Plato (we will return to Plato's life and times soon). Few people's thought had impact similar to Plato's, but by now we know how to understand such a claim: when we say that Plato's ideas were extremely influential, we are not describing some force they had within themselves. We are saying that the intellectual tools that Plato developed to meet the challenges of his time and place were reshaped and reapplied for different challenges in different places and different times – centuries, indeed millennia, later.

Plato and the Culture of Theory

Plato: Truth and Episteme

Compared to the work of Augustine and Plotinus, the relevance of Plato's work to the history of science is straightforward: his fundamental problem is exactly *how is knowledge of this world possible.* His answer, as his disciples may have made us expect, would come through the metaphor of light and vision: "the soul is like the eye: when resting upon that on which truth and being shine, the soul perceives and understands" (*Republic*, Book VI, http://classics.mit.edu/Plato/republic.7.vi.html).

To know, Plato reasons, is to get hold of Truth – this is the very meaning of what knowledge is. Truth, by its very essence, is what *does not* change; what is always and everywhere the very same; "becoming" is the opposite of "Truth and being" (*Republic*, Book VII, http://classics .mit.edu/Plato/republic.8.vii.html). Yet no object we perceive is ever *the same*: it looks different from every perspective and with every change of light or background; it feels different at different times and in different contexts. The same room looks wide from this angle and narrow from that; is warm if we come in from the cold, cold if we sit in it for a while. Worse: the impossibility of gaining knowledge from our surroundings may not only be the problem of our own fickle senses. The material world itself is constantly changing, and we change with it; what my body was a year ago is not what it is today; what could be said about this tree yesterday no longer holds today; even the mountain is no longer what it is while we point at it.

Knowledge, Plato concludes, simply cannot be such a relation between the changing and the changing – between our changing body and the changing world. It has to be a direct acquaintance between that part of us that does not change – our soul – and a reality which is also always one and the same. So if we know of this particular horse that it is indeed a horse, and of that horse over there that it is also a horse, and that it is in fact the same horse we saw just a little while ago in a different part of the field, there must be a real, ideal, exact and specific *Horse*, with which our soul is acquainted, and which it can then recognize in both these horses and in any one of their different and changing brethren we encounter is this world. There must be a realm of reality that is pure, eternal and always the same with itself, with which the soul can acquaint itself directly, a realm in which "the use of the pure intelligence [assures] the attainment of pure truth" (*Republic*, Book VII). Our world, the material world, can only be a reflection – an imperfect image or representation – of this realm of true reality. Only the pure forms – *eidos* – that populate this realm can be objects of true knowledge – *Epistêmê*. The most pure, general and exact of these forms are the mathematical ones and knowledge of them – mathematical knowledge – is the paradigm of episteme. The way we fumble about in the material world, our changing and unreliable familiarity with the ever-changing and unreliable objects which we commonly mistake for knowledge, is mere *doxa* – unwarranted opinion, vague approximation, grasping at shifting shadows.

Shadows and light are indeed the metaphors through which Plato explored his epistemology, or his ideas about knowledge. He presented these ideas in dialogues, in which his mentor Socrates is featured as the main interlocutor, and in his dialogue *The Republic* he has Socrates call upon his readers to imagine prisoners in a long, dark cave. The prisoners are chained to their seats since birth, their backs to the narrow entrance. Behind them is a great fire, in front of which objects are held and moved above a wall, as in a puppet show behind their backs. They are only able to look forward, at the cave's wall, where the shadows of these objects fall. For the prisoners, then, shadows are all there is. If one of them were released and allowed to turn around, he would not, at first, see anything else or any better; he'd be blinded by the light. As his eyes gradually adapt and he climbs out of the cave to see *real* objects, he will find returning even harder. Blinded again, this time by the darkness upon return, he would hardly be able to convey to his mates what he had seen. Worse – his new knowledge would be of little use, indeed an interruption for those who still dwell in the shadows. "Strange prisoners," Plato has the interlocutor remark; "no stranger than us," replies Socrates.

The brazen may say that only 'strange' is appropriate for Plato's ideas. Why should we think of ourselves as sitting in the dark, watching only shadows? The idea that the world we know so well is but a reflection of some remote and unreachable Real is not particularly appealing. The material things surrounding us give us our notion of hard facts (both literally and metaphorically). Horses, tables and mountains are recalcitrant; they resist our touch and are oblivious to our wishes, and this seems to be the very foundation of what we think of as real and true. And why should we accept this stark distinction between idealized 'episteme' and flawed 'doxa'? It's not clear what's to be gained from thinking that our acquaintance with our world is a mere approximation and Truth can be attained only by having our mind depart from it. Why should we even be interested in the strange ideas of a person who lived almost 2,500 years ago, in a culture so different from ours?

Estrangement

We should, because these are still our ideas. We also think that the table in front of us is hard only *to us*. That in *true* reality, it is actually almost a vacuum, comprising particles with no size, moving at unimaginable velocity. We think that this reality is beyond our daily experience and available only to a select few; that it can only be captured by mathematics because it is, in effect, a purely mathematical reality. Figure 2.1 provides a nice illustration of this approach. It's a diagram of the discovery – one of the most important empirical achievements in decades – of the so-called 'Higgs boson'; a particle named after the mathematical physicist who predicted its existence. It is perhaps not surprising that the prediction was based on a mathematical analysis of an abstract, mathematical theory. What *is* particularly telling is that the *empirical* discovery, made at the great CERN particle accelerator, is also mathematical: it is registered as a statistical difference on a graph plotted by a computer. Like Plato, we think it naïve to expect an elementary

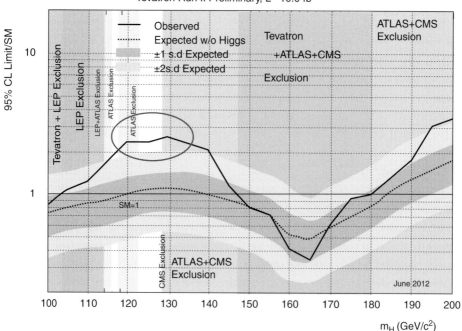

Figure 2.1 A diagram representing the discovery of the 'Higgs boson' at the CERN particle accelerator. The particles and the relations between them can only be perceived and described mathematically: in terms of frequencies, rates of decay, etc. Even the discovery – an emblematic empirical achievement – is fundamentally mathematical: it involves computing a very large amount of data to detect statistically significant differences between levels that the theory predicts in the case of the presence of the Higgs boson and in the case of its absence.

'particle' to have the same material existence of a sand particle: to have a distinct location, to be hard and impenetrable, to be accessible to the senses. Like Plato, we believe that the infrastructure of our world, the realm where elementary particles reside, is mathematical.

Plato's answer to the question of knowledge *is* strange. It is very *particular*. Both his worries and his solutions make sense only against the backdrop of a particular Greek culture, at a particular time in its history, reflected on by a member of a particular class. Yet it is still ours. Not because we are still troubled by the ancient Greeks' dilemmas or can truly sympathize with their solutions. Even less so because their questions or answers have some universal power by which they transcended their origins – like cathedrals, questions and answers do not travel by themselves; they have to be actively exported and imported. Rather, Plato's vision happens to be also ours because we are – albeit distant, indirect and well-modified – *products* of his line of thinking. We still believe in a version of his vision, because, for different reasons and in different and complex ways, people of remote times and places (remote from Plato as much as from us) adapted these ideas for their own purposes, producing, finally and contingently, us.

It is as crucial to realize that being *ours* does not make the ideas any less strange. This is one of the very important benefits of learning history: it allows what the great playwright Berthold Brecht called *estrangement* – the ability to look at ourselves from the outside, as if we were strangers. From the estranged perspective we can see that beliefs we take for granted are not self-evident at all: they have surprising origins and depend on surprising assumptions we didn't know we held. We can see that *we*, too, are completely particular. We are contingent products of historical processes that could have evolved in completely different ways; we are the effect of such processes, not their reason or their end. Understanding this about science, about what we know to be *true*, is all the more revealing.

Plato formulated the combination of ideas that we discussed in Chapter 1 in their later pagan and Christian versions: the abstract is superior to and 'more real' than the material; truth resides in an independent abstract realm, outside our world; and true knowledge is like light – flowing uninterrupted and forcing itself on us irresistibly. He also provided a strong clue as to his own resources, in claiming that at the apex of this pyramid of existence, most abstract and general, eternal and unchanging, are the pure mathematical forms, and hence that mathematical knowledge is our best example of certain knowledge; of true episteme. This reverence of mathematics was the teaching of a peculiar philosophical-religious sect, to which Plato – though he never explicitly said so – apparently belonged: the Pythagoreans.

The Pythagoreans and their Mathematical Reality

It's unclear whether Pythagoras was a real person; he was supposedly born and died on the island of Samos in the sixth century BCE, and led an itinerant life. But the story that the Pythagoreans told about the events that moved him to establish their religion nicely capture their fundamental beliefs. Curiously, their tale about the greatness of pure mathematical knowledge begins with a strictly empirical inquiry: Pythagoras was walking by a smithy, so the story goes, when he was struck by the beauty and harmony of the sounds coming from the metal being hammered inside. He decided to investigate the issue, so he took home some pieces of metal, and then other materials, and examined how harmonious sounds could be produced from them. The sounds were not pleasant (or unpleasant) in themselves, of course, but only in relation to others: two or more different sounds would sound beautiful together if they were in harmony, or in *consonance*. His great discovery was the one that would still enchant the cathedral builders millennia later: the sounds produced were harmonious, he discovered, if the length of the pieces producing them related to one another in the simple ratios between the first four digits – 1:2 produced what we call the eighth (*do* and *do*; *re* and *re*, etc.); 2:3 produced the fifth (*do* and *sol*; *sol* and *re*); 3:4 produced the fourth (*do* and *fa*; *fa* and *ti minor*). Most important: these harmonies were completely independent of the material used or the way the sound was produced: metal, wood or hair; beating, blowing or plucking; the same mathematical ratio always created the same harmonic consonance (the story and the ideas are beautifully represented in Figure 2.2).

For the Pythagoreans, this meant that mathematics exists prior to and independently of matter; that the beauty of Pythagoras' consonances spoke of matter 'partaking' in these ratios. Mathematics was not merely a means to measure or calculate parts of the material world – it was the true essence of all existence. More particularly, it meant that the world was structured by the whole numbers (1, 2, 3 and so on), and the ratios between them (1:2, 1:3, 2:3, 2:4, etc.). The simplicity was not accidental: it told of perfection. The consonances and their beauty captured this deep truth, a truth of metaphysical and religious significance. We don't know how the Pythagoreans worshipped and what their rituals were – they kept them secret – but we do know quite a lot about what they established in their mathematical inquiries, which in themselves clearly were of religious significance.

What particularly enchanted the Pythagoreans was the orderly structure and surprising symmetries between the whole numbers. Plato made one such structure famous in his *Timaeus*. This dialogue, very popular among scholars of the age of the cathedral, tells a Pythagorean, mathematical-magical story of creation, in which a workman-like god, whom Plato calls the *demiurge*,

Figure 2.2 The fable of Pythagoras' discovery of the mathematical laws of musical consonances: hearing the harmonies from the blacksmith (upper left), Pythagoras experiments and establishes that the musical consonances are completely determined by mathematical proportions, whether realized in the metal of the bell, the water in the glass, the thread of the string or the wood of the pipe. The illustration is from Franchino Gafori's *Theorica Musice* (Milan: de Lomatio, 1492).

constructs the world from numbers and figures. Plato presents a series of numbers, which came to be called 'the lambda,' after the triangular-shaped Greek letter (Figure 2.3). When one arranges the squares of two in sequence on the left-hand side of the lambda and the squares of three on the right, the musical consonances (1:2; 2:3; 3:4) appear in order, and the numbers on the right become interesting sums of the precious ones in the series:

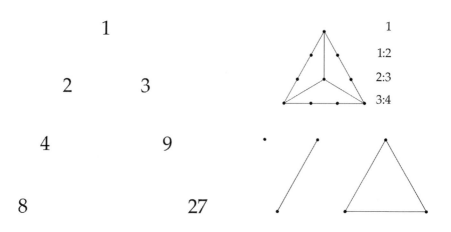

Figure 2.3 Cosmological significance assigned to mathematical regularities: in the *Lambda* on the left, the arrangement of the squares of two on the left and the squares of three on the right captures the musical consonances. In the *Tetractys* on the right the set of ten points represent the dimensions – from none (point) to three (pyramid), and the relations between them also conjure the consonances, another testimony to the musical harmony of the cosmos.

3=2+1; 9=4+2+3; and 27 is a sum of all the former numbers. The fact that these regularities seemed to emerge simply from the geometrical arrangement of the numbers captivated the Pythagoreans, and they moved to classify the numbers accordingly, as triangular, square and pentagonal numbers (Figure 2.4). These classes (which can continue indefinitely) had, again, surprising and exciting relations between them. For example: *every square number is the sum of two successive previous triangular numbers* (known as the Pythagoreans' first theorem); and *every square number is the sum of all the previous odd numbers* (third theorem). The truly thrilling point about these relations, however, was not that they existed, but that they could be *proven*.

The Concept of Proof

The proof is the greatest and most unique contribution of the Pythagoreans to the concepts and practices of knowledge that would come to make science as *we* know it. Not the particular proofs (which are easy to convey by drawing – see Figure 2.5), but the very concept of proof. We all learn proofs in high school and some of us get quite good at constructing them, but rarely do we stop to think what they are. A proof is a type of argument: a series of propositions, some assumed and the others following some laws of inference. What distinguishes *a proof* from all other forms of argument is that in a proof the inference is such that if the assumptions are true then the conclusion is *always and everywhere* true. So whether the dots in Figure

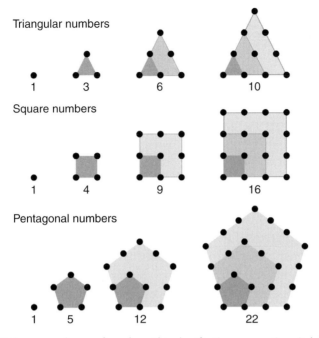

Figure 2.4 Pythagorean classes of numbers. This classification can continue indefinitely: hexagonal numbers; heptagonal; octagonal; and so forth, and the classes have surprising relations between them (see Figure 2.5).

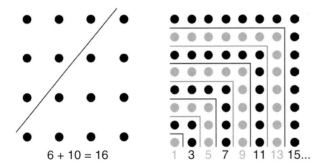

Figure 2.5 On the left – the proof of the first theorem: *every square number is the sum of two successive previous triangular numbers.* On the right – the proof of the third theorem: *every square number is the sum of all the previous odd numbers.* The power of the proofs is in their undeniable universality: the figures can be drawn in any means or style, and can be extended indefinitely, but will still convey the same relations.

2.5, for example, represent apples, people or galaxies – the proof guarantees that the relations claimed in the theorem will hold universally and eternally. Once we have proven that the sum of the angles of any triangle equals two right angles, it matters not at all that no triangle we may draw – indeed that no physical triangle ever – will be accurate enough to embody this truth. It is our drawing which is wrong, never the proven proposition.

Here again is this strange idea of perfect, certain, universal and eternal knowledge, perhaps at its most extreme. What does 'always and every-where' even mean? Obviously we can never check if a claim is 'always' true, so it seems spurious to worry about it. And it seems spurious to care about triangles which we can never draw. If we know, say, that the area of the square built on the hypotenuse of a right triangle is equal to the combined areas of the squares built on the other two sides (the famous theorem named after Pythagoras), what else do we gain by knowing that it will *always* be so? We don't seem to add anything to our knowledge by 'proving' this relation – this proof being the very achievement by which we remember Pythagoras' name. And what does it mean that Pythagoras' theorem is true even though no triangle in the world is *really* right-angled? The ancient Egyptians, for example, knew this theorem very well – we can see it embedded in their great engineering feats and it appears on their papyri – but it never occurred to them to *prove* it.

From any practical point of view, proofs are redundant. During Pythagoras and Plato's time (the sixth to fourth centuries BCE) – and well before and after – the Egyptians and the Babylonians could boast achievements of both knowing-how and knowing-that significantly more powerful than the Greeks' in medicine, metallurgy and what we might now call chemistry. Their engineering feats, from irrigation projects to pyramids, alongside their sur-veying, astronomy and astrology, are particularly interesting for us in the discussion of proofs. This is because they demonstrate *mathematical* capac-ities which were the envy and the model of the Greeks for centuries – all without proofs. The Egyptian Rhind Papyrus (Figure 2.6), for example, from the sixteenth century BCE (a millennium before Pythagoras!), contains a slew of arithmetical and geometrical problems and their solutions – but no proofs.

The uncompromising commitment to the absolute power of the proof had difficult implications for the Pythagoreans. It turned out that one can prove their fundamental assumption – that all quantities are either whole numbers or the ratios between them – false. Here is that proof, summarized in the textbox. The mathematically averse can skip, but the effort is worth-while – the magical way by which the unexpected conclusion enforces itself can illuminate the allure that proofs had on the Pythagoreans – and still have on us:

Let's assume a square whose sides are 1. (1) Since it's a square, ABC is a right triangle, so the square of its hypotenuse AC – the diagonal of the square – equals the sum of the squares of AB and BC. If $AC^2=1^2+1^2$, then $AC^2=2$ and $AC=\sqrt{2}$. (2) Let's assume that this number – the square root of 2 – is a ratio between two numbers, call them m and n: $\sqrt{2}=m/n$, and (3) reduce them until they have no common factor: $m \neq an$. (4)

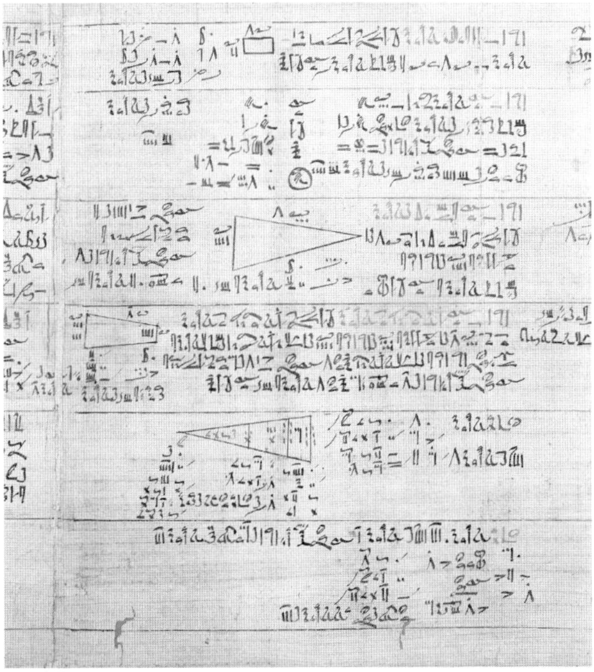

Figure 2.6 The Rhind Papyrus: an Egyptian mathematical textbook dating from c. 1550 BCE. Rather than proofs, this textbook comprises specific problems and general directions for how to solve them. The first part contains arithmetical problems – for example, how to divide a given number of loaves of bread among 10 men – as well as tables converting complex fractions to unit fractions, for example: 2/101 = 1/101+1/202+1/303+1/606. The image here is from the second part, which contains problems of geometry, like finding the volumes of cylinders and the slopes of pyramids. The last part comprises a collection of various exercises, such as multiplication of fractions. For Ahmose, the scribe who signed the papyrus, mathematics was an earthly, practical discipline.

Square the ratio: $2=m^2/n^2$, and you get (5), that the square of the nominator is double the square of the denominator: $m^2=2n^2$. This means that m^2 is even – it has 2 as a factor, hence (6) m is also even; the square root of an even is necessarily even. This is because an odd number multiplied by itself can't create an even number, since no multiplication by 2 occurred. If m is even, then it is double some other number – call it q (m=2q), so (7) the square of m is four times the square of q ($m^2=4q^2$). But the square of m is twice the square of n ($m^2=2n^2$), which means (8) that twice the square of n equals four times the square of q ($2n^2=4q^2$). Or, (9) the square of n equals twice the square of q ($n^2=2q^2$). This means that the square of n is even, so (10) n is also even. But we've shown that m has to be even, which means that *if* there is a ratio between m and n, *then* there is a common factor between them: 2. Or, to think of it slightly differently: if √2 is a rational number, then m and n share q as a factor. But from both perspectives this goes against what was assumed at the start: that **m** and **n** share *no* common factor. So √2≠m/n; the square root of 2 cannot be a ratio between two whole numbers, **m/n**. It is, literally, *irrational* – without ratio.

   This is a proof by way of negation: the assumptions (one assumption was Pythagoras' theorem and the other that √2 is rational) turned out to contradict each other, and the conclusion is that they *cannot* both be true. Because Pythagoras' theorem had been proven, then √2=m/n had to be false. Because it was *a proof*, this conclusion was necessary: the assumption that all numbers are rational *had* to be false. The Pythagoreans had to abide by it, and they indeed abandoned the investigation of numbers to concentrate on geometry. According to the third-century Neo-Platonist Jamblichus,

ABCD is a square whose sides equal 1.

(1)   **Diagonal AC=√2 (by Pythagoras' theorem)**

(2)   **Assume AC is a ratio between two whole numbers m and n: √2=m/n.**

(3)   **Reduce m and n so they have no common factor: m≠an.**

$$1^2 + 1^2 = 2$$

(4)   **Since √2=m/n, then 2=m²/n²;**

(5)   **hence m²=2n².**

(6)   **So m² is an even number, hence m is also even: m=2q (because a root of an even number is necessarily even).**

(7)   **Hence m²=4q².**

(8)   **According to (5) m²=2n², so: 2n²=4q²,**

(9)   **hence n²=2q².**

(10)   **So n is also even, and m and n share a common factor: 2.**

But this contradicts assumption (3), from which follows that (2) can't be right, or: √2≠m/n. √2 isn't a ratio between two whole numbers; it is 'irrational'.

the Pythagoreans acted swiftly – by killing the person who constructed the proof. Macabre humor aside, the fable delivers a powerful insight: that the idea of knowledge which is limitless and certain is not only both strange and compelling – it's also costly. It makes difficult, religious-like demands. The concept of *Real* knowledge, *episteme*, which we inherited from Plato the Pythagorean – abstract, eternal knowledge, detached from the changing world – this concept only makes sense against the background of a very peculiar understanding of the world and of us, humans, within it. This understanding was rooted in the very particular culture and politics which were home to Plato – those of the Greek *polis*.

Plato in Athens

Plato's Athens, like the other Greek towns on the Greek peninsula and the neighboring islands, as well as on the coasts of Asia Minor, Sicily and North Africa, was an independent city-state. It called itself a *democracy*, but its *demos* – 'the people,' the citizens – was very limited; perhaps 5 to 10 percent of its population. It comprised the heads of households: men, of property, old enough to have inherited their fathers', and wealthy enough to own slaves. Manual labor – menial as well as skilled – was performed by these slaves. Their lot was not that of the African slaves forced into the horrors of American cotton fields and cocoa plantations – they were protected by law and custom, could own property and were allowed into most facets of life. Yet they were *bound*. They belonged to a household, and within it they had to stay. They did possess knowledge – among them were carpenters and shoemakers, foremen and even teachers and physicians – but for their masters, this knowledge could never be more than *techne* – organized, tradition-supported practical *know-how*, carried and encoded in the body. It was knowledge that bound a man to his place, his tools, his body. Women and children, even of the higher classes, were also bound, if somewhat differently: to the transformations of their bodies, which they could not control.

For this reason – so the Greek citizens explained to themselves – *politics*, the public realm of the polis, was reserved for them: they, and only they, were *free*. They were not bound by the immediate, the bodily and the material. Their freedom allowed them to rule their passions by their reason; to manage their household with prudence, to weigh all things by good measure. Public matters were decided by these free men through public arguments, because free men could aspire to the type of knowledge required to run the polis: to consider general principles; to infer the future on the basis of the past; to reach decisions on the basis of abstract arguments. They could transcend the here and now; they, and only they, could aspire to *episteme*.

It was, quite probably, not the way *all* Greeks explained their world to themselves; it is only what we have learned from those Greeks who left written records, and those would have belonged to the class of the free. As with the medieval masons, it is hard to tell what Greek slaves and women thought (though not always impossible – we have, for example, poetry written by educated Greek women, such as the great Sappho of Lesbos); whether they contemplated their human condition and whether, in doing so, presented it to themselves differently than their masters did. For these masters of the *demos*, however, this was the cultural-political context in which *proof* carried such force: an abstract, pure argument, whose power stems from its structure and is therefore free of the bounds of the here and now. And this is the context in which Plato's idea of knowledge made sense; in which truth could be envisioned as residing in a realm of its own, outside our world. A reader who finds it hard to believe that such abstract considerations as Plato's philosophy can stem from such practical roots may consider Plato's biography.

He was born circa 427 BCE to an aristocratic family in Athens. His father hailed from the early kings of Athens and his mother was related to Athens' legendary lawmaker Solon, and when his father died in his childhood, his mother married an associate of Athens' leader Pericles. Plato was in military service from 409 to 404 BCE, participating in the last rounds of the Peloponnesian War (431 to 404 BCE), which pitted Athens and its allies against Sparta and the other Peloponnesian poleis. After the war he turned to the political career expected of a man of his pedigree, joining the oligarchy of the Thirty Tyrants that came to rule Athens; but deterred by their violence, he left. In 403 BCE, Athens restored its democracy and Plato tried to return to politics, but failed. This was followed by perhaps the most significant encounter in the history of philosophy: Plato met Socrates and became his disciple. He documents Socrates' style of thought in his earlier dialogues: an unrelenting questioning, which attempts to take nothing for granted and leaves the interlocutor aware of the weak and contradictory assumptions underlying his (always *his*) deepest beliefs.

This relentless questioning of Athenian moral convictions cost Socrates his life. He was condemned to death in 399 BCE on charges of treason, and Plato, who was present at Socrates' execution (by poisoning), left Athens temporarily and traveled to Italy, Sicily and Egypt. When he returned, in 387 BCE, he founded a school, the famous *Academia*, which we will discuss below together with the other Greek philosophical schools. Twenty years later, at the height of his fame, Plato left again, this time for Sicily, on an invitation by Dionysius the tyrant of Syracuse, to instruct him on how to rule according to his (Plato's) philosophical principles. The political-philo-sophical experiment failed, with Plato barely escaping alive, but he wasn't deterred enough not to try the experiment twice again, with the next two Sicilian tyrants – Dionysius II and Dion (Dionysius' son and brother) – with the same results. He finally retired to teach and write at the Academia, pass-ing away at a ripe old age in 348 or 347 BCE.

What Plato's biography powerfully shows is that, even if his philosophy urged abstract detachment from matter, it was by no means itself detached from immediate material-political concerns. Plato, like his mentor Socrates, his student-rival Aristotle (we'll discuss Aristotle's times at the Academia below) and almost any other influential thinker, was a member of a cultural-political elite, who made sense of their world in ways that explained and justified their status and privileges. But this is *not* to say that their philoso-phy, or any philosophy, was insincere or 'just politics.' This was their genuine understanding of that world, and what we have seen earlier is how, through adoption of and adaptation for different uses in later times, this understand-ing has been made fundamental to *our* understanding of *our* world – even though it is so very different from theirs. So to understand this philosophy, we do need to look at their social and political culture from the outside, as it were, but also to attempt to understand their worries and follow their solu-tions in *their* terms. For the Greek culture from which we inherited our con-cept of abstract, mathematical knowledge, the greatest mystery was the one we already encountered with that late representative of this culture, Plotinus: the mystery of being and becoming. How can order reside in variety? How can unity allow diversity? How can one become many?

Parmenides' Problem and Its Import

Parmenides' Challenge

The mystery could be presented from different perspectives: there are so many horses in the world; what makes them all one thing – *a horse*? And if there is one thing – *the horse* – why are horses so very different from one another? Similarly, all things move in very many directions – is there one place to which they all belong? And if there is such a place – why do they need to move at all? The answers carried not only metaphysical and epistemological, but also ethical, political and, as we saw with Plotinus, religious implications. The problem of the one and the many, of unity and diversity, was so fundamental to Greek thought that one could sort the different Greek philosophical schools according to the way they tried to accommodate these challenges.

The most influential formulation of this quandary was provided by Parmenides, who lived in the Greek colony Elea in south-west Italy about a century before Plato (Plato honored him by making him the main interlocutor of one of his dialogs). Parmenides related it in a philosophical poem of which only fragments are left, as is usually the case with Greek thinkers before Plato. The poem is written as if dictated by a goddess, who begins with this declamation:

> Come and I shall tell you what paths of inquiry alone there are for thinking: The one: that *it is* and *it is impossible for it not to be*. This is the path of Conviction, for it leads to Truth. The other: that *it is not* and *it necessarily must not be*. That ... is a path wholly unthinkable, for neither could you know what-is-not (for that is impossible), nor could you point it out.
>
> *Parmenides, Fragments, David Gallup* (ed. and trans.) (University of Toronto Press, 1984), p. 55 (emphasis added)

The words of Parmenides' goddess seem odd only at first glimpse. On a second look they convey a very solid philosophical insight: that there cannot be any middle ground between existence and non-existence. What *is*, what exists, simply exists; and what does not exist is nothing at all. Indeed, even saying 'what does not exist' does not refer to anything, and is therefore an empty string of words, "wholly unthinkable"; we can only know, perceive, or discuss what exists. Nothing can 'almost exist' or be 'on the way to, or out of, existence,' so what *is* cannot have come into existence – it could not have been nothing and then become something, because when it was nothing, it simply wasn't. So what *is* necessarily is and "it is impossible for it not to be"; it has always been, and will never cease to be. Furthermore, what *is* can never change, because change always involves something that didn't exist coming into existence and vice versa and this, again, is "wholly unthinkable." What exists is *one and the same with itself*: if there is more than one existing thing – two or more separate entities – what exists between them? Emptiness? Non-existence? But non-existence "is not and it necessarily must not be." And for a similar reason, *what is* cannot move: in order to move there has to be some empty space into which it can move. But empty space is just nothingness and that, again, necessarily "must not be." If we follow the "path [that] leads to Truth," we realize that existence is one, eternal, unhanging – there is no reality in variety, in birth and demise, in change.

One may ask if it makes any sense at all to doubt the existence of change or motion – aren't these the most obvious, the most general of all facts? Can we even think without change and motion? Parmenides' disciples agreed that we were all accustomed to think this way, but still, they argued, these very fundamental assumptions are but illusions. The most well-known of them, Zeno (also from Elea), produced a series of arguments – the famous 'Zeno's paradoxes' – to demonstrate just that. Imagine, he suggested, a mundane episode of motion: two things moving in the same direction at different speeds. Let's say it's a race between Achilles and a tortoise. Achilles runs ten times faster than the tortoise, and starts 10 meters behind it. You may think that Achilles would overtake the tortoise in no time – he is so much faster. It will indeed take him only a step or two to cover the 10 meters – but by the time he has covered those initial 10 meters, the tortoise would have progressed a meter. Achilles will of course quickly cover that meter, but by then the tortoise will be 10 centimeters ahead – and so on. The distance between the two will keep diminishing, but never disappear – Achilles will never manage to catch up to the tortoise, so he will never overtake him. Or, suggested Zeno, think of another common experience of motion: an arrow shot from a bow towards a target. To reach its target the arrow will have to cover half of the distance, and then half of the second half, then half of that quarter, and so on – the list of halves will continue to infinity, so the arrow will never have enough time to cover the distance and make it to the target. In fact, to get to that halfway mark the arrow will need to cover half of that half, and before that half of that, and so on – the arrow, in fact, will never leave the bow. Motion is an illusion.

The Atomists

How should one understand Zeno's paradoxes? After all, we *do* see fast runners overtake slow ones and arrows hitting targets – what is the meaning of this discrepancy between what we experience and what reason tells us? The Parmenideans' answer was that what we are told by our senses, our experience of this world, is an illusion. We should derive consolation from knowing that all our pains and sufferings – as well as our hopes and passions – are such illusions, and as far as knowledge is concerned, we should strictly follow reason. Plato, we saw, concurred – we can understand his philosophy as a sophisticated development of the Parmenidean insight. In his version, this world, the world of variety, motion and change, is but a cave, and our experiences are but shadows. For him, the only things that *truly exist* are the pure forms, and they follow Parmenides' laws for what "*it is* and *it is impossible for it not to be*": the forms are eternal, unchanging and only known by reason.

But not all Greek thinkers were as impressed by Parmenides' crystal-sharp reasoning. A different answer to his challenge came from Leucippus (c. 480–460 BCE) and Democritus (c. 460–370 BCE), who lived about a generation before Plato, and Epicurus (c. 341–271 BCE), who was born a few years before Plato's death. For them, the founders of the Atomist school of thought, explaining the world as we experience it was more crucial than following the edicts of pure reason. They couldn't deny or refute Parmenides' insights and Zeno's arguments – these made an enormous impression on Greek thinkers – but they compromised with them. Motion and variety, in their thinking, were just undeniable truths about the world. Hence, if motion requires empty space for moving things to move into, then emptiness had to exist, even if reason did not approve. Moreover, if we cannot understand how from the oneness of existence comes the variety of existing things, we simply have to assume that many things exist, separated by empty space. But what the title 'Atomists' actually shows is that they desperately tried to maintain as much as they could of Parmenides' teachings. The many things that *existed*, according to them, satisfied all the conditions that the goddess in Parmenides' poem set for what *is*: they were uniform, eternal and unchangeable. Each one was an atom: *a-tom*. indivisible, *in-dividuum* in Latin.

Some 400 years after the first Atomists and in a very different context, Lucretius (99–55 BCE), writing in Latin in the waning days of the Roman Republic (not much more is known about him), provided a sophisticated elaboration of ancient Atomism in a philosophical poem *On the Nature of Things* (*De rerum natura*). Here is an excerpt from Book Two, translated by William Ellery Leonard:

> Now come: I will untangle for thy steps
> Now by what motions the begetting bodies
> Of the world-stuff beget the varied world,
> And then forever resolve it when begot,
> And by what force they are constrained to this,
> And what the speed appointed unto them
> Wherewith to travel down the vast inane:
>
> ...
>
> For, when, in their incessancy so oft
> They meet and clash, it comes to pass amain
> They leap asunder, face to face: not strange –
> Being most hard, and solid in their weights,
> And naught opposing motion, from behind.
> And that more clearly thou perceive how all
> These mites of matter are darted round about,
> Recall to mind how nowhere in the sum
> Of all exists a bottom, – nowhere is
> A realm of rest for primal bodies; since
> (As amply shown and proved by reason sure)

Space has no bound nor measure, and extends
Unmetered forth in all directions round.

...

And of this fact (as I record it here)
An image, a type goes on before our eyes
Present each moment; for behold whenever
The sun's light and the rays, let in, pour down
Across dark halls of houses: thou wilt see
The many mites in many a manner mixed
Amid a void in the very light of the rays,
And battling on, as in eternal strife,
And in battalions contending without halt,
In meetings, partings, harried up and down.
From this thou mayest conjecture of what sort
The ceaseless tossing of primordial seeds
Amid the mightier void – at least so far
As small affair can for a vaster serve,
And by example put thee on the spoor
Of knowledge. For this reason too 'tis fit
Thou turn thy mind the more unto these bodies
Which here are witnessed tumbling in the light:
Namely, because such tumblings are a sign
That motions also of the primal stuff
Secret and viewless lurk beneath, behind.
For thou wilt mark here many a speck, impelled
By viewless blows, to change its little course,
And beaten backwards to return again,
Hither and thither in all directions round.
Lo, all their shifting movement is of old,
From the primeval atoms; for the same
Primordial seeds of things first move of self,
And then those bodies built of unions small
And nearest, as it were, unto the powers
Of the primeval atoms, are stirred up
By impulse of those atoms' unseen blows,
And these thereafter goad the next in size;
Thus motion ascends from the primevals on,
And stage by stage emerges to our sense,
Until those objects also move which we
Can mark in sunbeams, though it not appears
What blows do urge them.

http://classics.mit.edu/Carus/nature_things.html

Inane: emptiness, void

Amain: forcefully, violently

Asunder: apart, away

Spoor: track, trail

Lucretius may sound very ancient to some – 'how strange it is to argue scientific views in a poem!' – and very modern to others – 'isn't he talking about these particles the way we do?' But if we remember the lessons of the cathedral, we can instead look at what troubled Lucretius himself, in *his* time and place, and how he attempted to resolve his dilemmas; what he may have thought *his* audience would find difficult and how he thought he could sway them.

Lucretius confronted Parmenides' challenge by having order emerge from chaos and by disengaging uniformity from singularity. There is not only one, single *is* in his world, but very many particles – the 'begetting stuff.' The particles are identical – this is where the uniformity of existence lies – yet they move chaotically, and from their accidental collisions, bit by bit, the varied but orderly world of our experience comes to be.

Lucretius' challenge was twofold. First, he needed to convince his readers of this novel and strange story – the image of minute particles in chaotic motion creating the world of large, orderly things we know. Secondly, he needed to make them come to terms with the difficult idea that what they are familiar with is an outcome of something they can hardly understand. In particular, they needed to accept that what they can see (and hear and feel and taste) is an expression of things they cannot. To convince them that both ideas are not mere fantasy, he offers an analogy from a common experience: a ray of light enters into a darkened room, and little particles of dust are seen bouncing around. The readers can now visualize the motions Lucretius has concocted – the hardly visible (motes of dust) represent the invisible (atoms). With the analogy established, he offers a bolder move; the visible motion is not only a *representation*, but an *expression* of the invisible one. The chaotic motion of the minuscule dust particles becomes an argument for the motion of the even more minute atoms: how can the former be explained if not by the latter? Lucretius turns the dust motes into the top step of a ladder leading down, across the threshold of visibility and on to the tiny atoms, through their aggregates and aggregates-of-aggregates; so small they are.[1]

---

[1] I owe the first part of this analysis to my late, venerable teacher Amos Funkenstein, may he rest in peace, and the last part to my former tutor and PhD student Kiran Krishna.

Greek Philosophies and Parmenides' Challenge

The reason Parmenides' challenge is so crucial to our story is that it is still at the core of every scientific inquiry, whether it is about the motion of sub-atomic particles, the creation of inorganic crystals, the evolution of new species or the behavior of financial markets: how to understand variety in terms of unity? How to understand change in terms of stability and continuity? How to understand apparent disorder in terms of fundamental order? Plato's answer was to find unity in ideal forms in a different realm, outside this world; and to take variety and change as 'failures,' reflecting the deficiencies of the material imitations of the abstract ideal. The Atomists' answer was to find unity in the uniformity of matter, and to claim that order emerges spontaneously from chaos.

Lucretius' beautiful intellectual maneuvers are so agreeable to our ear that historians often take the Atomists' answers to these questions as more 'modern,' 'empirical' or 'scientific' than those of their contemporaries. This is a mistake. The Atomists were no less rooted in the culture of their times than any of the other philosophical 'schools' of antiquity (which we'll discuss in Chapter 4). Their philosophy did not aim at what we may think of as scientific knowledge, but was intended to provide metaphysical foundations for the *good life* worthy of a free man. What makes the Atomists' answer to Parmenides' challenge particularly interesting is rather the thoughtful *compromise* by which they handled the mystery of order and change. Compromises like that are particularly telling to the historian of science: they reveal what appeared to the historical agents as crucial; what they were willing to dismiss and what they wanted to maintain at all costs. One can see in Lucretius' poem that the Atomists *were* impressed by Parmenides' arguments. But they also had a strong conviction that experience should be explained, not denied, and therefore could accept *neither* the Parmenidean conclusions from these arguments, nor the Platonists'. They approached this discrepancy by adopting Parmenides' insights concerning existence, while simply ignoring the arguments against variety and motion. They could not explain how empty space, *a no-thing*, can still exist, how it can be *a thing*, but since motion requires this existence, they just accepted it. They could not explain how *the many* can arise from *the one*, so they just accepted that the many – which they interpreted as the many elementary particles of matter – exist. And they could not explain how existence changes, so they just assumed the incessant motion of these particles. Otherwise, they tried to maintain the uniformity, eternity and unchanging nature of existence in their atoms.

Atomist philosophy became an important source for the so-called *Scientific Revolution* of the seventeenth century, and Platonism, in its many guises, had an important following through the ages. But the most influential answer to the dilemmas introduced by Parmenides was different from both and was provided by Plato's prize student, Aristotle, for whom all philosophical positions had to pass the scrutiny of open-eyed observation and common-sense logical analysis.

Aristotle and the Science of Common Sense

Aristotle's Life and Times

Aristotle's philosophy in general and his philosophy of nature in particular were in fact so influential that the useful and respectable *Stanford Encyclopedia of Philosophy* (ironically called 'Plato': https://plato.stanford.edu) has, as of mid-2020, fifteen entries dealing directly with aspects of it, and over 800 (!) in which it is featured. His philosophy, as we shall see, has become the foundation of the university curriculum from its very inception and remained so into the eighteenth century. Muslim intellectuals and astronomers brought it to India and China, where it was integrated into indigenous thought. Most of what we know about Greek philosophy before Socrates is through Aristotle's summaries, and in some fields – such as rhetoric and poetics – his writings are still the starting point of any curriculum. In other areas, such as logic, it is an essential ingredient. There is no other thinker, apart from perhaps the founders of the great religions, whose concepts and modes of thinking and argumentation have held sway over such expanses of time and space, but just as we stressed and shall continue to stress, this does not mean his thought was not deeply rooted in the very specific culture in which it was conceived.

Like his mentor Plato, Aristotle's life was shaped by the tumultuous politics of the Aegean Peninsula of the fourth century BCE. He was born in 384 BCE in Stagira, northern Greece, to a physician father, and was almost certainly also educated as a physician. The tradition was that medical skills were kept secret and handed down from father to son, and biological and practical interests are very clear in his writings. When both his parents died, he was brought up by an uncle at the Court of Macedonia, and the medical education was replaced by a more liberal teaching, worthy of a nobleman: Greek, rhetoric and poetry. In 367 BCE, aged 17, he became a student at Plato's Academia – which by then had been operating in Athens for some twenty years – during Plato's first political visit to Syracuse. One has to assume they met upon Plato's return and many times in the decades to follow, but there is no record of these meetings.

The politics of the Academia, Athens and the Hellenic Realm would play a major role in Aristotle's life. He soon became a teacher at the Academia and was to remain there for twenty years, towards the end of which his political standing became tenuous. His Macedonian patron, King Philip, was moving to conquer Greece, and when his hometown of Stagira fell in 347 BCE at about the same time as Plato's death, Aristotle had to forsake his ambition of becoming Plato's successor as the head of Academia and leave Athens. He spent the next twenty years as a court philosopher, first in Assos (today Behram in Turkey) and then back in Macedonia, where he became the center of a group of philosophers with serious empirical interests in anatomy, zoology and biology. According to the great first-century chronicler Plutarch, Philip hired him as a tutor to his son, young Alexander, soon to be known as 'the Great.' It's at least as likely that Philip's interest was the political influence he could exert through Aristotle on the Academia, because in 340 BCE, when Philip began preparing to take Athens by force, Aristotle was forced to move again, returning to his home in Stagira, taking with him his intellectual circle. When Alexander finally subdued Athens in 336 BCE, the Academia was allowed to carry on with its work, but a year later Aristotle was sent to Athens to found a rival school – the Lyceum – much broader in its interests and more empirically oriented than the Academia. It is assumed that most of what we know as Aristotle's works are his students' notes from his teachings there, which were conducted while walking around – hence the nickname *peripatetics* for the Aristotelians throughout history (from the Greek περπατάειν *peripatein* – to walk up and down). After Alexander's death in 323 BCE, Aristotle had to leave yet again, retiring to a family estate in Chalcis, where he died the following year at the age of 62.

Aristotle vs. his Predecessors

Aristotle's philosophy is as earthly and involved as his life was, and so was his approach to Parmenides' challenge. Like the Atomists, he was not impressed by any denial of common experience: the philosopher's challenge is to clarify reality, not reject it; to explain how it works, not argue that it's just an illusion. But neither was Aristotle persuaded by the Atomists' account of the underlying reality, and especially by their obsession with motion in empty space. For the various Greek philosophical schools, 'motion' was synonymous with change of all kinds – water on the stove, for example, moved from cold to hot. But locomotion – motion from one place to another – Aristotle argued, is much too limited a concept of change with which to account for all the complexity of nature. Even more generally, he could not accept that order could emerge from disorder, or unity out of variety: it is absurd, he argued, to think that plants and animals, houses and cities, could somehow come about from the chaotic motion of uniform particles, as the Atomists would have us believe.

Nor was Aristotle impressed with the solutions provided by his great master, Plato. The concept of pure forms introduced, in his view, more mysteries than it solved. It was completely unclear how the ideal form relates to all its material embodiments – how *The Horse*, allegedly existing eternally and unchanging in another realm, relates to all the horses in the world. Furthermore, argued Aristotle, Plato cannot explain how the forms relate to one another: is there an ideal form of four-legged animals, a kind of super-form existing independently of the forms of horses and sheep? And then a super-super-form of an animal relating these forms to the forms of fishes and fowl? There seemed to be no end to it – Plato was taking similarities between real, worldly objects and treating them as if they were objects themselves.[2]

Aristotle's fundamental approach to the gap between reason and experience that Parmenides' and Zeno's arguments introduced was to submit philosophical argumentation to common sense. Truth had to be found in the real objects or our world, rather than in some ideal realm, and knowledge had to stand up to the scrutiny of empirical evidence and the views of all people, or at least most, or at least the wise. So Aristotle had no patience for Plato's enthusiasm for mathematics, in which he found an unhealthy and unhelpful fascination with abstract idealization. For him, *logic* was the means to knowledge, because it captured real relations between real objects (the main logical tool was the syllogism, an example of which is in Chapter 8). And in real objects, form and matter can be *distinguished*, but not *separated*: we can consider the form of a horse – being a horse rather than a mouse for example – independently of its flesh and blood. But this doesn't mean that this 'horsiness' exists someplace separate from the horse.

Plato and Aristotle represent fundamentally competing conceptions of knowledge, and the difference between them is reflected, in a nutshell, in what they have to say about *truth*. For Plato, truth is a relation to the eternal and transcendental: "the soul is like the eye," he says in his *Republic*, "when resting upon that on which truth and being shine, the soul perceives and understands and is radiant with intelligence; but when turned towards the twilight of becoming and perishing, then she has opinion only, and goes blinking about, and is first of one opinion and then of

---

[2] 'Reifying' is the philosophical term.

another, and seems to have no intelligence" (http://classics.mit.edu/Plato/republic.7.vi.html). Aristotle has little respect for such reveries: "To say that what is is not, or that what is not is, is false," he puts it in stubborn simplicity in his *Metaphysics*, "but to say that what is is, and what is not is not, is true" (www.perseus.tufts.edu/hopper/text?doc=Perseus%3Atext%3A1999.01.0052%3Abook%3D4%3Asection%3D1011b).

Aristotle's Alternative

Aristotle's own answer to Parmenides' challenge is shaped by this unrelenting demand to steer away from over-speculation; "to say that what is – is." Nature *is* change, *is* the coming to be and passing away of things around us. So for us to understand nature as we know it, the *is* and the *none*, what exists and what does not exist, cannot be kept so dramatically apart. There must be something between them – and that is Aristotle's great innovation: what almost-exists, or potentially-exists. The oak (his favorite example) does not fully exist until the acorn had sprouted, and the trunk grew, and the branches are covered with leaves. But the oak also, in a sense, already exists in the acorn – it exists *in potentia*. Given the appropriate conditions, the acorn *will* develop into an oak. It is not necessary that the acorn *actually* fully grows into an oak – a cow might eat it or a fire might destroy it. But it *is* necessary that it will never become a monkey or a maple tree. *By its own powers* (and given the appropriate conditions) the acorn will become an oak and nothing but an oak – this is what it means for the *actual* acorn to be a *potential* oak.

For Aristotle, there is no dichotomy between order and change. Potential can only be realized through change, and the realization of potential – the acorn's turning into an oak – is the paradigm of how nature changes in an orderly way. Nature is *orderly change*, and the task of the natural philosopher is to decipher this order.

How then are order and change balanced? What changes and what maintains orderliness and uniformity? *Matter* changes. Obviously, the matter of the oak is not that of the acorn – there is much more of it, and it's different in hardness, color and so on. What guarantees the orderliness of this change? *Form* does. It is the form of the acorn in which the potentiality of the oak lies. Matter is passive, uniform and changeable. It carries all the potentialities, but has no distinguishing properties of its own. Form is active and orderly: it bears the properties of the individual substance and guides its change.

If this sounds like antique and arcane jargon, consider that we still think of ourselves in a very similar way: most of the cells of our bodies have a life span much shorter than ours (between a few days and a few years – neurons and cardiomyocyte heart cells are the long-living exception), so it's fair to say that every seven to ten years almost all the matter of our body has changed. But we don't think of ourselves as being someone else or something other than who and what we were a few years ago. We think: '*I am changing.*' It is the order of this change, the development, the continuity that is 'I.'

We may think of this underlying unity as our *self,* or our *soul.* For Plato, we saw, this strict distinction between our changing body and our unified, consistent soul meant that the soul was independent of the body – it just resided in it, joined when it was born and departed when it perished. Christianity was happy to adopt this idea that the soul is immortal because it allowed it to solve the mystery of the injustice of life on Earth: souls are getting properly rewarded or punished after the death of their body, according to the way they navigated it during its life. For Aristotle, this was a fundamental mistake: that we can distinguish between matter and form – or body and soul – does not mean that they can be separated. Were the acorn to be eaten or burnt, both its matter and its form would be extinguished, even if we can understand very well the difference between them. When we die, both body and soul cease to exist. There is no realm in which forms dwell independently of matter, and matter has no existence if not shaped and determined by form.

Aristotle's World

Unlike Plato, then, Aristotle allows for only one world – the one we dwell in. And unlike the Atomists' world – which we to a large degree adopted – Aristotle's does not comprise moving bits of uniform matter. Its fundamental components are *substances*: definite individuals, like a person, a horse or an acorn. These individuals are inseparable compounds of matter and form, but their *essence* – what it is that they are: person, horse or acorn – is determined by their form. There are no substances without matter, but without form – *amorphous* – matter is not any particular thing; it's not a substance. Natural substances, like a person, horse, acorn or stone, have their own *essential nature*. There are also other objects in the world, like cups or desks, which are not natural but artificial – produced by art, their form is a reflection of the artisan's soul.

The nature of a substance is the sum of its properties and is determined primarily by its form and secondarily by the matter comprising it. This essential nature disposes every object to certain kinds of behavior – its 'natural' behavior. Thus fire naturally communicates warmth, rocks naturally fall, acorns naturally develop into oak trees and babies naturally grow into persons. If in *our* ('modern') world objects are ruled externally, by laws that pertain to all objects everywhere, in Aristotle's they are ruled from within, by their own natures. To know this world, therefore, is not only to know its overall regularities – which are studied mostly by abstract logic. It is also, and even more so, to know the particularities of substances, their essential natures and the behavior that follows from these natures, and how it differs from that of other substances.

Aristotle's world, then, is populated by substances, not 'things.' Its most fundamental elements are not simply material, but a combination of fundamental qualities (Figure 2.7). These fundamental qualities are ordered according to basic oppositions: warm vs. cold; wet vs. dry. Warmth and coldness are active: heat diffuses and separates; coldness draws things together. Wetness and dryness are passive: the wet receives impressions; the dry preserves them. The material elements of the world (of which all material things are composed) are combinations of these qualities: cold and dry is earth; cold and wet – water; warm and wet – air; and warm and dry – fire. Change the qualities, and the elements transform from one to another: heat up water, and it becomes air (such is vapor); cool water down and it becomes earth (ice); heat earth up and it becomes fire. Aristotle seems

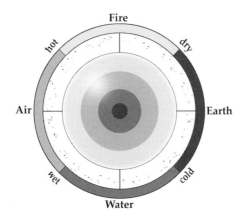

Figure 2.7 Aristotle's elements and the properties and oppositions comprising them. Note that although the origins of this analysis may have been common wisdom, the conclusions are not. For example: perhaps because he reasons that heating water produces air, Aristotle sets air to be wet and hot (like vapors), which isn't the way we usually experience wind and its effects.

to have simply picked the elements from common Greek lore (in Chapter 6 you'll find a poetic version of their transformations from the Platonic tradition), but one can see how they stand for the fundamental aspects of our daily experience: solidity and fluidity; heat and mobility. Experience also tells us that left to their own powers – following their own nature – earth and water will move downwards, and air and fire upwards. Earth will sink in water and fire will ascend through air, so we can infer what their natural places are: earth – at the center of the world; water – surrounding it; next – air; and at the periphery, furthest from the center, fire.

The Cosmos

The elements, it must be stressed, are not barren, uniform matter, like the atoms of Epicurus and Lucretius. They are expressions of primary principles and have distinguishing properties and active forms. So Aristotle's world is also not a uniform, empty space full of uniform, moving matter. It is an ordered whole, where every substance has its own proper place: a *cosmos* (Figure 1.12).

Since it has a clear center and a clear periphery, Aristotle's cosmos is necessarily limited in size – there are no places in infinite space. In fact, it was expected to be rather compact (more on that later). But it is not limited in time: it has neither beginning nor end. To the cosmos as a whole Aristotle applies Parmenides' insight about existence: nothing can come into existence out of non-existence, nor vice versa. There is no sense in asking 'what was before (or will be after) the cosmos,' nor 'what is outside it.' The cosmos is all there is.

And within Aristotle's cosmos things do change – they move. Motion is not simply what happens – it defines happening: time just *is* the measure of motion, and space comprises the places to and from which things are moving. Still, motion is *change*, so it requires a cause. This is a fundamental metaphysical principle for Aristotle and throughout history: every change requires a cause. This cause can be of two different sorts: substances can change according to their essential nature – a change directed by their own form – or they can change because of some external cause. In the case of motion, when a substance moves towards its proper place, the cause of its motion is its own nature, so the motion is natural. A stone, for example, is a piece of earth, so its proper place is at the center of the cosmos, and falling downwards is therefore its *natural motion* (it is clear to Aristotle that on the other side of Earth stones fall towards the same center). When something forces a substance away from its proper place – like a hand throwing the stone up – this is forced or *violent motion*.

Further, the intensity of change depends on the power of its cause. In the case of motion this means that the speed of a moving body is proportional to the force moving it, whether this force is natural – like the weight driving the stone down – or violent, like the force delivered by the hand. The heavier the body, the faster the fall; the stronger the arm, the faster the flight. A change is also affected by contrary causes, so the speed of the moving body is inversely proportional to the resistance that the medium offers to its motion: we have all witnessed a falling stone slowing down as it passes from the rare air to the denser water.

This combination of common-sense metaphysics and basic empirical knowledge – a fundamental trust in human reason and the human senses – is the hallmark of Aristotle's natural philosophy and what made Aristotelianism such an attractive intellectual framework for so many thinkers for so long. But this down-to-earth approach does not limit Aristotle to trivial conclusions – quite the opposite. His considerations of resistance, for example, allow him to argue that the Atomists' assumption – namely, that all motion requires empty space – is fundamentally wrong. In empty space, explains Aristotle, there is no resistance, and since speed is inversely proportional to resistance (in modern notation we can write it like this: $V \propto F/R$), zero resistance means infinite speed ($V \propto F/0 \propto \infty$). But infinite speed is impossible: a body moving infinitely fast is in more than one place at the same time, which is absurd. This means not only that motion does *not* require empty space – as the Atomists thought – it actually *necessitates* a medium: bodies have to move through *something*, not nothing.

Behind this nifty argument hides a very fundamental assumption about motion, an assumption that had a crucial effect on the philosophy of nature for two millennia. The assumption is that motion cannot be understood on its own; it is what happens to a body when it changes its *place*. So motion makes no sense without a place from which and a place to which the body moves. But in empty space there are no places, so motion in empty space makes no sense. The cosmos is full.

The part of the cosmos we dwell in is full of the four elements, whose proper places, we saw, are ordered in enveloping spheres: earth at the center, water around it, above it air and then fire. This is what it means for earth and water to be *heavy* and air and fire to be *light*, not just in comparison and in relation to one another, but in and of themselves. Earth and water belong in the center and their natural motion is down; air and fire belong in the periphery and their natural motion is up. But rarely do we come across the elements in their proper places; the incessant circular motion of the heavens mixes them up. Thus a piece of coal is heavy, so we can infer that it's made mostly of earth; but it will catch flame when heated, so it comprises also some fire; and fire being light, this flame will rise. The realm of the elements reaches up to the Moon, but above it is a completely different realm – the heavens – operating according to fundamentally different laws.

Even though we cannot observe the heavens from close range, Aristotle's considerations remain based on this combination of empirical and philosophical common sense. It is befitting of the noble bodies of the upper realm to be at rest, to not change, but we know that they do move. We don't see them move (their motions are too slow to observe), but we do see the heavenly bodies at different places at different times. Yet we also know that they always follow the same path and return to the same place, and as far as human recollection and records go, these have always been the same paths and same places. Metaphysical and empirical considerations lead to the same conclusion: that the motion of the celestial bodies is as close to rest as any motion can be – a motion *within* place, rather than from one place to another. A body can only move without changing its place if it moves in a circle, and the circle – equal in all dimensions, hence the most perfect of shapes – is indeed proper of the celestial motions. This circular motion has to have the Earth as its center, or it would approach and retreat from us, and no longer be *within* its place. It also must be of uniform (circular) velocity, because if it slowed down it would finally come to a halt, and if it accelerated, it would end up moving at infinite speed. Time, you remember, is unbound.

The natural motions of the celestial bodies are circular, different from the linear motions of the terrestrial substances, and their proper places are different. They must, therefore, be made of a something different from the four terrestrial elements – the fifth essence, the *quintessence*, as it would be called in the Middle Ages, or the *aether*, to adopt Plato's term (which Aristotle didn't actually use). The aether, unlike the terrestrial elements, isn't composed of oppositions, so it doesn't possess their penchant for transformation: it is unchanging, and is so subtle that it offers no resistance that could slow down the celestial motions. Here is one of the places where the power mill of Aristotelian thought seems to grind to somewhat of a halt: it is not completely clear how to reconcile the insights about motion and medium with the supposed eternity of the celestial motions. Thinkers in Europe and the Middle East would spend much of their intellectual energy for many centuries throughout the Middle Ages and into early modern times on problems like this.

Natural Philosophy and the Causes of Change

The celestial realm in the Aristotelian cosmos is thus as different as can be from the terrestrial one. Whereas under the boundary marked by the Moon's orbit all bodies are made of the four elements, above it they are made of a fifth; whereas on Earth bodies move in straight lines, in the heavens they move in circles; and in general, whereas in the realm of the elements everything is coming to being and passing away, in the heavenly realm all is eternal and unchanging. This *sounds* like Plato's distinction between the world of matter and the world of pure forms, and the Christian thinkers of the age of the cathedral indeed dedicated much effort to assimilating these two pagan philosophies together with their own monotheism, reading the Christian relations between the Creator and His creation into these distinctions between the eternal and the changing. But for our interests in the origins of science, it is important to pay attention to the deep difference between Aristotle and his great master. For Aristotle, even though the rules operating above and below the Moon are so different, both realms are parts of the same cosmos, and need to be studied by the same methods. Aristotle's cosmos, with its upper and lower realms, comprises substances and essences, connected to one another in causal relations, which the philosopher should decipher: 'to say of what is that it is.'

The Aristotelian natural philosopher, from Antiquity through the Middle Ages, is therefore particularly interested in causes, because nature is change, and change, as we said, requires *cause*. The philosopher's task is to find the order in this change; to find the regular causes of motion: *why* it is that the Sun moves around the Earth, that the tiger moves towards its prey or that the acorn moves to become an oak? It is also part of his task to reveal the causes of strange phenomena, which seemingly break these regularities: why, for example, does wood, made mostly of earth, float on water?

When *we* think about a cause of one event, we think about the immediate event which preceded it: the window broke *because* it was hit by the stone; we know this because when the stone hit the window, the window changed from a flat surface to a pile of shards. Aristotle would call the stone the *efficient cause* of the breaking – "the primary source of the change or rest" – but his concept of cause is much richer than that. To fully understand why the window was broken, we of course need to understand that it is made of glass. For him, this is part of the cause – the *material cause* – "that out of which a thing comes to be" (all quotes here and below from Aristotle, *Physics*, in Jonathan Barnes (ed.), *Complete Works (Aristotle)* (Princeton University Press, 1991)).

Matter in Aristotle's philosophy, we said, is what is subject to change, and the order, direction and end of the change is determined by the *form*. So to know why the encounter between the stone and the window ended up in a pile of shards, it is not enough to know that it's made of glass. We also need to know the form, or the essence of the window. This is the *formal cause*, "what it is to be." It makes sense: if the glass was in a different form – say of fibers – it wouldn't have been broken. Finally, Aristotle requires that we understand the reason for the change – "that for the sake of which" what happened, happened. Why did the stone hit the window pane? For what purpose? This is the *final cause*, which represents a very fundamental assumption in Aristotle's natural philosophy: things in the world happen for a reason. They don't only happen *because* something else happened before, but also *in order for* something to happen later. The final cause, then, is the most revealing of the four.

Aristotle knew of neither glass windows nor fiberglass, but the fact that his reasoning can so smoothly be applied to them may elucidate the enchantment it held on scholars for so long. This is especially true regarding the four causes, or four aspects of causality. From Aristotle's point of view, his predecessors were wrong in limiting their attention to only one of the causes: the Atomists to the material; Plato to the formal. We moderns, he'd have said, are too hung up on efficient causes.

In fact, our window example would have probably sounded to Aristotle rather uninteresting exactly for that reason. We, perhaps like the Atomists, think that in order to understand nature we need to concentrate on its simplest events. Aristotle, on the other hand, insisted that the cosmos needs to be understood in all its rich complexity. This is why he mostly disapproved of mathematics within natural philosophy: mathematics requires and provides abstractions and idealizations. Aristotle was interested in causal explanations, not mere calculations, so his natural philosophy is qualitative. He would also not pick up simple collisions as examples for these four layers of a complete account of a natural phenomenon. He may invite us to look at the acorn changing into an oak: the material cause of the change is the matter of the acorn, of course; the efficient cause perhaps the water and the warmth of the soil. The formal cause is the form of the oak which the acorn embeds, and the final cause – the oak. The acorn changes *in order for* the oak to come into being. The oak is the *telos* of the acorn – its end and goal – just as the center of the Earth is the telos of the stone's natural motion: Aristotelian natural philosophy is *teleological*.

This causal, qualitative, teleological analysis is a powerful tool. Consider, as another example, the generation of a man: the matter is the mother's womb and the efficient cause is the copulation. The formal and final cause seem to be merging here: it is the adult man who is both the origin of the process and its end. Aristotle even applies this mode of analysis to artifacts – things brought into being by the human hand. The material cause of the statue, his favorite example, is bronze; the efficient cause is the art in the artisan's hands and tools; the formal cause is the shape of Artemis or Apollo; and the final – the completed statue. Again, one might say that the final cause is the most explanatory of them all: one cannot understand any of the other three without considering their *telos*: the statue to be.

Conclusion

In the seventeenth century, when the New Science was defining itself in opposition to the Aristotelianism of the medieval universities, the idea of a final cause became a focus of ridicule. We have inherited this attitude, and probably think it's strange to claim, like Aristotle, that "nature belongs to the class of causes which act for the sake of something." But how else do we answer Aristotle's question on why it is that "our front teeth [are] sharp, fitted for tearing, the molars broad and useful for grinding," if not by saying '*in order* that we can properly digest our food?' Some readers may suggest that we do know how to explain the usefulness of teeth without supposing they are made for purpose: they developed by a series of accidents, at each stage those whose teeth were better survived longer and produced more offspring. But Aristotle, who's aware that the assumption of final causes is problematic, is also familiar with this line of thought. He's simply not convinced by it:

> A difficulty presents itself: why should not nature work, not for the sake of something, nor because it is better so, but just as the sky rains, not in order to make the corn grow, but of necessity? ... Why then should it not be the same with the parts in nature, e.g. that our teeth should come up of necessity-the front teeth sharp, fitted for tearing, the molars broad and useful for grinding down the food-since they did not arise for this end, but it was merely a coincident result; and so with all other parts in which we suppose that there is purpose? ... [and then] such things survived, being organized spontaneously in a fitting way; whereas those which grew otherwise perished and continue to perish, as Empedocles says his 'man-faced ox-progeny' did.
>
> Aristotle, *Physics*, Book II, ch. 8

Empedocles lived a century earlier in Sicily and his work was an important resource for Aristotle. He credits him with the doctrine of the four elements, and one may find clear roots of Aristotle's approach to natural philosophy in the fragments we have from Empedocles' poem *On Nature*. So Aristotle gives careful attention to the argument Empedocles suggested: that legendary creatures, like the half-ox-half-man and other marvels and monstrosities of Greek mythology, do not exist because they can't survive. Empedocles was offering a way to understand how things in the world fit their use without assuming final causes. The reason all creatures are orderly is because only orderly creatures survived – only those whose teeth allow handling food properly; there's no need to assume that the cause of the way teeth *are* is the way they *should be*. We think this an excellent argument – it is the basis of the theory of evolution. Aristotle doesn't. He agrees that "Such ... arguments (and others of the kind) ... may cause difficulty on this point," but insists that "yet it is impossible that this should be the true view." He doesn't seem to have a good argument why. At the end of the day, as we said concerning his view of the Atomists, Aristotle simply can't accept that order can arise from chaos, just by chance: "For teeth and all other natural things ... normally come about in a given way; but of not one of the results of chance or spontaneity is this true." Things that come up the same way again and again don't do so by coincidence, Aristotle insists. Even if the survival argument can explain why disorderly things perish, it cannot explain why orderly things keep reproducing themselves regularly:[3]

> We do not ascribe to chance or mere coincidence the frequency of rain in winter, but frequent rain in summer we do; nor heat in the dog-days, but only if we have it in winter. If then, it is agreed that things are either the result of coincidence or for an end, and these cannot be the result of coincidence or spontaneity, it follows that they must be for an end.
>
> Aristotle, *Physics*, Book II, ch. 8

It was important for us to look a little more closely at Aristotle's writing, to get a glimpse of the tantalizing breadth, thoroughness and explanatory power that made him so influential for so long. The force of his arguments often seems to come across even – or perhaps especially – when the positions they support are no longer what we are taught by our science. We mentioned that the New Science of the seventeenth century defined itself in opposition to Aristotelian natural philosophy. But as we will see in Chapter 10, even its greatest hero – Isaac Newton – could not to shake off the stubborn common-sense and direct empiricism that gave this philosophy its power and also marked its boundaries.

---

[3] It's worth noting that Charles Darwin, from whom *we* learned the idea of the survival of the fittest, also couldn't explain how it is that the same set of teeth keeps appearing in the same species generation after generation.

Discussion Questions

1. Which ancient Greek assumptions can we still find within our ideas about knowledge? Where do we clearly differ? What's the significance of these 'archeological remains' of one culture in another?
2. Does the idea of the mathematical structure of the universe make sense? If so – what does it say about the material world around us? If not – how can one explain the inevitability and great success of mathematics in our knowledge of this world?
3. Do Parmenides' dilemma and Zeno's paradoxes seem daunting? If not, what has changed in our thought to take their edge away?
4. Can you conceive of mathematics without proofs? What would be lost? Can anything be gained?
5. Do the biographies of Plato and Aristotle make a difference in understanding their respective philosophies?

Suggested Readings

Primary Texts

Plato, *Meno*, 81–86, in: Plato, *Complete Works*, John M. Cooper and D. S. Hutchinson (eds.) (Indianapolis: Hacket, 1997), pp. 880–887 (also: http://classics.mit.edu/Plato/meno.html).

Plato, *Timaeus*, 27–40, in: Plato, *Complete Works*, John M. Cooper and D. S. Hutchinson (eds.) (Indianapolis: Hacket, 1997), pp. 1234–1244 (also: http://classics.mit.edu//Plato/timaeus.html).

Aristotle, *On the Heavens*, Book I. 8, in: Aristotle, *On the Heavens*, W. K. C. Guthrie (trans.) (Harvard University Press, 1939), pp. 68–81 (also: http://classics.mit.edu/Aristotle/heavens.1.i.html).

Lucretius, *On the Nature of Things*, Book II, On Atomic Motions (http://classics.mit.edu/Carus/nature_things.2.ii.html).

Secondary Sources

On classical Greek science:

Lloyd, G. E. R., *Early Greek Science: Thales to Aristotle* (London: Chatto & Windus, 1970).

On Greek culture and its epistemology:

Arendt, Hannah, *The Human Condition*, 2nd edn. (University of Chicago Press, 1958).

On Parmenides and his import in Greek philosophy:
Palmer, John, *Parmenides and Presocratic Philosophy* (Oxford University Press, 2009).

On Plato's life and work:
Friedlander, Paul, *Plato: An Introduction*, Hans Meyerhoff (trans.) (Princeton University Press, 2015 [1970]).

On Plato's metaphysics and cosmology in the Timaeus:
Broadie, Sarah, *Nature and Divinity in Plato's Timaeus* (Cambridge University Press, 2011).

On Aristotle's thought in biographical context:
Natali, Carlo, *Aristotle: His Life and School*. D. S. Hutchinson (ed.) (Princeton University Press, 2013).

On Aristotle's epistemology, philosophy of nature and their relations:
Falcon, Andrea, *Aristotle and the Science of Nature: Unity without Uniformity* (Cambridge University Press, 2005).

3 The Birth of Astronomy

Looking Up

Disregard, for a moment, all you have been taught and consider the following question, relying only on what you have actually observed: *when you look up at the sky, what do you see?* The answer is not as straightforward as one might expect.

You may start by answering: 'it depends.' During the day, we see the Sun. During the night, we see the stars. Most of us live in cities, so we don't see many of them, but it only takes a short ride out of town and a bright, moonless night to observe the sky as the ancients did: full of literally countless stars. The Moon is a bit of a mystery: although it usually appears during the night, we have all, on occasion, seen it by day. We probably wouldn't be assuming too much if we said that both the day and the night skies are perched above us like a dome (Figure 3.1 shows what this dome looked like above Jerusalem the day Titus burnt the Temple). It really looks this way, meeting the horizon at what seems like a segment of a great circle.

What else can we say? Most of us – not much. Some of us – sailors or navigators by hobby or by trade – can recognize the stars and the constellations (Figure 3.9). And what do the bodies in the sky – the Sun, Moon and stars – do? We all know that the Sun moves: it rises in the east and moves to the west, where it sets – that is, disappears under the horizon – and if we are in the Northern Hemisphere (where the heroes of the following story resided), this journey goes through the south.[1] The Moon obviously moves as well; though none but the true expert can tell where it will show up. With a little observation over a night one may notice that the stars also move, in the same direction as the Sun, and some of them also rise and set. But it is crucial to stress that we don't really see any of these heavenly bodies *in motion* – we infer that they move because we see them in different positions at different times.

It seems obvious (though we'll see that it isn't) that the rising and setting of the Sun is what determines day and night. Considering it a little further, we may recall that the Sun's daily motion is not completely regular. It always

---

[1] In the Southern Hemisphere, the Sun's trip through the sky has a northerly bend. Anyone who grew up in one hemisphere and has moved to the other will attest how deeply these directions are ingrained in us: even after many years, it is still unsettling to find that the direction of the Sun is the opposite of what you'd expect.

Figure 3.1 The sky dome, as it would have looked above Jerusalem on August 3, 70 CE, the day the Jewish Temple was set on fire by the Roman legions, quashing the Jewish rebellion. You will find other dates in the literature: the Jewish tradition commemorates the day as Nine of Av – the ninth month of the lunar calendar. For reasons to be clarified in this chapter and Chapter 7, synchronizing these calendars isn't trivial.

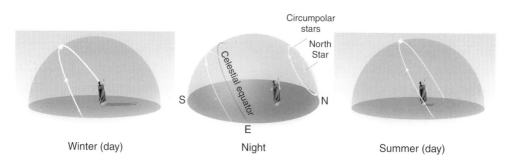

Figure 3.2 The Appearances (Northern Hemisphere). The Sun rises in the east, travels through the south and sets in the west. In the winter, it rises at a more southern point on the horizon, reaches a lower point and spends fewer hours above it; in the summer – vice versa. The stars always rise and fall in the same place. Some of them – the circumpolar stars – rotate around the North Star and never fall under the horizon.

travels in the same direction, but it doesn't always take the same time. The days get longer and shorter with the seasons, and the Sun's trajectory also changes. In the summer it's higher in the sky, and the days are longer; in the winter, the other way around. The keener observers among us may notice that both kinds of change are caused by the changing position of the rising and setting Sun. In the Northern Hemisphere, the farther up north it is on the horizon at dawn, the longer its trajectory, the higher its ascent in the sky, and the longer the day (Figure 3.2). This is inverted in the Southern Hemisphere – the farther south the Sun is on the horizon at dawn, the higher its ascent and the longer the day.

There are a couple of lessons to be learnt from this short exercise. The first is how little every one of us knows *personally*. Modern science, as a whole, produces knowledge whose depth and quantity would have been hardly imaginable to our ancestors. But this doesn't mean that each one of us knows more than an ancient sailor or a medieval peasant did. The opposite is more likely to be true. Knowing how to tell times and directions by looking at the sky was, for them, a matter of personal survival, so they had to make a point of learning it. It is exactly thanks to the great success of science that we have been relieved of this need, but at this very price: that we, personally, do *not* know. We have delegated much of our knowledge to other people – and to objects.

Another lesson is how surprisingly unyielding all these details are. Although we are submerged in these motions, they only present a very rudimentary order to our immediate experience. Nature does not speak to us. Not only do the mechanisms behind the phenomena remain hidden, the phenomena themselves are hidden behind our experiences, and the experiences are confusing and do not seem in and of themselves interesting or significant. It takes concentrated, organized observation to find enough order in these experiences to turn them into phenomena that we can attempt to systematize and understand. And it takes some basic assumptions – an image, a storyline, a hypothesis about how these endless details relate to one another – for us to recognize any pattern about which questions can be asked.

Making the Phenomena

The Two Spheres Model

Astronomy begins, then, not with innocent star gazing, but with the introduction of a basic model to guide observation. For the astronomers of ancient Mesopotamia and the Aegean region, that model was of two spheres: the image of our Earth, a sphere, nestled inside the bigger sphere of the heavens. The heavens and Earth share the same center, and the heavens – the celestial sphere – rotate around this center from east to west, carrying the Sun and the stars (Figure 3.3). When this rotation takes the Sun above the horizon, it's daytime; when below – it's night time (Figure 3.4).

This is not an obvious idea. The sky may look like a dome and the horizon does look circular, but otherwise nature does not declare its shape anymore than its motions. Egyptian cosmology describes a world that is by no means spherical and the author of *Genesis* doesn't seem committed to any particular shape for the firmament that God placed between the waters above and below. In Homer's epics and Hesiod's mythology of the eighth to seventh centuries BCE, the heaven *is* apparently a sphere – it's not completely clear,

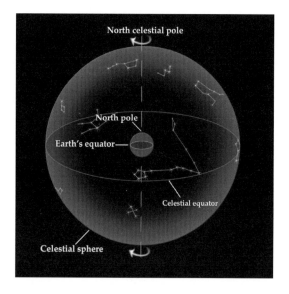

Figure 3.3 The Two Spheres model. The big sphere rotates as a whole around the smaller sphere which is Earth. The big sphere carries the Sun, the other planets and the stars in their constellations. It rotates from east to west, around an axis that extends north to south, and whose northern pole always points at one star – the North Star, or Polaris.

but Hades, the realm of the dead, is below and supposedly as deep as the Heavens above are high – yet Earth is a circular disc. Similar considerations of symmetry led some of the philosophers of the next three centuries (the so-called 'pre-Socratics') to argue for a spherical heaven, but many kept the Earth flat. Aristotle actually explains why the combination of a spherical (or hemispherical) heaven and a flat Earth is appealing yet wrong: the celestial horizon appears to be just a continuation of the terrestrial one, and because the celestial horizon is so far away, the curvature of the Earth becomes negligible. So it seems like we're on a flat surface, with a perfect hemisphere above us, he continues, but in fact the terrestrial horizon is very different from the celestial one. It's determined by the Earth's curvature and how high we are, and is rather close by (in modern terms: about 5 kilometers standing at sea level, and about 20 kilometers on a hill 100 meters high). Even those who considered Earth spherical didn't necessarily think of the Sun as completing a circular orbit around it. The idea that it might be spending the night under the Earth didn't occur to them all, and when it did, it seemed to go against common experience. Springs, for example, come from below ground, but they are colder early in the morning, so they couldn't have been warmed from below by the Sun during the night.

The Pythagorean fascination with the power of geometry and the Parmenidean arguments about the necessary oneness and perfection of being seem to have gradually outweighed these empirical considerations. Through the fifth century BCE, the two-sphere model became the favorite of Greek philosophers of

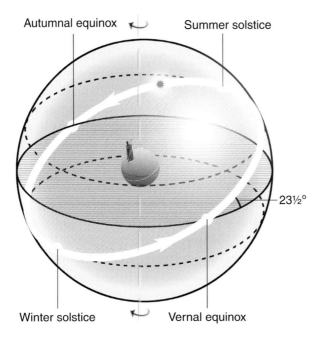

Autumnal equinox　　Summer solstice

23½°

Winter solstice　　Vernal equinox

Figure 3.4 The celestial motions according to the two spheres model. The celestial – outer – sphere rotates around the axis from east to west, represented by the clockwise arrows at the top and bottom. One such rotation is a day. For the observer on Earth – the inner sphere – it's daytime when the Sun, carried by the celestial sphere, appears above *their* horizon, represented by the thin bright line. The other stars also rotate with the celestial sphere, although their appearance above the horizon is not as dramatic and conspicuous as the Sun's. The Sun also travels on the ecliptic, represented by the bright band on the celestial sphere, from west to east, represented by the anti-clockwise arrows. It completes a full trip from one point on the ecliptic to the same point in a little more than 365 rotations of the celestial sphere around Earth. One such trip is a year. The length of daytime depends on the place of the observer, which determines their horizon, and the place of the Sun on the ecliptic. For example, when the Sun is in the summer solstice, it travels during one full day along the top dotted line, and spends more time above the horizon than below it. When in the winter solstice, it travels along the bottom dotted line and spends more time under the horizon. When the Sun is at one of the equinoxes, it travels with the celestial equator, which is a projection of the terrestrial equator on the celestial sphere, so it spends as much time above the horizon as under it. The other planets, not represented in this diagram, also travel on the ecliptic, though at slightly different angles, which is why the ecliptic is a band rather than a line. Each planet has its own pace and its own year.

nature, and both Plato and Aristotle provided elaborated versions of it. Plato, as one might expect, supports the notion that the Earth is spherical with abstract metaphysical arguments. Aristotle, as is *his* wont, adds to his cosmological analysis two arguments from observation as to why it should be spherical.

The first is that "the horizon always changes with a change in our position, which proves that the Earth is convex and spherical" (Aristotle, Meteorology, II.7, http://classics.mit.edu/Aristotle/meteorology.2.ii.html). Observed from different places, the same stars set (and rise) at different times and rise to different heights in the sky. This is most evident with the Sun: the same day is longer or shorter the farther north or south one travels. Were we residing on a flat surface, we would all have one horizon; the stars would rise and set at the same time and climb to the same angle above the horizon. This means that we have a dome not only above our heads, but also under our feet. Aristotle's other argument is that "in eclipses, the outline [of the shadow on the Moon] is always curved: and, since it is the interposition of the earth that makes the eclipse, the form of this line will be caused by the form of the earth's surface, which is therefore spherical" (Aristotle, On the Heavens, II.14, http://classics.mit.edu/Aristotle/heavens.2.ii.html). Aristotle unmistakably has the two-sphere image clearly ingrained in his mind: the world is a sphere; the Sun is the source of light and is also lighting the Moon; an eclipse is the shadow of the Earth on the Moon – so since its shadow is circular, Earth is a sphere.

Aristotle's considerations demonstrate that even in this rudimentary form the two-sphere model already assumes at least two crucial insights about the cosmos, neither of them trivial: that the Sun and the stars move together as one rotating spherical whole; and that the Sun produces daytime, the light of day being nothing but sunlight. These insights are far from obvious. Consider, by way of comparison, the monotheistic tradition: according to *Genesis*, you may recall, God created light, and designated it 'day,' on the first day of creation. He only created the sky[2] on the third day, and only on the fourth day did he create the heavenly bodies, anointing the Sun to rule the day and the Moon and stars – the night. All heavenly bodies were created "to give light upon the Earth," and the Sun's light isn't unique – it isn't the cause of the day. For the writer of Genesis, day and night are different entities, different creations – not astronomical phenomena.

It is with the two-sphere model that the confusingly changing positions of the bodies we see in the sky become *the heavenly motions* that concern the astronomer. The model itself tells us nothing about *the causes* of these motions. The Babylonians, for example, left no visual or verbal account of this model. They gave no evidence that they were interested in heavenly causes or expected their astronomy – their mathematical theory of heavenly motions – to fit their cosmology – their general understanding of the making of the cosmos. Their tables of planetary positions (about the planets momentarily), an exemplary achievement of observational astronomy to be discussed later, are patently free of theory. But even in these strictly empirical tables the two-sphere model is embedded. One can hardly understand them without the categories introduced by the model (and which we will discuss below): ecliptic, equinoxes, solstices, oppositions, and so on.

---

[2] The Hebrew word is רָקִיעַ (Rakee'a), which connotes a thinly beaten sheet of metal.

How It Works

With the two-sphere model, the appearances become astronomical phenomena. Let's reiterate and expand (Figure 3.4).

The celestial sphere – the heavens – rolls from east to west at a regular pace. The axis of this rotation goes through the Earth (the yellow dotted line), from the North Pole to the South Pole, and the largest plane perpendicular to this axis cuts Earth at its equator and the celestial sphere at the celestial equator (the red circles). The axis points always in the same direction.[3] In the Northern Hemisphere it points at a particular star: the North Star or Polaris, which is not a very bright star, but usually visible, so very useful for navigation. In the Southern Hemisphere, it points (almost) at Sigma Octantis, which is not nearly as visible, and hence not as useful. One full rotation of the heavens around Earth we call a day. [4]

The Sun and the stars rotate around Earth with the celestial sphere. Because the axis of this circular rotation is not above our heads (unless we're stationed at one of the poles), the Sun spends some of its orbit under the horizon – night – and some over it – day (the yellow circles). The length of the day depends on the angle between our horizon and the axis of rotation; that is, it depends on our latitude – our position on the Earth relative to the equator and the poles. The same is true for the stars: carried by the celestial sphere, they rise above the horizon in the east, travel in the same direction as the Sun (through the south in the Northern Hemisphere and through the north in the Southern Hemisphere) and disappear under the horizon towards the west. Stars which are close to the North Star (or to Sigma Octantis in the Southern Hemisphere) – whose angular distance from the celestial pole is smaller than the pole's angular distance from the equator – rotate in the same direction, but always remain above the horizon. These are the Circumpolar stars (on the right in Figure 3.2). Conversely, we can assume that there are stars we never see, because they always remain under the horizon.

This is what happens daily – the diurnal motion. What about the seasons? And the shorter and longer days? Here the two-sphere model allows an even more interesting understanding.

The length of the day changes because the Sun changes where it rises and sets. We can think of this change as the Sun shuffling back and forth within a small segment of the horizon – moving north in the Northern Hemisphere as the summer comes and south with the winter (and the inverse in the Southern Hemisphere) – making its path get longer and shorter accordingly, as in

---

[3] The axis actually moves – we'll touch on this below – but very slowly, so from the point of view of initial observation we can treat it as stationary.

[4] In English there is an ambiguity between 'day' as the time of light and 'day' as the time from one sunrise to the next. In this chapter, I'll try to avoid the confusion by calling one complete rotation of the heavens an 'astronomical day.'

Figure 3.2. But if we think of the whole celestial sphere as rotating around us from east to west with the Sun on it, we can think of the Sun, instead of oscillating, as moving regularly, in a constant direction, *on* the rotating heavens. In other words: the Sun, according to the two-sphere model, has two motions: one with the celestial sphere (east to west), and one on it (west to east).

In a little more detail: the two-sphere model explains the changes in the length of the day by having the Sun travel east along a big circle that is at an angle (which turns out to be 23½°) to the celestial equator (Figure 3.4). The Sun's journey *through* the heavens is much slower than the Sun's journey *with* the heavens. This means that every time the daily rotation of the celestial sphere carries the Sun above the horizon, the Sun is at a different position on the sphere. So every day the Sun rises at a slightly different point on the horizon and travels at a slightly different trajectory. It takes the Sun a little more than 365 astronomical days – 365 revolutions of the heavens around Earth – to complete one circling of the heavens. This is what we call 'a year', or in more pedantic astronomical terms – 'a solar year.'

The Sun's annual path in the heavens – the circle it draws on the celestial sphere – is the *ecliptic*. When the Sun is at one of the two points where the ecliptic crosses the celestial equator, it spends as much time above the horizon as under it, making the day and night of equal length; Figure 3.4 shows why. This is the vernal (spring) or autumnal *equinox*. When the Sun is at the northernmost point of this path – the furthest from the equinoxes to the north – then in the Northern Hemisphere it spends the longest time above the horizon and the day is the longest. This is the summer solstice. The southernmost point is the *winter solstice* with the shortest day. When the Sun is under the horizon in the Northern Hemisphere, it's above the horizon in the Southern Hemisphere, so the solstices are exactly reversed. Our horizon, we have noted, changes according to our latitude – how far north or south we are from the equator – and the two-sphere model now explains why. So how far on the horizon the Sun will get from solstice to solstice, how high in the sky it will get on each day, and how long the day will be all depend on our latitude. On the equator, the angle between the point on the horizon in which the Sun will rise on the solstice and the point it will rise on the equinox is the angle between the celestial equator and the ecliptic – those 23½°. On the equator the Sun's daily path is always perpendicular to the horizon. The two-sphere model explains that this is because the equator is perpendicular to the axis on which the celestial sphere rotates, so the day is always the same length as the night. On the poles, the Sun rises once a year, on the vernal equinox. It spirals up to 23½° above the horizon then spirals back down, disappearing under the horizon on the autumnal equinox, not to reappear until the next vernal equinox.

As fundamental and pre-theoretical as the model is, it already allows crucial insights and discoveries. Here is a particularly important one: almost all the stars are fixed – they remain in the same place in the heavens relative to one another, rising and falling at the same spot on the horizon. But it turns out that the Sun is not alone in its journey through the great celestial sphere. With it are a few other heavenly bodies, which also change their position among the fixed stars, rising and setting at different points along the horizon. These have come to be called *planets*; 'wanderers,' πλανήτης, in Greek. Unlike the fixed stars, these wanderers don't twinkle, which, according to Aristotle, is a sign that they are closer. More importantly, although each planet travels at its own pace (each takes a different period to complete a whole revolution, meaning that each has its own year), all planets share a path and direction with the Sun, travelling along a narrow band within a few degrees of the ecliptic. When divided into parts which are given astrological significance, this band becomes the *Zodiac* (Figure 3.5), which we'll discuss in some detail in the chapter on magic (Chapter 6).

Figure 3.5 The Zodiac. A segment of a mosaic tiling the prayer hall of a sixth-century (Byzantine period) synagogue near Beit Alpha, Israel, with Hebrew names for the signs. In a way that nicely illustrates the different facets of knowledge discussed in previous chapters, the mosaic is inscribed with two dedications near the entrance. One, in Aramaic, attends to the political, cultural and financial side: it dates the mosaic to the reign of Emperor Justin, and states that it was paid for by donations from members of the community. The other, in Greek, attends to the know-how. It reads: "May the craftsmen who carried out this work, Marianos and his son Hanina, be held in remembrance."

Making Time

The Astronomer's Role

As we realized earlier with our stargazing exercise, these concepts are not easy to come by. Even in this very rudimentary form, knowledge of the heavenly motions already demands acute and concentrated observation; a model that requires mastering; and considerable leisure. Why should anyone invest this labor and expertise? What can be gained from studying the sky? What's the use of astronomy?

Astronomers, one might say, handle time. They provide calendars. From the point of view of astronomy, calendars are predictions of regularly occurring events – times and places where the main heavenly bodies should be expected. The rise and fall of the Sun is the most prominent of these events, but every other planetary motion can also serve as a calendar. And each presents challenges to this demand for regularity, especially the Moon, whose behavior may seem completely erratic. From a wider cultural point of view, calendars determine days of significance – Holy Days – and related considerations, such as the direction of prayers or times of rituals. Some occurrences, although they happen fairly regularly, are rarer and thus carry special significance: the occasions when the Sun spends the most or least time above the horizon (summer or winter solstice), or the occasions when the day and night are equal (equinoxes, as we've learned). Some occasions are so rare that they seem extraordinary: the appearance in the heavens of an unknown object (comet) or the occlusion of the Sun or the Moon (eclipses). Aristotle even speaks, or tells us of observing the eclipse of Mars by the Moon – the rarer the occasion, the more competence the astronomer predicting it can boast. The more spectacular the rare event is, the more significant it is. What the exact significance is; what counts as beneficent or maleficent; what kind of influence the different heavenly arrangements carry for the individual or the community – this of course changes from place to place; from culture to culture. The Greeks titled the discipline that assigns and deciphers this significance *astrologia*, and in this sense it is not unreasonable to say that astronomy is the handmaiden of astrology.

It turns out that the answer to 'who needs astronomy' is also not trivial. One might think that everyone would find timekeeping useful: the time of day to milk the cows or feed the horses and the time of year to plant or harvest. But this practical time is not what the astronomer provides. The peasant can't rely on clocks and calendars; she listens to her cows and checks the readiness of the soil or the ripeness of the fruits. Hers is natural time: local, contingent, changing from year to year. The motions of the heavenly bodies, which make the abstract and idealized astronomical time, are of no use to her.[5]

There are many reasons to think of astronomy as the paradigmatic science, and this is one of them: that astronomy, and science in general, is impractical. It requires significant resources. As we saw, astronomy does not develop naturally from curiously gazing at the sky. It requires *expertise*, and expertise is expensive: the expense of supporting the livelihood of people who devote their time to developing it instead of producing food and shelter for themselves and their families. A society where every person has to provide for their own subsistence cannot afford astronomers. And such a society also has no use for astronomy. Astronomy is of use only to a culture organized enough to apply the knowledge astronomers produce: a culture with priests who consult calendars concerning rituals and kings who consult astrologers concerning war. So here is the point in a nutshell: astronomy, and by extension science, is only of use to a culture organized enough to make use of astrology.

Useful or not, the *Time* provided by astronomers and recorded in their calendars is the record of regular, orderly motion of the heavenly bodies.[6] Yet these bodies don't seem to move all that regularly. The heavenly sphere as a whole is orderly enough in its rotation: the length of the astronomical day – from sunrise to sunrise (or from sunset to sunset, as in the Jewish tradition) – appears to be always the same. But the length of the day and the night – the time the Sun spends above versus under the horizon – changes from day to day, as discussed above. What we haven't considered yet is that the pace of the change is not regular. This is a fact we can easily recognize: the days shorten very quickly, then stabilize so to speak, then lengthen quickly and so forth. From an astronomical standpoint, this means that the motion of the Sun on the ecliptic is sometimes faster and sometimes slower. The motion of the other planets along the ecliptic is even less orderly; not only do they change their velocity: they sometimes seem to change their direction. This phenomenon, called 'retrograde motion' (Figure 3.6), has a very important role in the history of astronomy, and we will return to it below.

Positions and Regularities

Astronomers produce orderly time from their observations of the motions of the planets, so it is important to explain here what 'observations' and 'motions' are for the ancient and pre-modern astronomer. As we noted at the beginning of this chapter, we cannot see the heavenly bodies move. Even the daily motion of

---

[5] This would finally change when Europeans took to the oceans for trade and conquest, but this is not until the late sixteenth century at the earliest.

[6] Whether these motions measure time or time is nothing *but* these motions is a very interesting philosophical question, but out of our scope here.

the whole heavenly sphere is too slow to really see as it's happening, although the changes of the place of the Sun or the stars during the day are obvious enough that we can think of them as having just moved from the place we saw them a little while ago to the place they are now. The motion of any of the planets on the ecliptic – even that of mercury, the swiftest of the lot – is much slower still. All we can directly observe is the planet's position against the backdrop of the fixed stars. The change in that position from day to day is what we call 'motion', and what the traditional astronomers termed 'anomaly.'

This term illustrates an important point: that the very idea that the heavenly bodies move is already a bold theoretical assumption. Moreover: the idea that they are bodies is just as bold. Until the advent of telescopes in the seventeenth century, only the Sun and the Moon looked somewhat like bodies – to the naked eye, the rest of the planets and obviously all the stars are just points of light; as we learned in the exercise we started with.

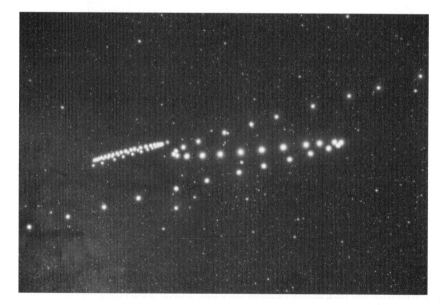

Figure 3.6 Retrograde motion. This is a sequence of photos from 31 January 2015 to 11 September 2016, showing Mars moving in an S-shape trajectory from upper right to lower left, and Saturn in a flattened compact loop behind it, against the background of Libra and Scorpio. The Milky Way is on the left and Antares is the bright star just under Mars in its fourth last position. This image is composed of numerous exposures spanning nearly a year, taken with a modern camera and lens. Such technology was obviously not available to the naked-eye-observing astronomers of yore. For them, distinguishing the planets from the backdrop of fixed stars and plotting the nightly changes of position as a motion over a long period demanded keen dedication and a powerful set of skills.

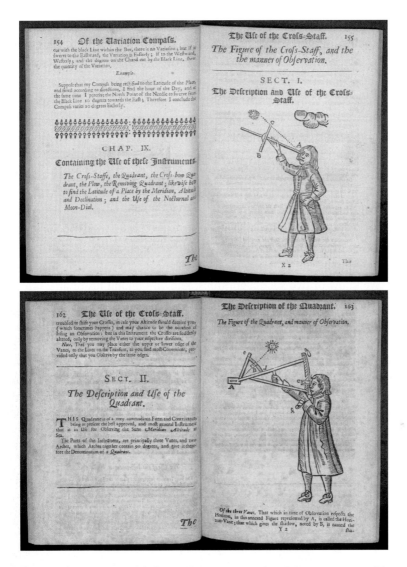

Figure 3.7 Naked-eye astronomical observation instruments. These instruments are used for measuring angles: between two stars; between a planet and a star; between a planet or a star and the horizon. The woodcuts are from John Seller's *Practical Navigation*, originally published in 1669 and in many editions since (including modern ones). The gentleman on the left is holding an advanced version of a very basic angle-measuring instrument: Cross Staff, also known as Jacob's Staff. The observer puts the end of the staff to their eye and moves one of the perpendicular rods so its ends 'touch' two celestial objects – for example, a known fixed star and a planet. The angle between one end of the rod, the eye-end of the staff and the other end of the rod is the empirical datum recorded. The gentleman on the right holds the most sophisticated instrument of that lineage – the sextant. The sextant is in a sense no longer a pure naked-eye instrument, because it comprises a lens and a mirror, but it is still used in a comparable way to capture an angle.

This is the subject matter of pre-modern astronomy: the relative positions of these points of light. These relative positions are angles: between two stars; between a planet and a known fixed star; between a star and the celestial equator (declination); between a star or a planet and the already calculated equinox (right ascension); and so forth. All observational instruments measure these angles, and as you can see in Figure 3.7, which is taken from a seventeenth-century manual, these instruments didn't change much – in a sense, until today. All 'motions' and 'velocities' are *angular* motions and velocities, or an angle covered in a given time: the astronomer's task is to find regularities in all these ever-changing velocities and directions.

How can this be done? The Babylonians, the keenest astronomical observers of antiquity, approached the challenge as a practical one: they tabulated their observations numerically (Figure 3.8) and developed algorithms that allowed them to predict the future positions of the planets as if they were moving regularly. We know of two such algorithms, known (among historians; otherwise the names mean sadly little) as 'system A' and 'system B.' Each, in a slightly different way, breaks the motion of the planet into a series of units, which approximate the motion as if its angular velocity changes in a uniform way. For a simplified example: system A is a kind of a 'step function,' according to which the Sun moves at a velocity of 30° in one mean lunar month and at a velocity of 27°7'30" in the next and so forth. In System B (a 'zig-zag function')

Figure 3.8 A seventh-century BCE Babylonian table of lunar longitudes, on a clay tablet from the library of Ashurbanipal in Nineveh (now near Mosul in Northern Iraq). The table records the daily change in the duration of the Moon's visibility during the month of the winter solstice.

the changes in the velocities are approximated as continuous. This provides good predictions of both the relations between the Sun's and the Moon's positions on the ecliptic and of the length of the solar year. What it does not give, or presume to give, is any explanation of why this is so. In a sense, it can be almost read without any physical image attached to it at all, let alone any theory: no motion or time, just changing relative positions.

For the Greeks, this type of approximation was not enough.

Saving the Phenomena

The Greeks were keenly aware of the quality of the Babylonian observations and calculations. Aristotle talks about "the Egyptians and Babylonians, whose observations have been kept for very many years past, and from whom much of our evidence about particular stars is derived" (Aristotle, *On the Heavens*, II.12, http://classics.mit.edu/Aristotle/heavens.2.ii.html). More than 400 years later the great Ptolemy (whom we'll discuss at length below) was still making use of what was by then 800-year-old Babylonian data, apparently still unsurpassable in accuracy and detail. But as we saw in Chapter 2, Greek classical philosophy and the culture of the ruling elite that it represented were hardly impressed by technical, practical solutions. Real knowledge, knowledge worthy of a free man – *episteme* – was, for the Greeks, knowledge of the general and the abstract, the simple and ideal, that lies behind the mundane and changing. The Greek astronomer could not fulfill his duties simply using efficient arithmetical methods predicting the positions of the Sun, the Moon and the planets. He was expected to provide a *theoria*: a geometrical model that would present these positions as the result of both *real* and *orderly* motions. He was required to make them intelligible. This was the meaning of the famous edict ascribed to Plato (in the sixth century, by Simplicius, one of the last great pagan philosophers and chroniclers): "save the phenomena."

Here is another very particular Hellenistic concept, which tellingly is still part of common philosophical parlance. The astronomical phenomena needed 'saving,' because they seemed irregular, yet couldn't be so. Nothing real was disorderly. The *real* had to be *orderly*, especially when it came to the heavenly bodies, the most noble inhabitants of the cosmos. The Babylonian tables were orderly, but they didn't show – or didn't presume to show – that their order was the real order. For the Greeks, imposed human order was mere *techne*, and of no interest to the philosopher. The task the philosopher set to the astronomer was to demonstrate that the apparent irregularity of the heavenly bodies is only apparent – that the heavenly bodies actually move in an ordered way, and it is only *to us* – as *phenomena* – that they appear irregular. The Babylonian tables

Figure 3.9 The constellations as they could have been observed at the same time and place as Figure 3.1. Similarly to Figure 3.1 and Figure 3.6, the patterns captured here by modern technology required superb skills and imagination from the ancient astronomer.

didn't discharge this task. Only a geometrical *theoria* could save the phenomena, because only geometry represented an order noble enough to capture the real orderliness of the heavenly motions. Even Aristotle, for all his usual suspicion of mathematics, admitted that geometry was the only proper instrument for astronomy. Some things in the cosmos, like light and musical consonants, were mathematical in their very nature and therefore needed to be studied mathematically. The heavenly bodies were of this sort.

If 'save the phenomena' survived as a philosophical concept, 'theoria' is still embedded in our concept of 'theory,' so it's interesting to briefly consider its meaning and to see what light it sheds on its modern namesake. What makes *theoria* a proper tool to save the phenomena? It's mathematical, which guarantees that it captures the order behind the phenomena. But this is not enough, because the Babylonian tables were also mathematical (though arithmetical rather than geometrical) and they did not satisfy the Greeks. The difference is that, unlike the tables, *theoria* is not a mere practical tool of prediction. It tells a story – it provides an account of the phenomena that explains *why* the phenomena are the way they are. In the astronomical case, the explanation is not necessarily causal, as Aristotle would have preferred, but it's still a 'story': it allows following bodies in motion; it tells of real changes in the world; it shows that things are the way they should be by following where they were and where they're going to.

Stories that explain why things are as they are belong, in most cultures, to the realm of myth, so it may be tempting to think of theory as a kind of myth: a narrative that makes sense of the mysterious and often frightening natural phenomena. But there is a fundamental difference between *theoria* and myth. Myths explain phenomena by domesticating them; by narrating the strange in terms of the familiar. Great natural events become, in the myth, effects of human-like reason and emotions of human-like gods and heroes. The painful and the frightening – thunder and lightning; storm or earthquake – are a sign of gods' anger and vengeance; the bountiful – sunlight and fertile rain – are a sign of their love and care. A *theoria* – theoretical – way of explaining is diametrically opposed to that. If the myth alleviates the mysterious by relating it to the familiar, the theory takes the familiar and explains it by the mysterious: the daily and common natural happenings become expressions of esoteric forces and motions – forms; essential nature; miniscule particles – and in Greek astronomy: perfect spheres.

Spheres

The import of spheres in astronomy is one crucial point at which Plato and Aristotle's philosophies converge. Plato's call to submit the heavenly motions to a strict spherical model is a general application of the Parmenidean and Pythagorean fascination with geometrical perfection. The detailed requirements for such a model are provided by Aristotle. The real heavenly motions, he dictates, must be *circular, concentric* and in *uniform velocity,* and the *theoria* should explain all apparent motions as configurations of these real motions. These requirements are embedded in Aristotle's cosmology (Figure 3.10): all heavy matter moves towards the center of the Earth and all light matter away from it. This means that this center is also the center of the cosmos and that the Earth is spherical. The planets travel around this common center of the Earth and the cosmos. Their trajectories are circular, and these circular motions are always at the same (angular) velocity.

To better understand the intellectual motivations behind these cosmological assumptions, we need to think again about Aristotle's (and the ancients') concept of motion. For them, as we discussed in Chapter 2, motion is change – in fact the paradigm of change – hence it requires a cause, a beginning and

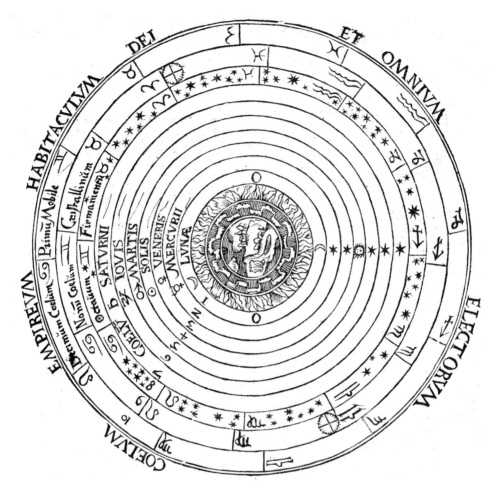

Figure 3.10 Aristotle's Cosmology as the basis for astronomy. The Earth at the center as the realm of elements and the planets surrounding it in order – the Moon, Mercury, Venus, the Sun, Mars, Jupiter, Saturn. Around them is the sphere of the fixed stars, which in this Christian text is identified with the Biblical Firmament (see Figure 1.12); around it the Ninth Heaven – "the Crystalline Sphere"; then the Tenth Heaven – the *Primum Mobile*; and around it the rest of the Christian context as in Figure 1.12. This woodcut is from Peter Apian's *Cosmographia* (Antwerp: Arnold Berckmann, 1539 [1524]) – an introductory book on astronomy, geography, cartography, navigation and instrument-making, which went through no fewer than thirty editions in fourteen languages during the sixteenth century.

an end, and a medium. The heavenly bodies rotate around us, but it doesn't seem right to think of them as truly changing. Like many of Aristotle's insights, this one is both a metaphysical and an empirical one. First, it is not appropriate for these noble, ethereal bodies to change. If they change, it also means that they come to be and pass away, like the mundane objects around us. We also know from our forefathers, Aristotle points out, that in general, the *motions* of the heavenly bodies *do not* change. The Sun has always risen in the east and set in the west; the constellations have always been the ones we know (Figure 3.9); no heavenly body has appeared or disappeared (the irregular appearance and disappearance of comets, meteors and their like attests that they belong under the Moon, within the earthly realm).

By all these considerations, the heavenly bodies should not change, and since motion is change, they should not be moving; they should be at rest. So how can one explain that the heavenly bodies *do* change their places in the heavens? Aristotle's rules for astronomy aim to resolve this tension; the sphere, in these rules, is not just an abstract symbol of perfection, but a working concept. It allows Aristotle to develop an idea of motion which is as close to rest as it can be.

The motion of the celestial bodies according to Aristotle's cosmology – circular, concentric motion of uniform velocity – is almost rest; it evades all the limitations that make terrestrial motion a paradigm of change. Moving in circles, the stars and planets move *within* a place, rather than from one place to the other. Moving in uniform velocity, their motion is eternal: never slowing, which may bring them to a stop; never accelerating, which may eventually make them reach infinite velocity. Moving in concentric orbits, the heavenly bodies move without ever approaching the center of the cosmos, from whence they could no longer move; nor recede to the periphery, beyond which there is nothing.

This is why the task of the Greek astronomer was to assign to the heavenly bodies these orderly motions – circular, concentric, uniform – and explain how the seemingly chaotic phenomena – the seemingly irregular motions of the planets – arise from them. This is what *to save the phenomena* meant, and it would remain so until Johannes Kepler changed astronomy's challenge in the seventeenth century.

Eudoxus' Nested Spheres

It is not entirely clear whether Aristotle invented these cosmological rules or whether he was giving philosophical formulations to ideas which were already evolving among the astronomers of his time. The first complete attempt that we are aware of to save the phenomena in this way is of Aristotle's contemporary Eudoxus, who lived in Cnidus in Asia Minor from c. 390 to c. 340 BCE, and we know about it indirectly, from Aristotle's own testimony and a later commentary. In Eudoxus' astronomy the heavenly bodies are set in concentric spheres, exactly as decreed by Aristotelian cosmology (Figure 3.11, left). At the center of all the spheres is the Earth, and at the periphery is the sphere of the fixed stars, which rotates from east to west in the regular *daily* motion we are familiar with. In between, Eudoxus placed (according to Aristotle's account) no fewer than fifty-five additional

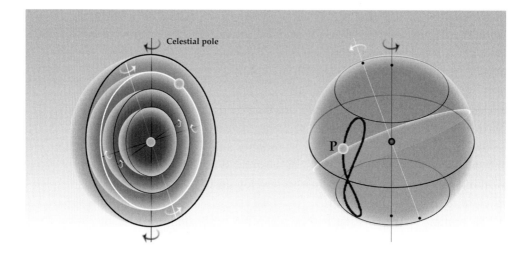

Figure 3.11 Eudoxus' system (on the left) and the way it explains retrograde motion (on the right). Each planet is carried by a number of rotating spheres. All the spheres are concentric and each one rotates in a uniform angular velocity, but each rotation is around a different axis and at a different velocity. The planets' apparent changes of velocity and direction, of which the retrograde, 'hippopede' shape is the most pronounced, are produced by the combination of these different motions.

spheres. Each sphere moved *regularly*, but each in its own direction and its own angular velocity, and every planet was carried by three or four spheres, producing its own motion while canceling the motions of the others. With this combination of separate motions Eudoxus could apparently reconstruct not only the observed changes of angular velocity in the annual path of every planet, but even their strange retrograde motions (Figure 3.11, right). This account of retrograde motion was particularly impressive; according to Eudoxus' model, it was a 'hippopede' – a figure-of-eight trajectory created by two spheres whose axes of motion were at an angle to each other (the name comes from ἵππος – horse – because the eight-shape was similar to the knot for hobbling horses' legs). Eudoxus turned the two-sphere model into a multiple-sphere one, but succeeded in saving the phenomena: the planets' seemingly irregular annual motions became an expression of circular, uniform, concentric motions.

  Was Eudoxus' model real or just an idea? Could it really produce the heavenly motions or did it just suggest how this might be done? One cannot calculate the motions of fifty-some spheres moving simultaneously in different directions and velocities using only pen and paper. But a modern computer can – so it can also be done by a mechanical calculating machine doing what we've learned to do electronically, namely: to turn the continuous motions into discrete computations. This is actually less mysterious than it sounds: perhaps Eudoxus built an armillary sphere, with rings that could be moved point by point to scratch a trajectory on the surface of some soft sphere – a fruit, perhaps – placed at its center?[7] This was not beyond the ancient Greeks' technological capacities, as the famous 'Antikythera mechanism' testifies (Figure 3.12). Named after the island near whose shore it was salvaged from a Roman shipwreck in 1901, this is second- or first-century BCE clockwork of at least thirty gears, perhaps even thirty-seven – not that many fewer than what Eudoxus would have needed.

But this is already more speculation than befits the serious historian. What is important to note – what is really telling as a crucial early moment

Figure 3.12 The Antikythera Mechanism: a second- to first-century BCE mechanical device for calculating relative positions of the planets and, in particular, past and future eclipses, both solar and lunar. It was discovered by sponge divers in 1901 in a 70–60 BCE wreckage of a Roman ship, apparently on its way from Rhodes to Rome, but the device's full complexity was only revealed in this century by X-ray tomography and computer-imaging technologies. It is made of bronze, about 20 cm in height, with dials on both sides. The front dial represents the ecliptic, with an outer ring divided into 365 days and an inner ring divided into 360 degrees and the signs of the Zodiac. The rear face had five dials, the most important following the Metonic cycle – the nineteen-year cycle in which the solar and lunar calendars synchronize. The dials were moved by a handle, through thirty-seven interlocked geared wheels of different sizes, whose accumulated ratios create the proportions between the dials' motions, mimicking the different cycles of planetary motions. If recent reconstructions (many available on the web) are indeed good approximations, the device represents mechanical and precision technology which was only redeveloped in Europe in the late Middle Ages. The mechanism was part of a trove comprising marble and bronze statues, jewelry, glasswork and coins, which recapitulate our discussion of the cultural import of such artifacts and the theories they embody: objects of wonder, whose superb precision fulfills aesthetic rather than practical values.

---

[7] I owe this idea to Ido Yavetz.

in the history of science – is the task Eudoxus set himself and the way he approached it. He had a set of phenomena which appear irregular, and he believed that they cannot really be irregular. There had to be an order they obeyed, and their irregularities had to be only *apparent* – only *to us*. Perfect order meant rest, but the phenomena were motions, and he could not explain motions through rest, so he had to compromise. He constructed a model out of motions which were as close to rest as possible: concentric, circular, of uniform (angular) velocity. The model could not provide a cause for these motions: he had to assume that circular motion was simply in the very nature of the heavenly bodies. But it did make physical sense: it put the Earth where it should be and arranged the stars and planets around it. Eudoxus' model *saved the phenomena*: it presented apparent disorder as the consequence of real order.

The Empirical Side of *Theoria*

Astronomers, as we saw, produce calendars – predictions of the future positions of the heavenly bodies – and in this empirical task the Babylonians were particularly successful in their arithmetical, approximating approach. The unique property of Greek astronomy, which makes it a founding moment in the history of what we now call science, is the addition of theoretical requirements to these predictions. Yet it would be wrong to think that Greek astronomy had no place for the empirical. Looking at the heavenly bodies with an overall image of their positions in mind and a conviction about the fundamental properties of their motions allowed the Greeks to make bold empirical claims – claims about sizes, distances and other properties of the heavenly bodies and the Earth. One of the most interesting features of these calculations is that they are completely irrelevant to that main task of the astronomer – to produce calendars. They serve curiosity alone.

Aristarchus of Samos (310–230 BCE), in his only extant work – *On the Sizes and Distances* – provides a very early example (Figure 3.13). Equipped with the image of spheres within spheres and the assumption that the Earth and the Moon are lit by the Sun (so, *pace*[8] Genesis, the Moon does *not* have its own light), he measured the relative sizes of the Earth and the Moon. What Aristarchus knew from observations is that eclipses happen only when the Moon is full. This correlation should have made perfect sense to Aristarchus: if the Moon's light comes from the Sun, it can only appear to us fully lit when the two are lined up – namely, when the Moon and

---

[8] Pronounced 'pa-che' or 'pa-ke'. From the ablative case of the Latin *pax*, literally meaning 'in peace' or 'without insult', and used to mean something like 'in respectful contradiction to ...'.

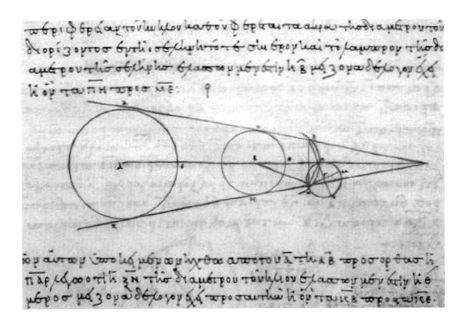

Figure 3.13 Aristarchus' calculations of the relative sizes of the Moon and the Earth from a tenth-century Greek copy of his *On the Sizes and Distances,* now in the Vatican Library. The Moon (the smallest circle on the right) comes in and out of the shadow of the Earth (in the middle). The angle between the Moon's position at the beginning of the eclipse and its position at the end is the Earth's angular size as observed from the Moon. The ratio between this angle to the angular size of the Moon as observed from the Earth is the ratio between their real sizes.

the Sun are on opposite sides of us, on the Earth. And it is in the same relative position that Earth can enter directly between the two and cast its shadow on the Moon. Eclipses differ in duration, so it also makes sense that the longest ones occur when the Moon passes through the whole radius of Earth's shadow. The angle between the Moon's position when the eclipse starts and its position when it ends – when the Moon enters and when it exits the shadow of the Earth – is the angular size of the Earth as perceived from the Moon. The angular size of the Moon is easily measured, and obviously, the distance from the Moon to the Earth is the same as the distance from the Earth to the Moon. So the ratio between the two angular sizes – 3.7, as it turns out – is the ratio between the diameters of the Earth and the Moon. All their other relative sizes (surface area, volume) can be easily calculated from this. Using the complementary insight that at half-Moon there is a right angle between the Earth, the Moon and the Sun, Aristarchus could also calculate the relative distances between the three bodies on the basis of the observed angle to the Sun (87°): another spectacular empirical achievement whose only purpose was to satisfy curiosity.

Aristarchus' younger contemporary, Eratosthenes of Alexandria (276–194 BCE), may have one-upped him by using even simpler observations and calculations to establish an even more exciting and impractical piece of empirical knowledge. With the aid of a gnomon – a simple sundial – he measured the Sun's angle at its zenith (its highest point in the sky). The length of the gnomon's shadow provides the angle, and it's easy to determine when the Sun is at the zenith; this is when the gnomon's shadow is shortest. It was important that the measure be taken at that particular point – it allowed him to compare angles taken at the same time in Alexandria and at Syene (today's Aswan). He found that the difference between the angles measured was 7°, which meant that the circumference of the Earth was 360/7ths of the distance between Alexandria and Syene, or 250,000 *stadia*. There are several guesses as to what a *stadion* is, and one of them – 185 meters, based on the dimensions of the stadium in Eratosthenes' Athens – gives 46,000 kilometers, astonishingly close to the modern number, which is about 40,000 kilometers (although we no longer think of Earth as perfectly spherical). But to do this would be to miss a crucial point. *Stadion* stands for a horse's pace. It is not a standardized unit. Eratosthenes' measure of the distance between the two cities was a crude estimation – approximately 5,000 paces of a horse – and he didn't pretend it was anything else. The measurement of the size of Earth, like that of the size and distance of the Moon, was an exercise in extravagant curiosity, and accuracy had an aesthetic, playful value, with no practical significance (compare the captions to Figures 1.10 and 3.12). Similar points can be made about the breathtaking accuracy of modern science.

Similar insights can be offered concerning the work of the man who Ptolemy himself considered the greatest of all Greek astronomical observers: Hipparchus of Rhodes (c. 190–120 BCE). Hipparchus' successors attributed to him a catalog of some 850 stars; great advancements in geometry (including an invention of a theory of chords – a basic version of trigonometry – necessary for his calculations); the invention of new observational instruments such as the equatorial ring; great improvement of the calculation of the lunar and solar months, and more. But he was (and still is) most revered for establishing the precession of the equinoxes.

Comparing his own meticulous observations to those of the Babylonians from centuries earlier, Hipparchus noted that the equinox is shifting; the place where the ecliptic crosses the celestial equator, measured against the backdrop of the fixed stars, moves to the west at a rate of about 1° a century. Loyal to the Greek way, Hipparchus accounted for the precession by assigning the celestial axis a small circle on which it rotated very slowly, completing a full cycle – a period – in some 26,000 years. Of note, again, is how rarefied and remote these calculations are – empirical, systematic and precise as they are – from any practical considerations. This time span is far beyond any society's capacity or need for planning. Like the size of the Earth and distance of the Moon, the precession of the equinoxes is an object of cultivated and refined curiosity, befitting a free Greek citizen, in which precision is an abstract, aesthetic ideal.

The Moving Earth Hypothesis

Greek astronomy is distinguished by its theoretical aspirations, and its empirical inquiries were both directed and enabled by these aspirations. It is therefore very interesting to see the role that empirical concerns – concerns about what was known from observations – played in theoretical considerations.

Particularly interesting from this perspective is the question of whether it is the Sun that moves around the Earth or vice versa. It's interesting because there is no direct way to resolve this conundrum by observation from Earth, and because it's common to think that the ancients simply assumed that the Earth stood still. This is not the case. Aristotle deemed both options prima facie plausible:

> As to its position [of the Earth] there is some difference of opinion. Most people – all, in fact, who regard the whole heaven as finite – say it lies at the centre. But the Italian philosophers known as Pythagoreans take the contrary view. At the centre, they say, is fire, and the earth is one of the stars, creating night and day by its circular motion about the centre.
>
> Aristotle, *On the Heavens*, Book II, ch. 13 (http://classics
> .mit.edu/Aristotle/heavens.2.ii.html)

For himself, we saw, Aristotle took the position that the Earth is stationary at the center of the cosmos. But he admits that this – unlike the sphericity of Earth – is not a position he can support by observations, only by general considerations of the shape of the cosmos.

With even the authoritative Aristotle remaining somewhat reticent in the matter, both options seem to have remained respectable, because a century later Aristarchus took the opposite view: that Earth, together with the rest of the planets, moves about the Sun.

Although he is the most famous believer of a heliocentric cosmos (Helios is Greek for the Sun) of Antiquity, we only know about his claims indirectly.

The original work in which this hypothesis was presented was lost – we know about Aristarchus' hypothesis primarily from Archimedes, the great mathematician and inventor of Syracuse (287–c. 212 BCE) whom we'll discuss later. But it seems reasonable to assume that his arguments for the centrality of the Sun, which imply the motion of the Earth, were similar to those Aristotle ascribed to the Pythagoreans, though perhaps a little less general and metaphysical and more specific and astronomical. As we saw, Aristarchus had particular interest in the light coming from the Sun, and his analysis of eclipses gave the Earth only a secondary role – a cause for shadow.

What is most important about Aristarchus' suggestion is that it was seriously considered – and rejected. The main argument against it was evoked by Archimedes. If the Earth moves around the Sun (so was the reasoning) it has to cover very great distances. It should be very far away on, say, the vernal equinox, from where it was on the autumnal equinox. This means that we should have a very different line of sight to the heavens, and should be able to note what is called *parallax* – a difference in the angles we observe between the stars. But such annual stellar parallax was not observed. The stars are arranged in the heavens in the same way throughout the year – the constellations don't change with the seasons. If we were, as most agreed with Aristotle, at the center of the cosmos, this is exactly how it should be. If Aristarchus was right, then the only possible reason we do not observe any stellar parallax had to be that the stars were so far away that the size of our orbit around the Sun was completely negligible in comparison to this distance. In Greek parlance, it had to be as if the whole orbit of the Earth were like a point in proportion to the distance to the stars. One could even estimate this outrageously large proportion. The human eye, at its very best, can distinguish half a degree. The distance from one side of the orbit to the other is 180°, so if Aristarchus was right, then in order for us not to observe stellar parallax, the distance to the stars should be more than 360 times the diameter of the orbit. This in and of itself seemed ridiculously large to the Greeks. Moreover, with the calculations of Aristarchus himself, as well as Eratosthenes and others, they thought they had a fairly good estimation of our distance to the Sun, which is of course the radius of the Earth's orbit. This number, times 360, was so large they could hardly express it with their number system. In fact, Archimedes presents this argument in the context of introducing a new way in which even such ridiculously large numbers could be written. It was absurd, and the Greeks, in spite of being apparently somewhat attracted to the idea of a moving Earth (or they wouldn't have reintroduced and discussed it occasionally), had to reject it.

The Legacy of Greek Astronomy: Ptolemy's Orbs

Still, the main legacy of Greek astronomy was theoretical. It culminated in the work of the Alexandrian Claudius Ptolemy (c. 85–165) whose main book, known throughout the mediaeval world by its Arabic title *Almagest* (from the Arabic pronunciation of the Greek *magiste* – greatest), remained the most important work in astronomy well after the collapse of the Hellenic realm, until the second half of the sixteenth century (Figure 3.14). The significance of Ptolemy's astronomy, like that of Aristotle's philosophy and Galen's

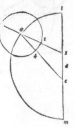

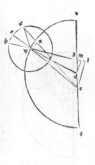

Figure 3.14 Saturn's epicycles in a Latin edition of Ptolemy's *Almagest* from 1496. Georg von Peuerbach (1423–1461) translated and edited the first four books, and his disciple Johannes Regiomontanus (1436–1476) completed this edition after his death. Note: as explained in Figure 3.15, the center of the deferent is d, but the points of reference are t – Earth (*terra*); and e – *equant*.

medicine (which we discuss in Chapter 8), lies much beyond any of its particular claims. The *Almagest* shaped the fundamental practices, criteria and methods of studying the heavenly motions everywhere, from Europe through the Muslim realm to China. For 1,500 years, to be an astronomer meant to employ and slightly modify the theoretical tools provided by Ptolemy in order to improve upon the calculations of previous generations working within essentially the same framework.

The *Almagest* begins with a careful explication of Aristotelian cosmology. Clearly, to Ptolemy and his disciples over some fifteen centuries, his work was a brave attempt to *save the phenomena* according to the rules set by Aristotle: to reduce the changes in planetary positions to circular, concentric motions of uniform angular velocity. All in all, it was a glorious success; Ptolemaic astronomers produced accurate circular models of the heavenly motions. But in order to do that, they had to compromise almost all of Aristotle's rules – except for circularity. In a sense, the essence of Ptolemaic astronomy is these bold compromises.

To begin with, Ptolemy gives up Eudoxus' three-dimensional model of moving spheres-within-spheres. Instead of attempting to handle the complex mathematics of all these interdependent motions, Ptolemy constructs an independent two-dimensional circular model like that on the bottom left of Figure 3.15 for each one of the planets. The three-dimensional model made sense physically: the moving, nested spheres were physical objects that carried the planets and moved one another. For the sake of simplicity and workability, Ptolemaic astronomy sacrifices the aspiration to make such physical sense.

Worse: every one of the computational tools introduced by Ptolemy is a crafty violation of one of the fundamental laws of cosmology that his astronomy is supposed to serve – concentricity and uniform velocity.

First, Ptolemy's models place the Earth *outside of* the geometrical center of each planetary orbit (Figure 3.15, top left). With this eccentricity – the planet is ex-centric, out of center – Ptolemy calculates the apparent changes in planets' angular velocity as if they are merely a matter of perspective. Examine the top right of Figure 3.15: if a given planet travels around the center **C** with uniform velocity, and takes the same time to get from **a** to **b** as it does to get from **g** to **f**, then the angles it will traverse in those times will be equal: ∠a**C**b=∠g**C**f. But if we, on the Earth, are out of the center, say at **E**, it will appear to us *as though* the planet traveled once a longer angle, ∠a**E**b, and once a shorter one: ∠g**E**f. By manipulating the eccentricity – the distance between **E** and **C** – the Ptolemaic astronomer could make the model produce for a given planet, from a circular orbit with uniform velocity, the non-uniform angular velocity that we observe from Earth.

Secondly, Ptolemy offers (following the work of Hipparchus and Apollonius) an ingenious way of presenting retrograde motion as an apparent outcome

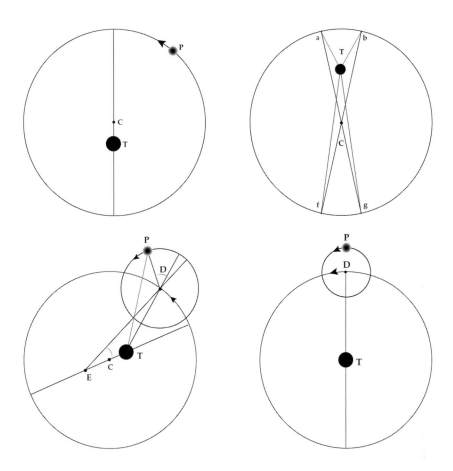

Figure 3.15 The Ptolemaic System and its geometrical tools: the planet is assumed to move in uniform angular velocity. Manipulating the center of the planet's motion and Earth's position in relation to it yields the non-uniform observed motions. Top left – eccentricity: Earth – **T** – is outside the geometrical center **c** of the deferent. Top right – the way eccentricity accounts for changes in observed angular velocity: if the planet moves in uniform angular velocity around the center C, it sweeps ∠**cab** and angle ∠**cfg** at the same times, but from **T** these angles will look different. Bottom right – the epicycle and the way it explains apparent retrograde motion: the planet P always moves counter-clockwise, but when it's on the inner part of the epicycle, it will appear to the observer at **T** that it moves in the opposite direction. Bottom left – the elements of a complete model. The center of uniform angular velocity of the deferent is neither **T** nor **C** but E – the equant point. **C** is at the same distance from **T** and from **E**. The line **TP** should conform to observations, and the rest is to be constructed to yield uniform circular planetary motions.

of circular motion with uniform velocity. The model places the planet on a small circle – the epicycle – which is carried by a bigger circle – the deferent (Figure 3.15, bottom right).[9]

---

[9] Over the centuries, Ptolemaic astronomers, especially in the Muslim realm, developed models with small deferents and large epicycles, and many other variations.

The planet **P** moves along the epicycle around its center **A**, which in turn moves along the deferent **ABD** around its vacant center **E** (the Earth, remember, is at its calculated eccentricity). Both circles move in the same direction, so the apparent angular velocity changes with the planet's position on the epicycle – another way to account for the non-uniformity. More crucial, however, is that when the epicycle carries the planet inside the deferent, it appears to us as if it has changed its direction and is moving backwards: retrograding. Setting the relative sizes and velocities of the two circles and adding more epicycles on top of epicycles, the astronomer could make the model produce any complex apparent trajectory while maintaining circular orbits and uniform velocities. The power of this tool is such that a thirteenth-century Persian astronomer and polymath, Nasir ad-Din al-Tusi, even showed how a combination of two uniform circular motions (known as 'the Tusi couple') can generate an apparent rectilinear motion (Figure 4.12).

Sometimes, however, eccentricity and epicycles were not enough to account for the planets' changes in direction and velocity. At other times, they actually worked against each other – for instance, the eccentricity and epicycle required could be too large to fit into the deferent together. For such cases Ptolemy introduced perhaps the boldest of his tools – the *equant* point (E in Figure 3.15, bottom left). The equant is a point inside the deferent – neither the geometrical center nor the eccentric Earth – around which the angular velocity is uniform.

The procedure can be understood thus: the astronomer would try to construct a model under the assumption that the planet's motion around the geometrical center of its orbit (the deferent) is in uniform angular velocity. Since from the Earth the velocity doesn't look uniform, he would calculate how far away the Earth should be from the center – both of the orbit and of uniform motion – to create this non-uniformity, which he'd assume is only apparent, as the illustrations at the top of Figure 3.15 demonstrate. If this distance turned out to be too large, or the epicycle would get in the way, he would place the Earth at a convenient point, and set the center of uniform motion at the calculated distance – no longer the geometrical center of the orbit. This new point is the equant point – E in Figure 3.15, bottom left.

Usually the two tools were used in tandem: eccentricity was simply halved, and the models placed the Earth and the equant point at the same distance and on opposite sides of the geometrical center of the deferent, as seen in Figure 3.15, bottom left. Yet the equant point was quite different from all other Ptolemaic tools. Unlike eccentricity, which made the planet revolve uniformly around a particular point with a clear meaning – the real geometrical center of its orbit (even though this center was no longer populated by the Earth) – the equant was a point with no distinction of its own; it was only there to make the model work. The astronomer had no physical or geometrical constraints in choosing the placement of the equant; he simply chose the point which would make his model best fit the phenomena and be the most convenient for calculation.

Conclusion

Some readers may be familiar with the celebrated work of the philosopher Thomas Kuhn on scientific revolutions. Reflecting on the above, it should come as no surprise that Ptolemaic astronomy served Kuhn as the inspiration and model for his idea of 'normal science': science working in a smooth and systematic way, according to clear rules and criteria, to solve well-defined questions. Using Kuhn's terminology, we can say that the Ptolemaic astron-omer had a clear 'paradigm' – Ptolemy's *Almagest* – that he could emulate. He had clearly defined 'puzzles' to solve: sets of planetary positions which he had to reduce to models like that in Figure 3.15, bottom left. He had a set of tools with which to construct these models as well as clear criteria for successful models. He was also aware of a number of nagging 'anomalies': unsolvable problems which he conveniently ignored while meticulously constructing his models.

But as we'll see in later chapters, Ptolemaic astronomers of centuries to come, most notably Copernicus, were acutely aware that the marvelous power of their tools was also their weakness. They could produce excellent calendars because these tools were so flexible, but this flexibility came at the price of the very ideas they were supposed to serve. The equant point in particular seemed like outrageous technical self-indulgence; if a planet moves with uniform veloc-ity around a point which is *not* the center, it obviously *does not* move with uniform velocity around the center. In using an equant, the astronomer was breaking the very rules of heavenly orderliness which were supposed to direct his practice. Ptolemaic astronomy made no physical sense – no Earth to move around, no spheres to carry the planets. In fact, it was treading dangerously close to relinquishing the very ideal of saving the phenomena and returning, in practice, to the Babylonian approach of providing a theory-free algorithm for calculating planetary positions (even if theirs was a geometrical algorithm, unlike the Babylonian arithmetical table).

For the fifteen centuries following Ptolemy, his system was the astronomers' 'normal science.' By the end of this long period, we'll see later in the book, it was the attempt to rescue the old ideals of concentric, uniform, circular motion that brought about the complete upheaval of astronomy, and with it of much else that was believed about the cosmos and our place in it.

Discussion Questions

1. Is there something interesting to be learned from the fact that most of us know so little about the most basic phenomena – the daily motions of the planets?
2. Does the idea that superlative precision primarily fulfills aesthetic values seem to you relevant to modern science? Can you provide examples for or against this idea?
3. What can be learned from the fact that Hellenistic astronomers and philosophers considered the possibility that the Earth moves and rejected this for empirical reasons?
4. Which of the Hellenistic empirical astronomical findings – the size of the Earth, the cyclical deviations of the calendar, the relative sizes of the Earth and the Moon, etc. – do you find most impressive and/or most telling?
5. Is it fair to say that the most impressive achievement of Hellenistic astronomy came at the expense of its most fundamental premises, and what is the significance of that?

Suggested Readings

Primary Texts

Aristotle, *On the Heavens*, Book II, 13–14 (http://classics.mit.edu/Aristotle/heavens .2.ii.html).

Ptolemy, *Almagest*, Book I, 1; 3; 5; 12; Book III, 3. In: Toomer, J. G., *Ptolemy's Almagest* (London: Duckworth, 1984), pp. 35–37, 38–40, 41–42, 141–153.

Secondary Sources

Aaboe, Asger, *Episodes from the Early History of Astronomy* (New York and Berlin: Springer, 2001).

Dreyer, J. L. E., *History of Astronomy from Thales to Kepler*, 2nd edn. (New York: Dover (1953).

Evans, James, *The History and Practice of Ancient Astronomy* (Oxford University Press, 1998).

Goldstein, Bernard R., "Saving the phenomena: the background to Ptolemy's planetary theory" (1997) 28 *Journal for the History of Astronomy* 2–12.

Kuhn, Thomas S., *The Copernican Revolution: Planetary Astronomy in the Development of Western Thought* (Cambridge, MA: Harvard University Press, 1992).

Neugebauer, Otto, *The Exact Sciences in Antiquity*, 2nd edn. (Dover, 1969).

Ptolemy, Claudius, *Ptolemy's Almagest*, G. J. Toomer (trans.) (Princeton University Press, 1998).

Van Helden, Albert, *Measuring the Universe: Cosmic Dimensions from Aristarchus to Halley* (University of Chicago Press, 1986).

4 | Medieval Learning

The Decline of Greek Knowledge

What happened to Greek knowledge? Not just the astronomy of Eudoxus, Eratosthenes and Hipparchus, but the natural philosophy of Aristotle and his disciples with its forays into what we'd call botany and zoology; the medicine of Hippocrates and Galen (to whom we'll return in Chapter 8); the geography of Ptolemy; the mathematics of Euclid and Archimedes and much more – all that knowledge, the longing memory of which Raphael captures so marvelously at the beginning of the sixteenth century (Figure 4.1) – and of which we now have only fragments or recollections of.

The Burning of the Library

Folk history offers an attractive answer: it all burned down with the library in Alexandria. The library was indeed a marvelous edifice of culture. It was established in 283 BCE by Demetrius Phaeleron – a student of Aristotle's Lyceum (see Chapter 2) and an ousted dictator of Athens. It was part of the *mouseion* (Μουσεῖον), a temple of the muses, financed and protected by Ptolemy, the Hellenic king of Egypt.[1] Ptolemy provided the library with a mandate to collect *all* knowledge, supported by aggressive instructions and means to purchase, confiscate and borrow (with no intention to return) every text within reach. At its prime, so ancient sources claim, the library comprised some 500,000 papyri, and when one realizes that it would only hold one copy of every text, and that the best modern university library holds a few million books (Harvard library boasts 17,000,000), many of them in multiple copies and editions, the magnitude of this ancient cathedral of knowledge becomes overwhelming. And the library was indeed burned down. The water of the Nile, we're told, was for days dyed black with the ink.

---

[1] There was no relation between King Ptolemy and the astronomer Ptolemy, but medieval scholars did occasionally confuse them, drawing the astronomer with a crown on his head.

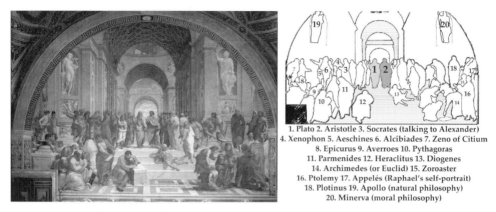

1. Plato 2. Aristotle 3. Socrates (talking to Alexander)
4. Xenophon 5. Aeschines 6. Alcibiades 7. Zeno of Citium
8. Epicurus 9. Averroes 10. Pythagoras
11. Parmenides 12. Heraclitus 13. Diogenes
14. Archimedes (or Euclid) 15. Zoroaster
16. Ptolemy 17. Appelés (Raphael's self-portrait)
18. Plotinus 19. Apollo (natural philosophy)
20. Minerva (moral philosophy)

Figure 4.1 Raphael's *The School of Athens* fresco from 1510 in the Apostolic Palace of the Vatican.

Yet the import of the catastrophe becomes less clear when considering that it occurred at least three times: in 48 BCE, with Caesar's conquest of Cleopatra's kingdom; in 391 (from now on, when not otherwise mentioned, all dates are CE), during Christian riots against pagan temples; and in 642, by the invading Muslim warriors of Caliph Omar. The fact that the story is so remarkable yet so unspecific suggests that it recalls more than a specific incident: it records a trauma; a cultural memory of a great treasure lost. The library was indeed destroyed, and more than once, but the loss of Greek knowledge cannot be attributed to a singular dramatic event. It was, rather, a long, gradual decline, during which the particular concept of knowledge that we described in the previous two chapters lost its cultural anchors.

The Schools of Athens

Here is another way to think of the particularity of Greek knowledge and learn some general lessons from its rise and decline about the construction of the cathedral of science:

Around 387 BCE, Plato begins to meet regularly with fellow scholars and disciples in an olive grove at the outskirts of Athens – the Hecademia, named after the mythical hero Hecademus – establishing the school that will come to be known by this name: Plato's *Academia*. Some fifty years later, around 335 BCE, Aristotle returns to Athens with the Macedonian conquerors and establishes the *Lyceum*, his school, in a gymnasium named after the neighboring temple of *Apollo Lyceus*. In 312 BCE, Zeno of Citium (not to be confused with Zeno of Elea, author of the famous Parmenidean paradoxes) starts teaching in the *Stoa poikilê* ('The Painted Colonnade') of the Athenian agora, founding the Stoic school. Just a few years later, in 306 BCE, Epicurus buys a house and a garden, in which he establishes the Epicurean school.

It is very telling that the great schools of Greek philosophy are named after their places of residence: knowledge requires a real, physical place, even if it aspires to be most general and abstract. A place allows for memory: a school would have a library, the most magnificent exemplar of which was of course the one in Alexandria. (We tend to think of memory as place-less, floating somewhere 'in the cloud.' But 'the cloud' has a very physical place: big server-farms in locations like the Nevada Desert.) It would have a dominant father-teacher figure, whose teachings comprise the core of the library and whose memory is related to the place, perhaps mythically. These teachings are the core of the doctrine which defines the school – Platonism or Aristotelianism; Stoic Skepticism or Epicurean Atomism – and consecutive generations of scholars would debate the interpretation of the master's ideas while developing them and arguing against the rival schools.

This was a very fragile network. It could only be maintained by that unique Greek culture which we described in Chapter 2, if in an admittedly idealized manner; a culture that assigns high value to impractical, free debate; that takes with utter seriousness the power of argument; that accepts the great speech as the prime political act. By the time the library was burned for the first time, this culture had already lost its vivacity.

From the Greek Polis to the Roman Empire

In the four centuries between the establishment of the Athenian philosophical schools and Caesar's invasion of Egypt, the role of Greek culture and language in the life of the people of the Eastern Mediterranean basin dramatically changed. When Alexander the Macedonian (356–323 BCE) – allegedly Aristotle's tutee – left the Greek Peninsula in 336 BCE to embark on the heady journey of conquest that took him through Mesopotamia and Persia and into India (Figure 4.2), he left in his trail at least a half-dozen Greek-style poleis (city-states) named after him, from Egypt to Afghanistan; a lasting impression of Hellenic military prowess; and monuments of Greek material culture, from theaters to public baths. When Alexander died in 323 BCE (in the palace of Nebuchadnezzar II, King of Babylon), at only 30 years of age, his generals were left to divide between them (not peacefully) the enormous areas of the ancient kingdoms of Egypt and Mesopotamia as booty. The rather tight-knit Greek archipelago of poleis thus turned into a vast, loose galaxy whose language of commerce and governance as well as the culture and religion of its courts, was Greek, whereas local languages and cultures remained intact. Jews spoke Hebrew and practiced Judaism;

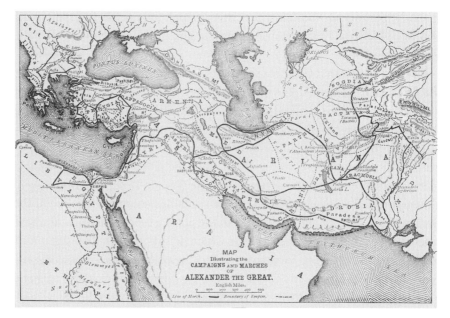

Figure 4.2 Alexander's conquests (336–323 BCE), which marked the boundaries of the Hellenistic realm. From Cassell's 1890 *Illustrated Universal History*, Vol. I: *Early and Greek History* by Edmund Ollier. Artist unknown.

Syrians spoke Aramaic and worshiped Canaanite gods; Cleopatra, from whom Caesar seized Alexandria and Egypt, was a Hellenistic queen of a Hellenistic kingdom – the first of her dynasty to learn the local language.

The great Hellenic conquest had, then, a paradoxical consequence. 'High' Hellenic culture, the culture of the ruling class, masters of abstract knowledge – especially, as we saw, in the Greek context – became detached from the 'low' culture of everyday life. The Roman takeover of the Hellenic realm in the two centuries before the Christian era deepened this alienation. Quite contrary to the Greek aristocracy's ideal of freedom from worldly constraints, the Romans stylized themselves as simple and industrious. The Roman noblemen, unlike their Greek predecessors, prided themselves on their practical prowess, rather than an abstract freedom from the vicissitudes of this world. Leaders of a nation of warriors – suckled by the she-wolf – rather than scholars or artists, they happily adopted many facets of Greek high culture, which they acknowledged as superior to theirs. They took on Greek mythology and poetic structures, philosophy and rhetoric, which meant that Greek remained and even deepened its hold as the learned language of the evolving Roman Empire. At the same time, however, it was deprived of its role as the language of traders and politicians – this role had now been taken over by Latin, the conquerors' tongue.

Removed from most facets of life, the fragile network of Greek learnedness had little to sustain it, and required no grand catastrophes to decay.

The Encyclopedic Tradition

The First Roman Encyclopedists

This is also how the state of scholarship was perceived by the great *Latin*
scholars of the time: Marcus Terentius Varro (116–27 BCE) and Marcus
Tullius Cicero (106–43 BCE). Both were minor Roman nobility and politicians by trade. Both were witnesses and participants in the mayhem of the
final days of the republic and early decades of dictatorship – Varro was
rather flexible with his loyalties, whereas Cicero's staunch republicanism
resulted in his execution as 'enemy of the state.' Both were well versed in
Greek high culture by virtue of their intellectual ambitions and both even
traveled to Athens and studied in the Academia for a while. So for Cicero
and Varro saving Greek knowledge for future generations had the urgency
of rescuing a civilization from oblivion, and they attempted to do so by
collecting every part of it they could lay their hands on, translating it into
Latin and then organizing and condensing it into what will later be called
encyclopedias – Cicero in a series of dialogues and Varro in his now-lost
Nine Books of Disciplines (*Disciplinarum libri IX*). The consequences of
their efforts were complex. On the one hand, they had undoubtedly saved
many Greek ideas from oblivion, and for centuries to come Europeans
would owe almost all their knowledge of those ideas to the encyclopedic
tradition that Varro and Cicero originated. On the other hand, this very
tradition may have pushed many of the original texts into extinction: as
encyclopedias came to define the canon of knowledge, those texts which
were not summarized, or at least mentioned in them, fell out of the pur-
view of scholars and were forgotten. But it might have had a similar effect
on those texts which *were* included and preserved. Available in digested
version, there was less incentive to go through the expensive and laborious effort of reproducing them, and the full, original version went extinct.
(Recall that each and every text had to copied by hand. This will only
change with the introduction of the movable press in the fifteenth century
that we'll discuss in Chapter 5.)

Moreover, in digesting them for their Latin readers, Varro and Cicero
were unwittingly dismantling both the particularly Greek ideas they lab-
ored to save and the intellectual infrastructure that gave them their meaning. Greek learning with its plethora of schools, as we saw, was a highly

combative field of competing positions. The encyclopedists did not try to emulate this arena of fierce criticism and nuanced argumentation. Instead, they presented a reservoir of sterilized concepts, ready to be put into rhetorical use and organized according to 'disciplines.' These disciplines became important in shaping knowledge and we'll return to them below.

Pliny's Natural History

But the Latin encyclopedic tradition did not merely rehash old Greek ideas. One of the most important troves of real and new knowledge for many centuries to come was compiled by another Roman minor nobleman-turned-politician: the *Natural History* (*Historia Naturalis*) of Gaius Plinius Secundus (23–79), better known as Pliny the Elder, who lived about a century after Varro and Cicero. Born in northern Italy, Pliny worked his way up the military ranks as the Roman republic, so dear to those earlier Roman scholars, was fading into distant memory and a new tyranny was forming itself under the deranged rule of Caligula and Nero. When the military finally took control of the empire and his old comrade Vespasian became emperor, Pliny was called into service and sent on administrative duties as far as *Hispania*, *Africa* and *Gallia Belgica*. He died heroically while trying to save a friend trapped by the famous eruption of Vesuvius which destroyed Pompeii and Herculaneum.

Throughout his adventurous and exciting life, Pliny wrote history, biographies and works on grammar and rhetoric. He would do it at night, after and before his military and administrative chores, and *Natural History*, the pinnacle of his work and his only surviving text, carries the marks of the urgency and verve of its production. It's a work befitting a young and quickly expanding empire. Much less interested in disciplining and categorizing knowledge as it is in capturing its vastness, it attempts to display as much as possible of the enormity of the world in all its breathtaking variety. Pliny's introduction to his third book captures this mood:

> Thus far have I treated of the position and the wonders of the earth, of the waters, the stars, and the proportion of the universe and its dimensions. I shall now proceed to describe its individual parts; although indeed we may with reason look upon the task as of an infinite nature, and one not to be rashly commenced upon without incurring censure. And yet, on the other hand, there is nothing which ought less to require an apology, if it is only considered how far from surprising it is that a mere mortal cannot be acquainted with everything. I shall therefore not follow any single author, but shall employ, in relation to each subject, such writers as I shall look upon as most worthy of credit. For, indeed, it is the characteristic of nearly all of them that they display the greatest care and accuracy in the description of the countries in which they respectively

flourished; so that by doing this, I shall neither have to blame nor contradict any one.

Pliny, *Natural History*, Book III, Intro., V. 1: 151
(http://resource.nlm.nih.gov/57011150RX1)

There are thirty-seven of these books, comprising no less than 117 chapters – a massive work, befitting this "task as of an infinite nature." Pliny offers no fundamental philosophical principles to lead his work, but it is ordered: from the world as a whole, through the stars and the planets, to the marvels of meteorological phenomena – wind and rain; lightning and thunder; and even "showers of milk, blood, flesh, iron, wool, and baked tiles." Next comes the Earth and its various parts: "the countries, nations, seas, towns, havens, mountains, rivers, distances, and peoples who now exist or formerly existed" and then "to Man, for whose sake all other things appear to have been produced by Nature" (Pliny, *Natural History*, Book VII, ch. 1). From there Pliny goes on to describe birds, fish, trees, remedies, metals, precious stones – every fact that he can gather from "writers ... most worthy of credit." He grants this trust to hundreds of them, whose names he respectfully acknowledges, but has little place for their evidence or arguments – it is facts that he collects, many of them quite wondrous, like the "wild beast which kills with its eye," the unicorn of Figure 4.3, or the dragons of distant shores:

Heb. Reem, Unicornis. *Eenhoorn,*

Figure 4.3 A unicorn from Pliny's *Natural History* in a 1644 Dutch edition (*Handelene van de Natuere*, Amsterdam). His report on it is quite reserved – he clearly doesn't pretend to have witnessed one: "There are in India oxen also with solid hoofs and a single horn; and a wild beast called the axis, which has a skin like that of a fawn, but with numerous spots on it, and whiter; this animal is looked upon as sacred to Bacchus" (*Natural History*, Book VIII, Ch. 31, V. 2: 280–1, http://resource.nlm.nih.gov/57011150RX2).

Æthiopia produces dragons, not so large as those of India, but still, twenty cubits in length. The only thing that surprises me is, how Juba came to believe that they have crests. The [Æthiopian dragons] are known as the Asachæi ... and we are told, that on those coasts four or five of them are found twisted and interlaced together like so many osiers in a hurdle, and thus setting sail, with their heads erect, they are borne along upon the waves, to find better sources of nourishment in Arabia.

Pliny, *Natural History*, Book VIII, ch.13, V.2: 231

(http://resource.nlm.nih.gov/57011150RX2)

Pliny finds most things trustworthy, and every detail as important as the other: name and size; behavior and historical anecdotes; legends, practices and use.

The Medieval Encyclopedists

This is not to say that Roman culture produced no sophisticated knowledge beyond *Natural History*. The opposite is true. Mostly, it is know-how that this culture is credited with: the masons' arch, discussed in Chapter 1, along with the aqueducts, roads and theatres in which it was deployed, as well as the managerial skills to distribute water, dispatch soldiers and organize spectacles all are aspects of this know-how. But even this kind knowledge had a theoretical, abstract component to it, like the analysis by Vitruvius we mentioned in Chapter 1. Moreover, Cicero himself was not just a rhetorician and opportunistic compiler of Greek thought: his Stoic philosophy, like that of another Roman politician-turned-scholar a couple generations later – Lucius Annaeus Seneca (c. 4 BCE–65 CE) – was original and very influential. The work of some of the great heroes of Greek learning of the early centuries of the Christian era was so steeped in Roman culture that, even when it was conducted in Greek, it may just as accurately be considered Roman Knowledge. An excellent example is Galen (Claudius Galenus, c. 130–c. 200), the most influential writer on medicine for 1,500 years, whom we'll discuss at length in Chapter 8. A Greek aristocrat by descent who moved in the Hellenistic realm of Asia Minor, Greece and Alexandria, he first made a name for himself in the distinctly Roman practice of physician to gladiators, and spent the peak of his career in Rome.

Yet by the end of Galen's life most of the writings of his Hellenistic predecessors were hardly available, and of the grand edifice of Greek learning only scattered fragments and wistful memory were left. This memory is what is expressed in the myth of the burning of the library in Alexandria and what turned the encyclopedic writings of Varro, Cicero, and especially

Pliny into a long-lasting tradition. The last and very influential pagan member of this tradition was Martianus Capella, who (like Augustine) flourished in Roman North Africa in the fifth century and whose *On the Marriage of Mercury and Philology* (*De Nuptiis Philologiæ et Mercurii*) was so popular there that some 250 manuscript copies of it still survive and those from the sixth century had already gone through many earlier hands. The very title of the book reveals the concept of knowledge embedded in it: philosophy (represented by philology) and rhetoric (represented by Mercury, the emissary god) are being joined together. Gone is the strict Greek distinction between pure *episteme* and practical *doxa* as well as Plato's hostility to rhetoric and its pretense to knowledge, interfering with philosophy's pursuit of *Truth*. The two are now to be wedded, and knowledge is served to them as gifts from the seven 'free' disciplines (from Varro's list, Martianus deleted medicine and architecture and maintained the seven liberal arts we'll discuss later), represented by their respective muses (Figure 4.4, left).

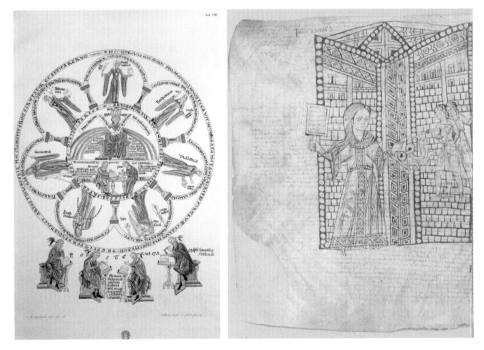

Figure 4.4 Medieval illustrations of Martianus Capella's *De Nuptiis Philologiæ et Mercurii* (*The Marriage or Philology and Mercury*), where philology stands for philosophy, and Mercury, the god of commerce and words – for rhetoric. On the left: a copper engraving illumination by Herrad of Landsberg (1125/1130–1195), "Hortus Deliciarum" (Garden of the Delights); an allegory of "Philosophy surrounded by the seven liberal arts," each represented by its muse. On the right: "Dialectic conquers the serpent of debate." This is an earlier illustration of a tenth-century manuscript.

As the image in Figure 4.4, right, powerfully reflects, debate – the lifeline of Greek learning – has become an anathema, and logic (termed 'dialectic') is not there to order and adjudicate argumentation, but to quash it.

Figure 4.4, right, is taken from a Christian edition of Martianus' encyclopedia. Indeed, the aversion to dispute and disagreement became especially apparent when Christian scholars adopted their Roman predecessors' longing for the lost treasures of Greek knowledge and took upon themselves the task of gathering them for future generations. Isidore (c. 560–636), the bishop of Seville, compiled an influential encyclopedia towards the end of his life which he titled *Etymologiae* and organized mostly, as the name suggests, around the origins of the meanings of words. A century later, The Venerable Bede (672–735) – an English monk who earned his title by being probably the most erudite person of his time – added a historical approach to the collection of *all* knowledge. For much of his acquaintance with the ancients, however, Bede was already indebted to Isidore, just as Isidore was to Martianus and just as Rabanus Maurus (c. 780–856), a German monk who became the bishop of Mainz and the great encyclopedist of the next century, would be indebted to him. With little access to the originals, the encyclopedists had to trust earlier encyclopedists on both matter and form: Isidore's *De Natura Rerum* (*On the Nature of Things*) – his summary of his knowledge of astronomy – became the model for Bede's text of the same name and of similar content. Bede's text was then emulated by Rabanus in his *De Rerum Naturis* just as his *De universe ... sive etymologiarum opus*, as its name reveals, mimics Isidore's *Etymologiae*.

Christianity and Learnedness

For Christian scholars of the early Middle Ages, Greek learning posed less of a threat when it was stripped of its highly contentious character. They were much more comfortable when presenting deeply antagonistic positions – say Plato and Aristotle's competing ideas of the relations between matter and form – as nothing worse than differences in wording. Yet it was far from obvious that Christians should devote such attention to Greek knowledge in the first place. As we saw in Chapter 1, early Christian thinkers were much more inclined to dismiss it for its pagan origins. And it is important to stress that people like Bede or Isidore were not just scholars who happened to be Christians. Their Christianity was the most fundamental part of their being – intellectually, personally and institutionally: Isidore was a bishop and Bede, as he attests, almost never left his monastery. But the social and cultural role of such people was dramatically different from that of the likes of Tertullian, who could contemptuously ask, a few centuries earlier, "what indeed has Athens to do with Jerusalem?" (see Chapter 1).

The Changing Cultural Role of Christianity

In the first decades of the fourth century – about half a century after Tertullian's death and just before Augustine's birth – Rome saw a period of new prosperity after a long decline. This came under the rule of Flavius Constantinus (272–337), better known as Constantine the Great. Constantine fought his way to the throne and ruled with an iron fist (from 306 to 337), but he was also a capable administrator and a fearless reformer. Two of his reforms would have a crucial, if indirect, impact on the way knowledge would take shape in Europe for many centuries.

The first, in 330, was to move his capital from Rome to the ancient city of Byzantium, which Constantine renamed, unsurprisingly, Constantinople. On the coast of the *Bósporos* straits (today's Bosphorus) and thus at the gate to Asia Minor, Byzantium represented a clear strategic and commercial vision; Constantine was consolidating the emperor's rule after decades of decentralization. But the long-term cultural effect of the move was the opposite of the unification he had in mind: it radically deepened the gap between the western parts of the empire, where Latin was the language of politics and commerce, and the eastern parts, which would remain Greek-speaking. A century later, Rome would be sacked – by the Visigoths in 410, by the Huns under Attila in 450 and finally by the Vandals in 455 – and the communication between scholars on the now-hostile border would become very strenuous. For those on the Latin-dominated west, the Hellenistic treasures of the east would fade into nostalgic memory.

The other reform, in 331, was to reverse the standing of Christianity. Succumbing to its growing popularity, Constantine lifted the restricting measures aimed at curbing its spread (and provided many opportunities for martyrdom that Christians seem to have been taking in at a rate that the authorities found alarming). He then went further, granting Christianity enough privileges to turn it, for all practical purposes, into something of a state religion. Whether Constantine himself converted, and if so how sincerely – questions that occupy historians – and even the fact that his acts would be reversed (and reversed again) by future emperors carry relatively little significance. What is crucial is that Christianity came to adopt the organized, centralized, hierarchical structure of the Empire with its common language – Latin – and its written law. When Rome fell and the Western Empire – already more of a federation of Germanic tribes than a coherent empire – collapsed, the Catholic Church found itself the custodian of this law and of the learnedness it required.

The Church, as discussed in Chapter 1, was primed for the challenge. Augustine, a colonial Roman for whom these vicissitudes were literally world-changing, dismissed Tertullian's contempt: "we ought not to give up music because of the superstition of the heathen ... we ought not to refuse to learn letters because they say that Mercury discovered them" (*De Doctrina Christiana*, II, XVIII). Learning – any learning, as long as belief was unshakable – elevated the soul.

The Church took to the task as it usually did – institutionally, but with a strong sense of its shortcomings: "Woe to our age, for scholarship has died out among us," wrote Gregory (538–594), bishop of Tours and a fine scholar himself, in his massive *Decem Libri Historiarum* (*Ten Books of Histories*,

Figure 4.5 The scriptorium. Usually the monk would not be alone in the scriptorium – like most aspects of monastic life, intellectual work takes place in the presence of others. But the image, from a fifteenth-century manuscript and depicting the translator and copyist Jean Mielot (died in 1472), captures well the devotional concentration of the work in the scriptorium, which by decree was conducted in complete silence.

or *History of the Franks*, as it's commonly known). Already a couple of generations before, Anicius Manlius Severinus Boëthius (480–c. 525) was keenly aware that his excellent Greek – he may have been educated in the East – was becoming a rarity. Boëthius was a high official in Rome, by then Christian and under Gothic rule, and much of his political rise and fall was predicated on the relations between Rome and Constantinople. So he was also self-serving when he took it upon himself to collect, translate and comment on as much of the Greek masters as he could, especially in logic and mathematics. It was still a worthy endeavor: his version of Aristotle's logical works, for example, became the only significant portion of the latter's corpus available in Europe until the twelfth century, and some Greek arithmetic we still know only through his translation and adaptation. Boëthius' aspirations were shared by his successor in the Ostrogoth court of Rome: the long-lived Flavius Magnus Aurelius Cassiodorus (c. 485–c. 585; Boëthius, by comparison, finished his life by execution, on whose details there are a few versions, all equally horrid). Cassiodorus went as far as trying to establish a Christian academy in Rome – the very idea that repelled Tertullian. When this failed, he founded a monastery with an adjacent learning center on his family estate in Vivarium, Southern Italy. The Vivarium monastery was dedicated to the study of the marvelous collection of texts, both sacred and secular, in its library. The collection was not Cassiodorus' own: it was the fruit of a spirited project of gathering, copying and translating manuscripts; the monks at Vivarium were scholars.

The Monastery and the Scriptorium

With Vivarium as a model, the monastery became Christianity's primary institute of knowledge. At the heart of many monasteries was a library, and at its heart – the *scriptorium*; the room of the scribes (Figure 4.5). In the scriptorium, manuscripts – usually on vellum, fine parchment made of calf skin – were produced and reproduced: copied, translated, annotated, commented on and illustrated. Almost no censorship and very little selection were practiced – the veneration of the wisdom of the ancients outweighed any dread of "heresies ... instigated by philosophy," as Tertullian had put it (see Chapter 1). The hunger for this wisdom was insatiable and any text was deemed worthy of fulfilling it: sacred or profane, heretical or obscene. A strict law of silence ruled the scriptorium: the work there was a spiritual exercise. Texts were not only reproduced materially; they were absorbed, committed to memory, elevating the soul. And this spiritual learnedness was institutionally sanctioned. The monk's vows bound him to a life devoted to prayer and work, strictly governed by a written "Rule of the Order," but if he was literate, he could discharge his work duties in the scriptorium. "My desire has always been either to study or to instruct or to write" said Bede, "in that I have always observed the rule of the order and the daily singing of the hours in church" (Bede, *Historia Ecclesia*, cited by Pedersen, 48).

Bede was a Benedictine, and the Rule he was following had been written some 150 years before his birth by St. Benedict (Benedict of Nursia, c. 480–550). But in the world surrounding the monastery, there was very little orderliness. The Roman Empire was first taken over by the Germanic tribes populating the areas it occupied to its north, and its final collapse left Europe divided into numerous kingdoms, princedoms and fiefdoms. The relations between and within these entities were strictly hierarchical, but they were ruled much more by loyalties than by law. At the bottom of the hierarchy stood the *serfs* – the peasants bound to the land they cultivated but didn't own. They paid for it in crops and labor to the lord of the *manor* – the fundamentally self-sufficient rural estate. The lord owed his allegiance to the king who granted him the estate in exchange for services rendered – usually military. Often the exchange happened many generations earlier, between ancestors of both, but the lord's first loyalty was still to his remote king rather than to his local serfs. The kings and princes within the Kingdom of the Franks – that is, the Holy Roman Empire – owed their allegiance to the emperor, elected by the most powerful of them every time the previous one passed away.

It befell the Church to offer some structure and stability to this amorphous cluster of local customs and long-distance loyalties – known as 'feudalism' – by which Europe was governed. With its official structure, its uniform language (Latin) and its universal code of law (the Roman), the Church provided the only institutional cross-European uniformity. Even more important was the Church's role we discussed in Chapter 1: to grant overall legitimacy to this political system. This legitimation it provided by embedding the political hierarchies in to the 'Great Chain of Being': the metaphysical and religious hierarchy stretching from profane matter up to the abstract divine (Figure 1.13). It was a legitimacy essential for ruling. Emperors who tried to defy the pope's decrees learned that their military might was no match to the power he could unleash through the symbolic act of excommunication. So much so, that Henry IV's willingness to take upon himself any measure of humiliation in order to relieve the pope's wrath – to 'walk to Canossa,' where the pope withdrew himself – became a proverb of surrender.

Medieval Know-How

It was only under the auspices of the Church that organized production of knowledge could take place. Declining commerce with the great cities of the Mediterranean and little indigenous urban culture meant that Europe of the centuries after the collapse of Rome had no cultural infrastructure to support anything similar to Hellenistic learning. This is not to say that the people of the manor or the budding medieval town developed no knowledge worthy of our story, only that this was almost exclusively the 'know-how' that by its very nature leaves behind artifacts, but few testimonies as to how they were produced.

Traces of this know-how, however, are to be had – sometimes embedded in the artifacts, sometimes more directly. Recipe books, for example, tell of elaborate knowledge of local flora and fauna and their nutritional and medicinal value. This was the domain of the witch, the rural physician and the herbalist; the recopies of the town apothecary show that his knowledge (unlike the rural 'medicine woman' or herbalist, the apothecary was always a man) spanned further: he was literate, imported *materia medica,* and mixed his own preparations. In the manor, as we learn from paintings and remains, natural energy was harnessed in increasingly efficient ways. The improved use of beasts of burden, which we discussed in Chapter 1, is perhaps the simplest of examples. Mills, marshalling the elements and comprising sophisticated mechanical devices such as cranks and geared wheels, provide a more elaborate example. Mills steered water and wind, moving grindstones to produce flour from grain, big saws to cut wood, and large bellows to stir up furnaces in which iron was smelted. The mills were a great classical invention whose use significantly declined with the collapse of the Western Roman Empire, but re-emerged in the eighth century and onwards: the *Domesday Book*, a 1086 survey of England and Wales prepared for William the Conqueror, reports 5,624 (!) water mills in England alone, and the number seems to have doubled every century. Many monasteries, as they aggressively transformed themselves into manors by appropriating the land around them, became centers of technological development of this kind. Brewing, distilling, dyeing and founding were ways to add value to the raw materials by which the serfs paid their tithes – their charges for using the land. The monasteries thus had the wherewithal and the motivation to develop practical disciplines like herbalism and metallurgy – and with them alchemy – as well as mill technology.

Within the monastic realm, 'practical' didn't only mean 'material.' Observance required observing time: both "the daily singing of the hours" (as Bede termed it above) and the yearly arrivals of the holy days. The daily task was greatly aided by the same mechanical technology that allowed for mills: geared wheels were essential for the great medieval invention of the mechanical clock. Wheels were not enough: the clock required the insight that in order to measure it, time needed to be broken down into segments, and thus too the mechanical contraption to perform the task: the verge-and-foliot escapement (Figure 4.6). The origins of the European escapement are unknown – a very different escapement was used in the great eleventh-century Chinese astronomical clock of Su Song. We'll also probably never know if and how the medieval application of geared wheels related to the technology of the Antikythera mechanism we discussed in Chapter 3, which also had a chronometric task. If there were such ancient resources from which this technological knowledge stemmed, the monastery would indeed be the place where they could be expected to surface.

Monastic daily life followed artificial time, ordered by the bell marking the hours for prayer (think of the children's rhyme *Frère Jacque*), in which the mechanical clock with the ticking of the foliot had a natural place. The annual time was kept with the help of the *computus*: an applied version of the

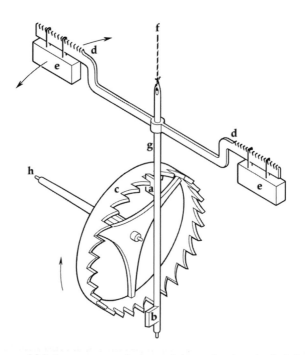

Figure 4.6 *Verge-and-foliot* escapement – the heart of the medieval mechanical clock. Illustration by David Penney ©. The crown wheel **c** is moved clockwise by weights – the clock's engine – and its axle **h** is connected to the dial, directly or through some more gears. The teeth **a** and **b**, in right angle to each other, are pushed by the wheel, causing verge **g**, hanging on the chord **f**, to oscillate left and right. The swing of the balance ('foliot' – fool) **d** alternately engages and disengages the teeth, allowing the crown wheel to 'escape' (hence the term) and leap only one tooth at a time. The pace of oscillation is regulated by moving the weights **e** inside or out. The further out the weights are, the higher is their 'positional weight' (see Chapter 9), the longer the oscillation, and the slower the pace of the clock.

Hellenistic astronomical calendar, by which the dates relevant to the Christian were determined (Figure 4.7). The *computus* was a 'paper tool,' rather than a mechanical one, but it was as much a piece of practical, technological know-how and the time it kept was as artificial as the clock's. It did not attempt to capture accurate heavenly positions (such as the equinox and the full Moon

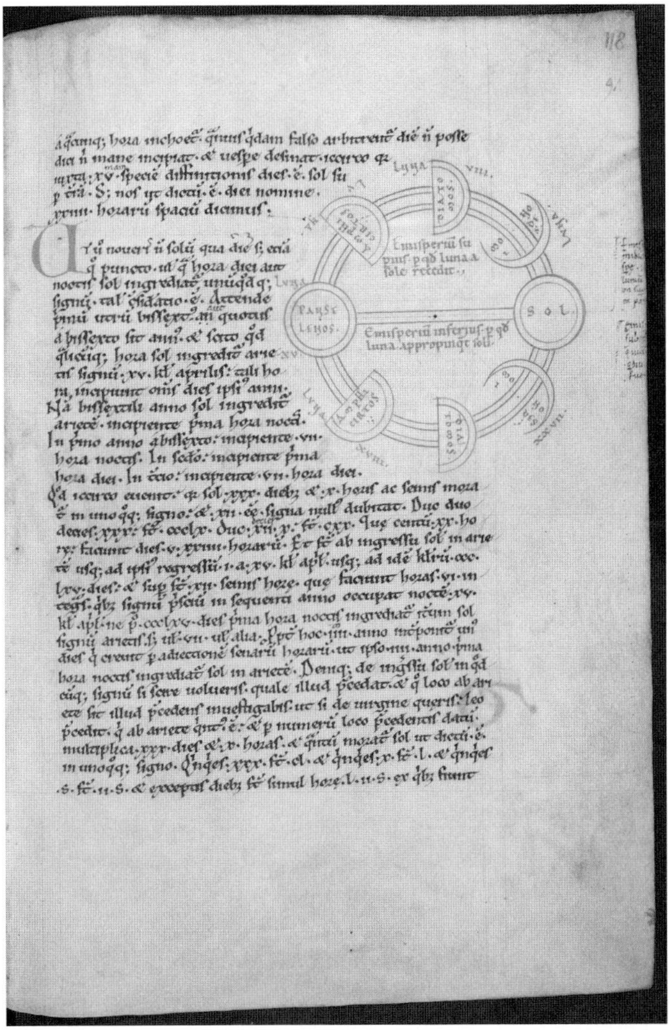

Figure 4.7 Medieval knowledge of Hellenistic astronomy. A diagram of the relation between the phases of the Moon and its position relative to the Sun (Sol, on the right) from Abbo of Fleury's (c. 945/950–1004) *Opinion concerning the System of the Spheres*. The text is part of a compilation of astronomical texts, written in Durham cathedral by a number of scribes during the second quarter of the twelfth century, now at the University of Glasgow Library (MS Hunter 85). The most important is Bede's *19 Year Cycles* (referring to the Metonic cycle – see Figure 3.12), a *computus* providing the dates for Easter for the years 1–1253 (so for some 500 years ahead of his time), showing powerful command of the practical sides of that astronomy.

which mark Easter) but to set regular dates, more or less around those astronomical events, by which rituals could be regulated and instituted.

The monastery was not confined to bookish knowing-that, knowing-how was not necessarily material, and both knowing-that and knowing-how had important social and institutional roles. In the town, a peculiar medieval institution of knowledge developed: the *guild*, an association of practitioners of a craft. There were many dozens of guilds in any town, and hundreds in Paris or London: for the masons, carpenters, tailors, surgeons-barbers, silk merchants, drapers – one for every trade. As the economy developed and with it the distribution of labor, the guilds splintered into specific domains of skill: The Worshipful Company of Carpenters is to this day separated from The Worshipful Company of Joiners and Ceilers (both going strong in the City of London since 1271 and 1375 respectively) because carpenters use nails while joiners use only glue. The guild provided to its members control over tools and materials, protection against competition, mutual financial and political support and even some religious solidarity, but most importantly – it sanctioned the knowledge of its members and monitored its distribution.

A boy would be taken by his father to a master's workshop or store at an early age and be put under his custody as an apprentice. He would do menial jobs in the household in exchange for room and board and would acquire the trade and its secrets through work; starting with the simple tasks, which gained in complexity as he gained in age, skill and his master's trust. It was to the guild that the master would have to answer if he relinquished his duties – say, by enslaving the boy without instructing him – and it would be the guild that would test the boy – now a young man – to decide if he had acquired the trade. Upon acquitting himself well in the guild's test, the apprentice would become a journeyman, taking to the roads and offering his services to other masters, now for salary but still with an eye to improve his skills. After the number of years set by the guild, the journeyman would be allowed to take the final examination, and if successful, would be 'set free by the guild' – which meant he had become a master himself, permitted to set up his own shop and take apprentices.

The guilds were very important in the development of medieval politics and economy – some even say stunting it with their in-built hierarchy and conservatism. For us the guilds provide two lessons beyond the particulars of medieval ways of knowing. First, they are another example of the essential relations between institutions and the knowledge they produce. Argumentation and the dialectic competition of ideas, for example, were built into the Greek philosophical school; careful maintenance and piecemeal evolution of material tools and bodily techniques were promoted by the guild. It turns out – and that's another lesson and somewhat of an antidote to our discussion in Chapter 1 – that pure 'know-how' can also be organized and systematically reproduced, even if not in written form.

Education and the Church

The know-how produced and maintained by the guilds was organized, but by its nature as trade knowledge it was local, particular and secretive. This fractured nature was representative of European culture in the millennium following the demise of the Roman Empire. The Church was unique in its commitment and wherewithal – institutional, linguistic and intellectual – to develop, sustain and distribute 'knowing-that,' and this fact did not escape the notice of Charlemagne – Karl der Grosse, or Charles the Great (742–814) – the king who united the Franks under his rule and established the Holy Roman Empire. The Franks were warriors – anything but scholars – yet Charlemagne was apparently attuned to the value of education. He already had a monk by the name of Alcuin of York (730–804) – scholar, theologian and poet – teaching the seven liberal arts in his court, and in 787 he decreed the following:

> [the King] has judged it to be of utility that, in the bishoprics and monasteries committed by Christ's favor to his charge, care should be taken that there should not only be a regular manner of life, but also the study of letters, each to teach and learn them according to his ability and the Divine assistance.
>
> Cited in F. V. N. Painter, *Great Pedagogical Essays* (Hawaii: University Press of the Pacific, 2003 [1905]), p. 156

Already educating its future clergy in cathedral schools and its monks in monastery schools, the Church was ready for the task. But beyond the education of Europe's elites, the Catholic Church had now been officially entrusted with the very creation of its *episteme*.

The University

Learnedness was the prerogative and duty of the literate monk, and learnedness requires conversing with other scholars – face to face or through their writings. An informal institution had thus developed: that of the wandering learned friar, making his way between monasteries, consulting their libraries, scriptoria and fellow scholars. He could travel alone or with a young scholar-apprentice or two, and if he gained a name for himself, a small group of disciples could have gathered around him. If this brotherhood of scholars – *collegium* in Latin – would grow more numerous than was convenient for travel and courteous to request other monasteries to host, it might opt to settle in a town, probably finding residence in the cathedral. When a number of such *collegia* assembled in a town, they might seek a charter from the king that, if granted, would turn them into a *universitas*.

Sovereignty and Its Bounds

By the twelfth century, the relative calm and prosperity (which we discussed in Chapter 1 in relation to the cathedral) allowed towns – many of them newly established – to sponsor such institutions. Like the guild, the *universitas* was an urban institution peculiar to the medieval fragmented society. It was a community that was granted partial sovereignty by the king, the pope or even the emperor, and thus made independent of the local authorities, such as the town's senate or the region's fief. It could be a religious body such as a Jewish community; a fundamentally economic one like the guild; or the assembly of colleges of monastic scholars for which we now reserve the title 'university.' The first university was chartered in Bologna in 1088; in Paris there was teaching already from the mid-eleventh century and the university was chartered in 1150; Oxford was teaching from 1096 and chartered in 1167, and Cambridge in 1209 and 1231, respectively. Over the next two centuries universities would be established throughout Christendom – from Portugal to Bohemia; from Sicily to Poland and Scandinavia; from Scotland to Germany.

The independence of the *universitas* meant that it conducted itself and judged its members according to its own rules – a very important privilege. In the case of our community of scholars, this privilege meant that, being a Church institution, students and faculty were protected from the gallows and the stake, since Church law didn't allow for capital punishment (when the Church wanted someone executed, it would deliver them to the secular authorities). But this independence of the university was carefully constrained. It had no swords and horses to defend it, only custom, and could only be maintained as long as those with real power did not feel threatened. We are well-acquainted with this dance of power: military or police stay off university campuses, until unrest on its lawns threatens to boil to the streets. But as long as the political authorities could be persuaded that all mischief was well contained, the privilege of *universitas* bound them to not interfere.

There was a formalized way for university scholars to declare that their deliberations were strictly within the bounds of the university and thus protected by its privilege of sovereignty. They would designate a claim as discussed *ex cathedra* – namely, within the context of teaching; or *ex hypothesi*, as mere hypothesis. This meant that they didn't commit to the truth of the claim; they neither believed in it nor intended to proselytize it. Under these designations all scandalous, not to mention heretical, positions could be discussed: could it be that God isn't good? Are there things He can't accomplish? Or even: could it be that He doesn't exist? *Ex cathedra* or *ex hypothesi*, the discussion was to be taken as a purely intellectual exercise, posing no threat to the world outside the university. We are familiar with these practices, if not as well formalized: a modern-day lecturer can discuss the most racist, sexist or otherwise offensive view she chooses, as long as she makes it clear that the view is discussed, not advocated.

Interlude: The Foundations and Decline of Academic Freedom

It is crucial to pause shortly and reflect on the lesson just learned. The freedoms and responsibilities we associate with the university and think of as essential for scholarship and education are not a modern achievement but remnants of an ancient institution. The academic freedom from political interference, the right to inquire into and to teach what one deems worthy, independently of the shifting preferences of public and politics – these are peculiar medieval privileges. They are not protected by civil liberties but tolerated as habit and custom. They don't reflect secular belief in human rights but religious trust in vocation: 'tenure' is not a labor protection granted to an employee, but a designation of an acceptance into a community of devoted learners. This is the reason it cannot be revoked for economic or organizational considerations, but only for severe breach of the community's code of conduct. The university has very quickly taken on the task of educating elites and training professionals, as we'll see below, but at its core it is not an institution providing a service to the wider society but a community of learners, whose reason for existence is its very scholarly pursuit. In this form, the university has been extremely long-lived and stable: its core structure, procedures and pedagogy have remained virtually the same over its 800 years of existence. Universities have been remarkable in producing and disseminating knowledge – humanist and scientific; abstract and practical; medical and technological. But these achievements, unparalleled by any other institution of knowledge, were but secondary consequences of their fundamental aim. This was, as we explained, to give home, independence and sovereignty to a community of scholars, for whom scholarship is a vocation.

The university has proven extremely resilient – or rather flexible and crafty – to external, hostile political pressure, but less so to the modern pressures caused by its own success. In the last decades of the twentieth century, universities were recruited as instruments of mass education and grand, expensive research. They became large institutions, requiring very large resources, provided from public funds and thus regulated by agents of the public realm. Almost unwittingly, universities relinquished the very independence that their title connotes. Indeed, by the first decades of the twenty-first century, they started succumbing to demands to reform their medieval ways so that they could be rationalized as a contemporary service provider: to consider their value as measured by its cost on the one hand and the satisfaction of an envisaged clientele on the other. For those who may assign this work as a textbook, these are formidable, if not traumatic changes. It is left to be seen if the culture of those to whom it may be assigned will still have a place for a sovereign *universitas* of learning.

Masters, Students and Pedagogy

Who populated the medieval university? They were all male. Quite early on the universities diversified and neither students nor masters necessarily belonged to a monastic order, but they were regarded as members of clergy by virtue of belonging to the university, and were hence regulated and protected by its laws. Because they mostly came from afar – anywhere within the vast realms of Christendom – these protections were not only privileges, but necessary for survival. The students were often very young – in their early to mid-teens – and for many of them this was their first formal education. They could be second sons, whose parents envisioned for them a career within the court or Church bureaucracy. The Church recruited its officials from the educated cadres produced by the university, so the basic Latin and mathematics of the first couple of years of study were necessary. They could also be, more rarely, Church-educated peasants; then as now, education was a good means of social mobility. Very quickly, it was to the university that a young man went to become a physician or a lawyer. A university could comprise something between 1,000 and 1,500 students at any one time – Paris would take in 500 a year, most of them leaving after a couple of years. Overall, it is estimated, European universities educated some 750,000 from the middle of the fourteenth century to the end of the sixteenth; not a small segment of a population averaging about 75 million.

They would study in two main ways we still use and one which we have regrettably lost. In the *lectura* (Figure 4.8, left), the master would stand at his podium, a manuscript in front of him, and would read, interpret and comment on the text to a presumably mostly passive (though hopefully attentive) audience. The comments gradually took the form of questions and answers and these *quaestiones*, copied and disseminated formally and informally, became the paradigmatic genre of medieval writing – their

Figure 4.8 Two of the main pedagogical institutions of the medieval university (the third being the *repetitio* or tutorial). On the left is the *lectura*, represented by the great fifteenth-century legal scholar Pietro da Unzola lecturing at the University of Bologna. Note that Unzola is depicted as reading and interpreting a text, but painted shortly after the invention of the movable press (see Chapter 5), the image also has some of the students looking at books, even though the text from which the miniature is taken – the *Liber iurium et privilegiorum notariorum Bononiae* – is itself in a manuscript form. On the right: the *disputatio*. The illustration is from the Statutes Book (*Statuenbuch*) of the *Collegium Sapientiae* – a theological seminary established in 1555 in Heidleberg. Note the hourglass behind the students, a testimony to the formalized and competitive nature of the debate.

structure and sophisticated language suggests that the questions were generated by the master, not the students. In the *repetitio*, a young master or a senior student reiterated and elucidated the material taught in class. Finally, in the *disputatio* (Figure 4.8, right), two senior students argued a point in front of the class.

The *disputatio* truly captures the spirit of high medieval scholarship and its complex of intellectual freedom protected by careful institutional constraint. The format of the *disputatio* was rigid. A question would be chosen by the master; one disputant would present a thesis, to which the other provided an anti-thesis; the first would then provide an argument, which would be answered by a counter argument, followed by an example and a counter example, and so forth, until the master would deliver a *determinatio* – a decision on who had won. Within this highly structured and clearly pedagogical context, however, the most explosive theses could be vehemently argued and even prevail, if supported by better arguments (although the master would also clarify the official Church position on the point, the position in which all must believe). The *disputatio* was a crucial part of the curriculum. To become a bachelor of arts, the student had to attend and 'determine' a disputation; to become a master, he had to both participate in one disputation and chair two more.

Curriculum

Tradition prescribed the seven liberal arts as the foundation of all learning, which made good sense: entering the university, students had to acquire the basic tools of scholarly language and reasoning. They therefore started with the *Trivium* – the three linguistic arts: grammar (of Latin); rhetoric; and logic (Figure 4.9, left). Logic was often called dialectic and was taught

Figure 4.9 The university curriculum, set as an allegory of the seven liberal arts, corresponding to the *Hortus Deliciarum* in Figure 4.4. On the left – the *Trivium*: Grammar taught by Priscianus, Rhetoric by Cicero and Dialectics by Aristotle. On the right – the *Quadrivium*: arithmetic taught by Pythagoras, Geometry by Euclid, Music by Milesius (the legendary king of the Irish) and Astronomy by Ptolemy. The image is a miniature manuscript illumination for the didactic poem *Der Wälsche Gast* (*The Welsh [meaning here – Romance] Guest*) by the German-writing Italian Thomasin von Zirclaere (1186–1235), lecturing to young nobles about chivalry.

according to Aristotle's logical-methodological texts, his *Organon*, especially once they became fully available (more on that below). Two years of this sufficed for most students; those who stayed and passed their exams would continue on to study the *Quadrivium*, or the four mathematical arts: arithmetic, geometry, astronomy and music (Figure 4.9, right). These particular disciplines, again, were the product of tradition. But this division of mathematics was also given good sense by the Platonist mathematician Proclus of Athens (412–485), the head of the Academia in his time, in his commentary on Euclid: arithmetic dealt with discrete static objects (represented by the numbers); geometry with extended static objects (lines and bodies); astronomy with discrete moving objects (the planets); and music – with extended moving ones (strings). These disciplines were hardly practical, but they trained the mind and prepared the student for going out into the world or continuing his education.

If the student chose to further his education, then after completing the requirements to get his *baccalaureate* – to become a bachelor of arts – he could decide whether to join the higher faculties – law, medicine or theology – or to continue in the arts faculty. The former two were fundamentally professional schools: with LLD (or Dr. Juris) one was certified as a lawyer; with MD as a physician. Though the teaching was mostly theoretical, periods of practical apprenticeship were included. The case of theology was somewhat more complex: it was the queen of disciplines, highly abstract and very difficult. Obtaining a ThD could take ten to fifteen years; however, it all but assured a good career in the service of the Church.

If the student decided to remain in the arts faculty and work towards the title *Magister* – master, that is, teacher – then he would move to study the three philosophies: metaphysics, ethics and, most importantly, philosophy of nature. This meant, primarily, studying Aristotle's works like *De Anima* (*On the Soul*), *De Caelo* (*On the Heaven*) and *De Generatione et Corruptione Animalium* (*On the Generation and Corruption of Animals*), and the commentaries on them. It took some time before the lower faculty was allowed to confer its own doctoral degree – the PhD – which then, as now, did not offer much by way of a career path outside the university.

Grammar was taught with a textbook composed by the obscure Priscian in sixth-century North Africa, and astronomy with the *Tractatus de Sphaera* (Figure 4.10) by the University of Paris teacher John of Hollywood, known as Johannes de Sacrobosco (c. 1195–c. 1256); a digested, non-technical version of Ptolemy's *Almagest*. For arithmetic and music there were treatises by Boëthius, and the basis for all theology was the *Sentences* (*Quatuor Libri Sententiarum*) of Peter Lombard (Petrus Lombardus, 1096–1160).

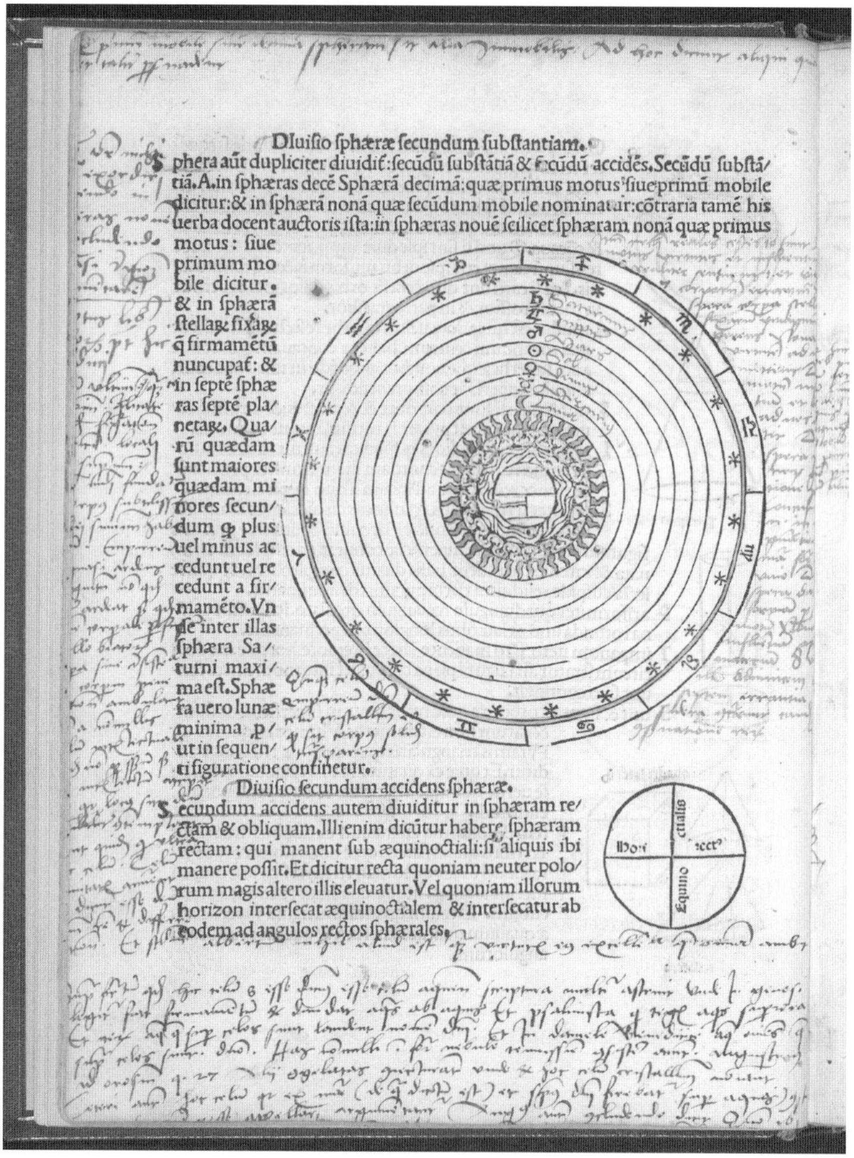

Figure 4.10 The primary astronomy textbook in the medieval university: *Tractatus De Sphaera* by Johannes de Sacrobosco (John of Hollywood, c. 1195–1256). As befitting a working book, this manuscript copy, from c. 1230 (currently at the Stillman Drake collection of the University of Toronto), is heavily annotated by at least two hands from two different times.

This was a massive compilation of Biblical and Church Fathers' thought, elaborated in Aristotelian modes of argumentation while carefully maintaining and stressing Augustine's hierarchy between pagan and Christian knowledge:

Aristotle said that (there are) two principles, namely, matter and species, and a third called the "operative (cause)"; also that the world always is and was. Refuting the error, therefore, of these and similar (men), the Holy Spirit, handing down the discipline of truth, signifies that God at the beginning of times created the world and before times eternally existed, commending (thereby) His Eternity and Omnipotence.

Lombard, Sentences, Book 2, Dis. 1, ch. 3.

The encyclopedic tradition provided the masters of the early universities with a wealth of such works, useful as textbooks. But they were scholars first, and teachers second. The intellectual urge that drew them together in the first place was enhanced by their assembly, and with it – the craving for the complete versions of the works they were teaching. These works, however – the original texts of Plato and Aristotle, Euclid and Ptolemy, Cicero and Galen – were not available to them.

The Great Translation Project

The original works were to be found close to where they were created – in the cities of the Hellenistic realm. The north-western part of this area – Greece and Asia Minor – still comprised the declining Byzantine Empire. The rest – Mesopotamia and North Africa – had been conquered by the Arabs in the seventh and eighth centuries, and was now part of great Muslim empires stretching from Persia to the Iberian Peninsula. The demand created some supply: from the places where Christianity and Islam rubbed shoulders – Constantinople, Sicily and especially Spain – samples of these texts were trickling in. Ambassadors, who were often scholars themselves, sometimes carried them, for their own use or as offerings. Merchants, sensing an opportunity, brought them as luxury goods. Constantinus Africanus (c. 1020–1087) is an interesting representative of the merger of travel, trade and scholarship which came to typify the age: a North African Muslim merchant, he used to sell Arabic medical manuscripts of the Galenic tradition in Salerno in South Italy, until he converted to Christianity, became a Benedictine monk in Monte Casino, and turned translation into his vocation. The texts collated and translated by Constantinus and his brethren in Salerno and Monte Casino would shape European medicine for centuries, and we'll return to them in Chapter 8.

Monte Casino was an early model of what was to come. Once Christian scholars took direct responsibility for the acquisition of these texts, an exhilarating project of intercultural exchange transpired. Its center was the city of Toledo, at the heart of the Iberian Peninsula, which Alfonso VI of

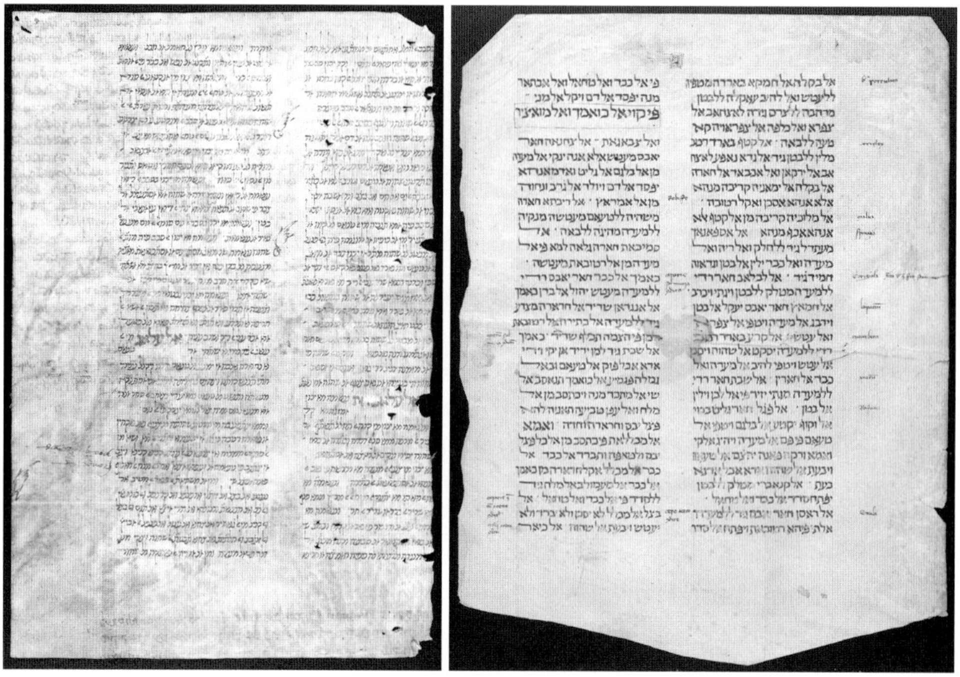

Figure 4.11 The multicultural context of the translation projects illustrated by two parchment leaves from the fourteenth century. Both are in the Arabic dialect of the Jews of the Western Muslim realm written in Hebrew script, but are not by the same scribe. The manuscript on the left is "On the causes of hiccupping" and "On the treatment for hiccupping." It is on a palimpsest (see Chapter 5) whose undertext is in Hebrew, which likely originated in the Maghreb or Andalusia, and carries later annotation in Latin and Castilian (note too the little drawings of hands pointing at presumably crucial points in the text). The manuscript on the right is a transliteration of *Kitab al-Mansuri* by al-Razi, originally composed in Arabic near Tehran or Baghdad in the tenth century. Its subtitle is "On the effect of pickles and sour substances" Originating from northern Iberia (probably Mallorca), it was later annotated in Latin. Together the manuscripts draw a beautiful itinerary of medieval knowledge traversing long distances and political, linguistic and religious barriers. One can surmise that after being penned by Jewish scribes, they made their separate ways from North and South Iberia to Castile and perhaps Catalonia, where they were read and annotated by physicians or scholars who could read them in their very particular dialect – or perhaps were collaborating with someone who could – but were more comfortable annotating in the Christian languages, so were probably university educated. Thanks to Rowan Dorin and Hagar Gal for the research and analysis.

Leon conquered from the Muslims in 1085 and turned into the capital of Castile. Toledo was an important Christian center under the Visigoths' rule in the sixth and seventh centuries, but for the Umayyad Caliphs who captured it at the beginning of the eighth century it was a remote provincial town, very far from their capital, Damascus. It became a focus of feuds between Arabs and Berbers, local and imperial soldiers and rulers, and by the time of the Christian conquest it was only a small city-state. Yet it was still an important center of learnedness, both Muslim and Christian, and on the foundations of this learnedness the Archbishop Raimundo (Raymond de Sauvetât, reign 1126–1151) turned the acquisition of these longed-for texts into a Church project by establishing the College of Translators at the library of the cathedral.

It was a college indeed: a scholastic brotherhood, whose members came from all over Europe. The most famous of those traveling translators was Gerard of Cremona (1114–1187), who came from his hometown in Northern Italy in his 20s and spent the rest of his life in Toledo. In the five decades of work, he translated some seventy books from all branches of science, among them works of Aristotle, Euclid, Archimedes and Ptolemy, as well as many great Muslim scholars like Ibn Sina, Al Kindi, Al Razi and Thabit ibn Qurra (more about whom in Chapters 5 and 8). In fact, even if the original work was in Greek, the translation was almost exclusively from Arabic. The Muslim scholars had long turned Greek knowledge into their own, and had first translated it to their own language of knowledge (we will return to that shortly). Gerard apparently mastered Arabic, but more often than not, the translation was a painstaking process by which a text would be rendered, word by word, from Arabic to the local Castilian, and from that to Latin. A few centuries later, the humanists of Renaissance Italy would smugly ridicule this process and the quality of its products, but it was a heroic effort, testifying to the true craving for this knowledge. It was also a true cross-cultural and cross-religious effort, in which people like John of Seville – a Spanish Jew converted to Christianity – would play a crucial role as mediators between cultures and languages.

Muslim Science

The First Translation Project

The Christian scholars traveling to Toledo were looking for the complete, original Greek texts they knew in fragments, but they were not retrieving their own lost cultural heritage. First, the Hellenistic culture that produced these texts became 'European' only in hindsight, in the Europeans' own mind. The ancient Greeks had little interest in the people to their north, whom they called 'barbarians' because the northern speech sounded to them like incomprehensible gibberish. They knew, admired, and sometimes feared and detested the great cultures to their east and south – Persia, Babylon and Egypt – and these were the directions in which they expanded their own culture, into the realms where Muslim culture would rise. Secondly, the Muslim scholars who inherited and adopted these texts were not holding them as a deposit. They had been using them to construct their own intellectual edifices for centuries.

When Mohamed united the Arabs, converted them to Islam, and led them to conquer the Fertile Crescent and the Eastern Mediterranean, they were mostly a nomadic culture. But it did not take long for the conquerors to get off their horses and camels and adopt much of the means and ways of governance of the empire they'd deposed. This required translating the documents of the Byzantine rulers, from whom the Arabs wrested the Hellenistic realm, and the skills for such a translation project were possessed by the Greek-literate bureaucracies they inherited from Byzantium. Translation Projects like that were a very long tradition in Mesopotamia: the Akkadians translated Sumerian documents already in the second millennium BCE. The Byzantine Christian-Orthodox intelligentsia, however, brought with it a disdain for pagan knowledge, similar to what we have read in Tertullian. During the first great Muslim empire – the Umayyad Caliphate, which reigned between 661 and 750 from its capital Damascus all the way from Persia to Iberia – not much Greek science was rendered into Arabic.

This would change dramatically when the Abbasids deposed the Umayyads. With their power base in the east – today's Iran and Iraq – they could take a much more neutral approach towards Hellenistic culture and had good political reasons to turn careful attention to its treasures. The notion of ancient knowledge was common to many of their constituents, who prided themselves on their own antiquity: the Jews remembered Moses, Abraham and Solomon, and the Syrians were convinced of their Babylonian origins. Most politically crucial, the newly Islamized Persians provided a convenient myth of antiquity: that all Greek knowledge was in fact of Persian origin, stolen by Alexander who had it translated and its sources destroyed. Under the savvy caliph al-Mansur (Abu Jaʿfar Abdallah ibn Muhammad – 714–775) the fable was gladly adopted, providing justification and drive for a two-centuries-long project of translating ancient knowledge into Arabic: from Indian and Persian, but first and foremost from Greek.

Court-sponsored and politically motivated, the knowledge sought and translated was, at first, mostly practical, with medicine taking precedence. Astrology was highly valued: for its place in medicine; as means for military and political prognostication; and as an ideological tool that enabled the Abbasids to demonstrate that their rule was predestined and inevitable. With astrology came astronomy, its necessary handmaiden, crucial for rites and prayer directions (Ptolemy's work was crucial to both). With astronomy came the necessary mathematics. Aristotle's logical works, including such difficult work as the *Topics*, were translated numerous times because they were deemed crucial for inter-faith disputations. These became increasingly important as the Abbasids' divergent political base drove them to fashion themselves as *Muslim* rulers, rather than exclusively Arab. With this political-cultural shift came a religious change: Islam became a prose-lytizing religion, bent on converting the infidels. This required developing a sophisticated theological stance, for which not only Aristotle's metaphys-ical works were translated, but also the Atomists'. Even Euclid's *Elements* may have been translated with an important practical application in mind: its geometry is inscribed into the marvelous new capital that Al-Mansur built for himself in Baghdad, on the banks of the Tigris and away from the Umayyads' Damascus.

Within a generation the translation project took on a life of its own. Like its Christian counterpart 300 years later, it was often an organized and cooperative project: the most famous of the translators, Hunayn ibn Ishaq (809–873), had a veritable workshop, employing his son Ishaq ibn Hunayn (830–910) and other members of their extremely erudite extended family. Hunayn was a cross-cultural mediator of the sort we've discussed: an Arab whose mother tongue was actually Syriac. As a Nestorian Christian, his knowledge of Greek – his liturgical religious language – was as thorough as his Syriac, so his collaborators would often render into Arabic what he first translated from Greek into Syriac.

This was an expensive undertaking, but translation became a source of pride and prestige to all involved. Sponsorship was provided by all parts of the Abbasid elite – political, intellectual and economic – who came to perceive the translation project as proof of their superiority over the Byzantines. Those Greek-speaking predecessors were so corrupted by Christianity, they maintained, that they lost their bond with and their claim over their magnificent heritage, whose genuine bearers were now the Muslims. The Caliphs would fashion themselves not merely as great warriors, but as great scholars. Harun al-Rashid (766–809), Al-Mansur's grandson, became legendary for his wisdom and erudition (his sobriquet means 'The Just'), and a subject of many folk stories. He's often credited with the establishment of *Bayt al Hikmah* (the 'House of Wisdom'), which was apparently more of an archive and library inherited from the Persians than the 'Arab Academy' it's sometimes purported to be. But the fable itself tells of the import erudition acquired in the public persona of the caliph – and especially erudition related to ancient knowledge (Greek, but also Persian and Indian). Harun's son, Al-Ma'mun (Abu Ja'far Abdallah, 786–833), used this erudition – the caliph's claim to personal command of logical and dialectical skills – to legitimize seizing supreme religious authority. These skills acquired such prestige, apparently, that they could be brandished in a strictly Muslim religious context despite their foreign and pagan origins.

Originality and Traditionality

The Muslim rule reunited the Hellenistic realm and opened the borders between central Asia, Asia Minor and North Africa, enabling the cross-fertilization of cultures and language – a flow of people, ideas, goods, agricultural species and techniques. It created great wealth and the urban culture necessary for the development of sophisticated knowledge, and by the end of the tenth century, when the whole Greek corpus had been essentially made available in Arabic, scholarly attention in the Muslim realm turned to critical inquiry of the potentials it presented.

The attitude they developed would frame scholarship across religious and cultural borders throughout the Middle Ages. It is, in fact, the core of what has come to be called 'Medieval Scholasticism': a venerating adoption of the Greek logical and cosmological framework, especially as shaped by Aristotle, with a careful critique of details, especially when these details failed to meet the criteria of rationality and evidence set by this very framework.

This attitude was most long lasting and influential in astronomy. Muslim astronomers eagerly translated and adopted all of Ptolemy's books, and especially the *Almagest* (the title with which they adorned it and is still in place), but they were unwilling to ignore the fact that it violated the very rules of Aristotelian cosmology on which it was supposedly founded (discussed in Chapter 3). The dissatisfaction was expressed already around the turn of the eleventh century, when the Greek-Arabic translation project was still going, by Abū ʿAlī ibn al-Haytham (965–1040):

> The epicyclic diameter is an imaginary line, and an imaginary line does not move by itself in any perceptible fashion that produces an existing entity in this world ... Nothing moves in any perceptible motion that produces an existing entity in this world except the body which [really] exists in this world ... no motion exists in this world in any perceptible fashion except the motion of [real] bodies.
>
> Cited by Saliba, *Islamic Science*, p. 332.

"The configuration, which Ptolemy had established for the motion of the five planets, was a false configuration," Ibn al-Haytham unhesitatingly concludes. "The motions of these planets must have a correct configuration, which includes bodies moving in a uniform, perpetual, and continuous motion, without having to suffer any contradiction" (cited by Saliba, *Islamic Science*, p. 343).

Ibn al-Haytham points here at two fundamental problems with Ptolemy. First, he is inconsistent, departing in practice from his own theoretical commitment to uniform concentric motion. And this implies, secondly, that his vaunted model is physically impossible. For Ibn al-Haytham that means that epicycles are as bad as equants, but this is not the way medieval astronomers – first the Muslims and later on Christians – understood his challenge. They didn't attempt to replace the Ptolemaic model with a physically possible one. Instead, they remained loyal to its methods and procedures and worked on solving the more disturbing inconsistency – the equant – by means of what they perceived as a less disturbing one – epicycles. We have already mentioned in Chapter 3 the most successful of these attempts: that of the Persian astronomer and polymath Muhammad ibn Muhammad ibn al-Hassan al-Tūsī (better known as Nasir al-Tūsī, d. 1274). Al-Tūsī has shown how a circle rolling inside a bigger circle can make a point on its circumference move along a straight line drawn through their common center (Figure 4.12). Setting both circles in motion in opposite directions, manipulating their sizes and velocities to achieve the desired relative (angular) velocity, astronomers could construct models in which

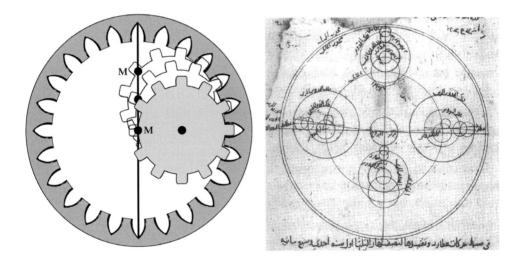

Figure 4.12 The Tusi Couple. On the left, a diagram of its modern version, materialized by geared wheels. As the smaller wheel roles counter-clockwise within the larger one – whose radius is double – point M moves up and down in a straight line through their common center. On the right, it's original introduction by Nasir al-Din Tūsī in his *Compendium of Astronomy* (this manuscript from 1289). The planet, carried by the small epicycle rolling between two deferents (the three reiterations of this small circle represent it at different positions), moves back and forth in a straight line in relation to their common center. This eliminates the need for eccentricity, let alone an equant point: Earth can be set at the geometrical center of all motions. On the right, another version: the model for Mercury developed by al-Tusi's disciple Ibn al-Shatir ('Abu al-Hasan al-Ansari) in his *La Nihaya al-sul* (*The Final Quest*) from 1304. Carried by the small double epicycles, the center of the big epicycle carrying Mercury continuously changes its distance from the center.

the planet constantly changed its distance from the Earth. They could thus identify the planet's center of motion with its center of uniform velocity, annulling the disturbing need for an equant point (and other such arbitrary devices of forcing order on planetary motion).

Tusi Couple is thus an emblematic achievement of medieval science as shaped by the Muslims and adopted by the Christians: ingenious manipulation of the elements of the Hellenistic tradition with only a very rare challenge to its principles. This doesn't mean at all that medieval scholars were satisfied with minor theoretical modifications of the science they inherited from the Greeks. Muslim astronomers in particular were very aware of the empirical shortcomings of that knowledge (the practical bent discussed above may have something to do with that), and made a concerted effort to ameliorate it. Well-organized astronomical observations are documented already from the time of Al-Ma'mun, and although they usually didn't last beyond the life and generosity of their patron, they've created a tradition that culminated in the eastern regions of the Muslim realm. Al-Tusi himself had the opportunity to head a well-equipped, well-manned observatory, founded for him by his Mongol patron Prince Hūlāgū, brother of the fearsome Kublai Khan, in Marāgha (south of Tabrīz in modern Iran). In the 1420s, an even more splendid observatory was established in Samarkand by another warlord-cum-scholar – Ulugh Beg (or Bey; 1394–1449), grandson of the fierce conqueror Tamerlane. These institutions provided astronomical observations which were not only many-fold more numerous than Ptolemy had access to (the *Almagest* is based on fewer than 100 observations over some 700 years), but of a different order of accuracy. The Samarkand observatory included, among many other splendid instruments, a 50-meter-tall (!) mural quadrant.

Even so, the instruments in Marāgha, Samarkand and later Istanbul were developments of the ones reported by Ptolemy: gnomons and sundials; quadrants and sextants, astrolabes and armillary spheres (Figure 4.13). In this sense, Ibn al-Haytham and his demand for a fundamental overhaul of Hellenistic science were an exception highlighting the rule. This is particularly telling because he was otherwise probably one of the two most influential scholars in the Middle Ages (the other – Ibn Sina, whose most important work was in medicine – we'll discuss in Chapter 8), not only in the Muslim realm, but also in Christian Europe, where he was known as Alhazen (or Alhacen, after his first name, Al-Hasan). His main legacy was in optics, where his book, *Kitāb al-Manāzir* (*Book of Aspects*), offered the same synthesis he was calling for in astronomy. It comprised a strict

Figure 4.13 The epitome and last vestige of Ptolemy-style observational astronomy: the observatory in Galata, Constantinople (by then, under Ottoman rule, and known as Istanbul). Established by Taqi al-Din Muhammad ibn Ma'ruf al-Shami al-Asadi (Turkish: Takiyüddin, 1526–1585) in 1570, the observatory featured many of the novel institutional characteristics that would mark Tycho Brahe's contemporary observatory (see Chapter 7). As one can clearly note, the instruments in use here (not unlike Tycho's, but with the important difference in size) are traditional angle-measuring devices, as the image clearly displays. Tellingly, the observatory was closed down and its purposefully constructed building demolished by 1580. The image is a copy of a 1581 illuminated manuscript.

mathematical analysis of lines of vision and their reflection and refraction; an empirical study of light, color and other visual phenomena, and a careful physiology of the eye. Its power was such that, when it was translated into Latin as *De Aspectibus* at the turn of the thirteenth century, it created a discipline named after it: *perspectiva*. The books written in that tradition, new in Christian Europe, by such luminaries as Roger Bacon (1214–1292), John Peckam (1230–1292) and Erazmus Ciołek Witelo (c. 1230–c. 1300 – see Figure 6.5), were fundamentally commentaries on Alhazen. Yet none of them would pick up his revolutionary project in astronomy. This would only happen about 400 years later in the work of Johannes Kepler.

Discussion Questions

· ·

1. If knowledge needs a place, can you think of types of knowledge that have lost their places? What has happened to them?
2. It is tempting to compare the encyclopedic tradition to its contemporary online version. Does the analogy hold? To what degree? Are the effects on knowledge and its culture similar? Different?
3. What is the importance of the unique continuity of the university as an institution of knowledge? Does the fact that its main pedagogical means are still fundamentally the same after so many centuries constitute an advantage or a drawback?
4. What kind of a university would you have liked to be a part of? What kind of relations would it have between its faculty (the academics)? Between faculty and students? Between the university as a whole and the surrounding world? Does your university come reasonably close to your ideal?
5. Are there lessons to be learnt from the translations projects, whether about knowledge in general or about scientific knowledge in particular? From the length of these projects? From the people and institutions involved? From their cultural and political import?

Suggested Readings

· ·

Primary Texts

Plinius, Gaius Secundus (Pliny the Elder), *Natural History (Historia Naturalis)*, John Bostock and Henry T. Riley (trans.) (London: Henry G. Bohn, 1855–1857), http://resource.nlm.nih.gov/57011150R, Dedication and Book I, chs. 1–2 (V.1, pp. 6–16); Book II, chs. 17–31 (V.1, pp. 50–63).

Abū ʿAlī ibn al-Haytham (Alhacen), "*Alhacen's Theory of Visual Perception*," Mark A. Smith (ed.) (2007) 91(4–5) *Transactions of the American Philosophical Society* 343–346, 561–577.

Grosseteste, Robert, *On light (De Luce)*, Clare C. Riedl (trans.) (Milwaukee, WI: Marquette University Press, 1978).

"The Reaction ... to Aristotelian ... Philosophy" in E. Grant (ed.), *A Source Book in Medieval Science* (Cambridge, MA: Harvard University Press, 1974), pp. 42–52.

Secondary Sources

On the fate of Greek science:

Lehoux, Daryn, *What Did The Romans Know? An Inquiry into Science and Worldmaking* (University of Chicago Press, 2012).

On the Muslim adoption of Hellenistic science:

Gutas, Dimitry, *Greek Thought, Arabic Culture: The Graeco-Arabic Translation Movement in Baghdad and Early 'Abbasid Society* (London: Routledge, 1999).

On Muslim science and philosophy:

Rashed, Roshdi (ed.), *Encyclopedia of the History of Arabic Science* (London: Routledge, 1996), Vols. 1–3.

Rashed, Roshdi, "The Ends Matters" (2003) 1(1) *Islam & Science* 153–160.

Sabra, Abdelhamid I., "The Appropriation and Subsequent Naturalization of Greek Science in Medieval Islam: A Preliminary Statement" (1987) XXV *Journal for History of Science* 223–245.

Saliba, George, *Islamic Science and the Making of the European Renaissance* (Cambridge, MA: MIT Press, 2007).

On European medieval science and philosophy:

Grant, Edward, *Physical Science in the Middle Ages*, Cambridge History of Science Series (Cambridge University Press, 1977).

Lindberg, David C., *Science in the Middle Ages* (University of Chicago Press, 1978).

Lindberg, David C., *The Beginnings of Western Science: The European Scientific Tradition in Philosophical, Religious, and Institutional Context, 600 B.C. to A.D. 1450* (University of Chicago Press, 1992).

On the rise of the university:

Grant, Edward, *The Foundations of Modern Science in the Middle Ages* (Cambridge University Press, 1996).

Pedersten, Olaf, *The First Universities*, Richard Nort (trans.) (Cambridge University Press, 1997).

5 | The Seeds of Revolution

Monotheism and Pagan Science

Historical periods are almost always introduced in hindsight, with some agenda. This is doubly true for 'Middle Ages,' a term we've thus far employed to designate the period after the fall of the Roman Empire. The millennium this term designates was far from being just 'in the middle,' and the people living through it experienced as many changes and upheavals as in any other era. But at least as regards the coming to being of science, one can hardly dispute that it was the ushering in of a new age, with the adoption of Hellenistic – *Pagan* – science by the dominant monotheistic cultures: first the Muslim, then the Christian.

We saw in Chapter 1 how early Christianity moved from Tertullian's strict rejection of "philosophy ... the rash interpreter of the nature" and its "heresies," to the compromise sketched by Augustine, for whom "pagan learning ... contains also some excellent teachings." But the late medieval scholars who thoroughly immersed themselves into these "excellent teachings" required much more detailed answers to the challenge of assimilating the pagan concepts and modes of thought into the demands of monotheistic belief. It was a major challenge, whose resolution would shape knowledge in this new era and for centuries to come.

The Fundamental Discrepancy

The Hellenistic approach to the study of nature, especially that evolving from Aristotle's clear and uncompromising formulation discussed in Chapter 2, is pagan not in the sense of suggesting *many* gods, but the opposite: in not allowing *any*. Most generally, it relies strictly on the collection of natural facts and a critical use of the human mind; it leaves no place for revelation, which, as we discussed in Chapter 1, is fundamental to the monotheistic religions. This is perhaps the most difficult for Judaism to accept, because the moment of God's direct communication with the Israelites at Mount Sinai defines it both as a religion and as a nation. Secondly, Aristotle's natural philosophy assumes strict causal determinism – every phenomenon has a clear natural cause – and leaves no place for miracles. For Christians this was particularly unacceptable, because miracles constitute their founding mythology – the virgin birth and the resurrection, as well as their basic rite, the Eucharist, in which the bread and the wine miraculously turn into the flesh and blood of Christ. Thirdly, Aristotle argues specifically for the eternity of the cosmos, leaving no place for creation, the most fundamental belief shared by all monotheistic religions. Finally, there is no place for an eternal soul in Aristotle: the soul can be distinguished from the body but not separated from it, and it perishes together with the body. This is difficult for both Christianity and Islam, with their concept of punishment and reward in the afterlife.

To the modern reader, it may seem that medieval scholars had a choice. On the one hand, they could – like many scientifically minded thinkers would do centuries later – abandon religion. But this was not an option that they even had the vocabulary to entertain: as we discussed in Chapters 1 and 4, religion provided the framework for every aspect of their life and thought. On the other hand, those scholars might have decided – like the early Church fathers and many monotheistic zealots since – that science needed to be rejected because it was too incompatible with religion. But they were too thrilled with the intellectual horizons opened up by Aristotle and the rest of Hellenistic thought to entertain this option either. Muslim, Christian and Jewish scholars strove to hold on to both horns of the dilemma – monotheistic belief and pagan science – and found two ways out of the contradictions. These were paved by two people who probably never met, although they were born within ten years and a 10-minute walk from each other, in Cordoba of Al-Andalus: the southern, Muslim region of medieval Spain.

Ibn Rushd

The elder of the two was ʾAbū l-Walīd Muḥammad Ibn ʾAḥmad Ibn Rushd (1126–1198), whose authority as Aristotle's interpreter in Christian Europe, where he was known as Averroes, can be gauged by the fact that whereas Aristotle was often referred to simply as 'the philosopher,' Averroes was known as 'the commentator.' He was a jurist from a jurists' family, and served as a judge and high-level courtier until the pressures of Christianity from the north and the invasion of North-African Berbers from the south brought about religious fundamentalism that forced him to flee to North Africa and he finished his life in Marrakesh.

The fundamental principle of Ibn Rushd's synthesis was the primacy of reason over belief – diametrically opposed to Augustine's approach which we discussed in Chapter 1. The Qur'an itself, he argued, commanded the devout Muslim to study philosophy. Even the very conviction of the existence of God, in Ibn Rushd's analysis, is mandated not by irresistible illumination but by philosophical consideration: it follows from the orderliness of

nature and the way it's so well-suited for humans to thrive in. The God arising from such considerations turns out to be rather different from the one
enunciated by religion. This philosophical God is an absolute and necessary entity, so His existence, His will and His knowledge are one. Creation,
therefore, could not have been an act at a certain moment in time: God's
will cannot change, and since He had by necessity always known that the
cosmos would be created, it was always by necessity true that the cosmos
would be created (in that Ibn Rushd followed Ibn Sina, whom we'll discuss
in Chapter 8 and whose thought he usually rejected as over-Platonist and
his interpretation of Aristotle as simply wrong). In other words: the cosmos
exists not because of what God does, but by virtue of what He is, hence it
is necessary and eternal.

Ibn Rushd was turning God into a completely abstract *ratio*; a pure intellect. This allowed him to express the fundamental Aristotelian principles
of necessary causation and the eternity of the cosmos as divine properties
and consequently to practice pagan science as if it were sanctioned by
monotheistic creed. In the scriptures, of course, God does change His will
and His mind, deciding to create and later to destroy the world, and since
the scriptures cannot be wrong, Ibn Rushd reasoned, their language must be
simple – sometimes allegorical, often personified – to convey to the vulgar
the truth the philosopher obtains through contemplation. This line of reasoning came to be called 'the two-truth doctrine' by the Christian scholars
deriding it.

It is not difficult to imagine how liberating Ibn Rushd's philosophy was
to monotheistic scholars seeking legitimacy for their enthusiasm for pagan
science, and how disturbing it must have been to the more conservatively
minded. In Christian Europe, in Latin translation, he immediately drew a
group of followers. The group was entered in the University of Paris around
Siger of Brabant (c. 1240–1280s), a rowdy fellow, who otherwise distinguished himself by heading a faction in a series of non-intellectual brawls,
which almost led to his execution. Most of what we know about the doctrines and arguments of this group comes from their fierce adversaries,
who called them 'Averroists,' but we can gauge their popularity from the
efforts directed at suppressing their teachings. We can understand that the
rising enthusiasm for Pagan science in the universities during the thirteenth
century clearly caused serious concern among religious authorities, both
within and around these institutions of learning and education, because
a series of regulations were issued to curb it. We can also learn from the
growing seriousness and details of these rulings that they were not particularly successful. They culminated in condemnations decreed in 1277
by Etienne Tempier, the Bishop of Paris, which include over 200 articles

of Aristotelian "Abominable errors" that no one should "defend or support" on penalty of excommunication. These articles are formulated in a way demonstrating sophisticated engagement with the pagan doctrines rather than the suppression they were purported to achieve. It is prohibited to believe, for example, (Article 6) "that when all celestial bodies have returned to the same point – which happens in 36,000 years – the same effects now in operation will be repeated"; or (Article 98) "that the world is eternal because that which has a nature by which it could exist through the whole future has a nature by which it has existed through the whole past" (quoted from Grant, *A Source Book in Medieval Science*, pp. 47–49).

Moshe ben Maimon

Christian scholars would not relinquish the study of pagan knowledge; nor could they afford the radical interpretation offered by Averroes. The compromise they would end up adopting was first formulated by Averroes' fellow Cordovan: Rabbi Moshe ben Maimon (1135–1204), known in Hebrew by the acronym Rambam and in Latin Europe as Moses Maimonides. Although Maimonides' family was forced into exile by the same strife that affected Ibn Rushd – they traveled all across North Africa to Egypt – he was still a product of the Jewish golden age of Muslim al Andalus and its magnificent cross-cultural wealth. He was still the most venerable figure of the two-millennia-long Jewish Rabbinical tradition; in Cairo he became the court physician and head of the Jewish community; and his great philosophical book, *More Nevokhim* (*Guide of the Perplexed*) of 1190 or thereabouts, was written in a Jewish dialect of Arabic using the Hebrew alphabet, explaining and commenting on Aristotelian thought and its Muslim interpretations to his "perplexed" Jewish flock.

Maimonides has no place for either Augustine's preference for faith or Ibn Rushd's for reason. God created both the world and the scriptures, and both are for human understanding and carry the same single truth. Yet since God put us in the world before giving us The Book – the Torah – understanding nature should precede and direct the interpretation of scriptures. Both these types of knowledge are directed by reason – fallible and limited human endeavors – and God himself is entirely beyond their bounds; we can only say what He's not – not corporeal, not in time, etc. The scriptures are an absolute authority on issues of faith and law, even if their reasons remain unknown to humans, but they don't contain clues about the make-up of the world. The study of this subject belongs to reason and observation, and in practice – to Aristotle and the philosophers. What to do, then, about the pagan lines of thought embedded in Aristotle's teachings? These need to be defeated on their own terms – using Aristotelian science to defeat Aristotle's arguments – in particular concerning the eternity and necessity of the cosmos. So, on the one hand, the orderliness of the cosmos reveals that it had to be by design, which entailed the existence of a designer, hence an all-knowing God:

> Do you hold that it could have happened by chance that a certain clear humor should be produced, and outside it another similar humor, and again outside it a certain membrane in which a hole happened to be bored, and in front of that hole a clear and hard membrane? ... Can someone endowed with intelligence conceive that the humors, mem-branes, and nerves of the eye – which ... are so well arranged and all of which have as their purpose the final end of this act of seeing – have come about fortuitously? Certainly not ... This is brought about of necessity through a purpose of nature. [And since] nature is not endowed with intellect and the capacity for governance [this] craftsman-like governance ... is the act of an intelligent being.
>
> Moses Maimonides, *The Guide of the Perplexed*, III.19, Shlomo Pines (trans. and ann.) (University of Chicago Press, 1963), p. 479

On the other hand, the many contingent elements of the cosmos show "that all things exist in virtue of a purpose and not of necessity, and that He who purposed them may change them and conceive another purpose" (Maimonides, *Guide of the Perplexed*, II.19, p. 303). God had willed the cosmos this way, and might have willed it differently, for reasons we cannot fathom:

> For in the same way as we do not know what was His wisdom in making it necessary that the spheres should be nine – neither more nor less – and the number of the stars equal to what it is – neither more nor less – and that they should be neither bigger nor smaller than they are, we do not know what was His wisdom in bringing into existence the universe at a recent period after its not having existed. The universe is consequent upon His perpetual and immutable wisdom. But we are completely ignorant of the rule of that wisdom and of the decision made by it.
>
> Maimonides, *Guide of the Perplexed*, II.18, pp. 301–302

Thomas Aquinas and Thomism

It took only a few decades for this compromise between pagan learning and monotheistic belief to take root in Latin Europe. Its powerful advocate was the Dominican friar Albert of Cologne (1200–1280), whose great erudition earned him the moniker Albertus Magnus. The eldest son of a

Count, Albert had an emblematicallly medieval scholarly career: he studied at Padua, Bologna and Paris; taught theology at Hildesheim, Freiburg, Ratisbon, Strasburg and Cologne, and became the pope's court intellectual, defending official orthodoxy. In this capacity, his preference for Aristotelianism over Augustinian Neo-Platonism had a resounding impact, especially coupled with his interpretation of and commentaries on the whole Aristotelian Corpus, by then available in Latin. A careful reading of Maimonides informed this interpretation: there is only one divine intellect, and hence only one truth, Albertus insisted in his clearly titled text *De Unitate Intellectus Contra Averroem* (*The Unity of the Intellect against the Averroists*) of 1256. The study of nature, like the interpretation of the scriptures, is therefore the study of the laws instituted by God, and carries its own religious value. This also means that philosophy of nature, even by the pious, should concentrate on the orderly and the regular, not the miraculous. Natural philosophy is the study of the causes God inscribed into nature, not of His reasons for doing so, so it has no reason to clash with religion.

This synthesis of pagan science and monotheistic creed, formulated in a Muslim context by a Jewish Rabbi, would be made into *the* framework of Christian scholarship by Albert's disciple and colleague, Thomas of Aquino (1225–1274).

Thomas, often called 'Aquinas,' had a similar background and career to his mentor, although it started with some more excitement. A son of a noble Italian family, he received his education at the ancient monastery school in Monte Casino and the relatively new university in Naples, but his decision to join the Dominican Order at an early age didn't sit well with his father. He was abducted by his brothers and imprisoned for a year in the family estate. After much failed persuasion and a heroic struggle with a prostitute (who was sent to seduce him), he was allowed to escape through the window and go to the University of Paris, where he spent years as both student and lecturer, mostly under Albert. He wrote vastly (from 1879 to 2014, only thirty-nine of the projected fifty volumes of his *Opera* have been published), his views moving in and out of the mainstream.

Yet for most of his life, Aquinas was not a teacher but a roving scholar for the pope, and in this position he had the authority and responsibility to develop the details of the synthesis between science and religion. He reasserted the Augustinian principle of the primacy of revelation in matters of faith, but stressed with it the Aristotelian principle of the primacy of experience and empirical knowledge in the matters of this world. He went as far as to agree that, philosophically, the eternity of the cosmos is the most probable hypothesis, although neither creation nor eternity can be conclusively demonstrated. Yet since faith dictates that the eternity of the world is false and heretical, creation is what the Christian has to wholeheartedly commit to.

Like Maimonides and Albert, however, Thomas found no contradiction between the two routes to knowledge: both lead to truth because God is the source of both. God operates in the world through its laws, which He instituted, so their study, through reason and experience, is a religious virtue. This route still only leads to limited knowledge – philosophy can only reveal the 'secondary causes' – the causal relations between things as God created them. The 'primary causes,' which are God's *reasons* for creating things the way He did – indeed for creating them at all – we could only know if God decided to reveal them to us – through revelation. Interpreting His words – the scriptures – is the task of theology. Treating "by the light of divine revelation ... the same things which the philosophical disciplines treat ... by the light of natural reason," theology is the highest of disciplines (*Summa theologiae*, Que. I, Art. 1, www3.nd.edu/~afreddos/courses/439/st1-ques1.htm).

Injecting content into these principles, Thomas turns to Plato's philosophy to bridge the gap between Aristotle and Christianity. He adopts and develops the distinction and relations between form and matter, mostly along Aristotelian lines: matter is one – uniform, passive, the seat of all potentialities; form is unique, active, the principle of development and change. In humans, the soul is 'the form of the body,' like in Aristotle; but it also can and does exist separately, like in Plato. Cosmologically, Thomas allows for pagan astronomy by identifying the Aristotelian heavens with the Biblical 'firmament' and Aristotle's 'unmoved mover' with God. In matters where Christianity is not necessarily implicated – whether it's epistemology or theory of representation – Thomas is as keen an Aristotelian as any of his Muslim predecessors.

There were other intellectual options: members of other educating orders, especially the Franciscans, didn't feel beholden to the Dominican philosophical positions and offered their own. But as the dust of contention settled, it was Thomas' synthesis that would come to define and structure scholarship in Latin Europe for at least four centuries. It wasn't simply a matter of popularity among scholars: it would be embedded into the university curriculum and its very structure of disciplines. 'Thomism' would become virtually synonymous with 'scholasticism' – the way Europeans of the High Middle Ages understood and investigated their world, their God and themselves from the Age of the Cathedral until early modern times.

The Renaissance

The relative calm and prosperity that allowed Europeans to build material and intellectual cathedrals was short lived. In 1337, a war over the succession to the French throne broke out between the Plantagenets, the rulers of England, and the Valois, who ruled France. It would last until 1453, modestly titled the 'Hundred Years' War,' and drew in most of Western Europe. The Holy Roman Empire, already battling Mongol pressures from the north-east since the middle of the thirteenth century, managed to engage itself in a series of doomed belated crusades to the south-east, which not only exhausted resources and lives, but left in their trail the destabilizing presence of military orders like the Templars and the Hospitallers. The religious-intellectual stability represented and buttressed by Thomism broke into a succession of ugly religious-cum-political conflicts, the worst of which saw the pope move from Rome to Avignon for most of the fourteenth century. By far the worst disaster, however, was neither political nor religious, but natural (although it most probably had been carried around by armies and traders). It was the bubonic plague, also known as the Black Death, which appears to have originated in Kyrgyzstan around 1338–1339. It disseminated to India and China, where it killed an estimated 25 million people, reached Constantinople in 1347 and from there it spread to the Middle East and Europe, where it proved most lethal. An estimated third of the population between Persia and Egypt was killed and no less than half of Europe's: Paris lost 50,000 out of its 100,000 inhabitants and Florence – 70,000 out of 120,000.

The New City-State and Its Prince

Yet rather amazingly, from the devastation emerged a prosperity the likes of which Europe hadn't experienced since the heydays of the Roman Empire. One reason may have been the brutal fact that the same resources were now shared by far fewer people; another was perhaps the energy unleashed in the process of rebuilding. More long-lasting was the opening of trade routes to the east by the Mongol unification of Asia – the same opening that allowed the spread of the plague – whose pioneers of a century earlier have been immortalized in the picturesque character of Marco Polo. The fantastic wealth that this cross-continental, cross-cultural commerce brought to trading cities in Italy like Genoa and Venice was augmented by an economic novelty: the bank. This institution embedded new and bold ideas, like the use of money not only as a currency but as a commodity; borrowing not only for immediate relief but for investment; monetizing risk; and lending – with the aid of bonds and promissory notes – more than is available in 'hard currency.' All these allowed for a multiplication of resources and made some families – notably the Medicis of Florence – even wealthier than the long-distance merchants (and later – explorers) whose expeditions they were financing.

Shielded by the Alps from the land warfare of the new empires, Italy was particularly poised to exploit these new opportunities. With the Papacy leaving and the Holy Roman Empire retreating to deal with its challenges from the north, a power vacuum was created in which these newly wealthy cities – Genoa and Venice; Florence and Milan; Pisa, Luca, Siena and others – could claim their independence and run their business to their own best (mostly pecuniary) interests. The city-state, reminiscent of that of antiquity, re-emerged, and with it – a new type of knowledge and a new type of knower.

Most of these cities became, initially, republics, or rather joint oligarchies, run by senates composed of representatives of the powerful families. More often than not, however, one of these families managed to wrest sole (though always contested) control, imposing their own head as the prince: the Este family in Ferrara; Sforza and Visconti in Milan; della Rovere in Urbino; and the greatest of them all, as we have noted, the Medici clan of Florence. These self-appointed princes were never assigned a role in the grand religious and cosmological theater which had imbued the rule of their predecessors with a sense of inevitability (see Chapter 1). They could not claim the bravery and loyalty of knights, the ancient lineage of kings, or the Church's blessing of emperors. They had acquired their rule of clan and city recently, by intrigue, wallet and stiletto, and their subjects knew it. They had to acquire legitimacy, and so they did – by building a court and surrounding themselves with painters and sculptors, poets, rhetoricians and philosophers. The task of these artists, writers and scholars was to glorify the prince; to reproduce his image, literally and figuratively, as a grand persona, to whom governance belonged by right. The patronage of Lorenzo de' Medici, 'il Magnifico' (1449–1492) – who personified more than anyone else the character of the Renaissance prince – extended to painters like Andrea del Verrocchio, Sandro Botticelli, Leonardo da Vinci and Michelangelo Buonarroti and to scholars like Marsilio Ficino and Giovanni Pico della Mirandola. Niccolò Machiavelli's *The Prince* of 1513, analyzing the personal traits and political strategies befitting such a prince, was dedicated to Lorenzo's grandson, Lorenzo Piero (1492–1519). The Sforzas of Milan competed with the Medicis for the services of da Vinci, and also had a particular taste for music: the glory of Lorenzo's contemporary Galeazzo Maria Sforza (1444–1476) was sung by composers the likes of Josquin des Prez, Alexander Agricola and Gaspar van Weerbeke. The glorification didn't always work: all this divine music didn't make the Milanese forget Galeazzo's cruelty nor the rival families forget Sforza's transgressions, and the stiletto claimed his life on Christmas day at the church entrance.

The Humanists

The artists and scholars congregating around the Renaissance court represented a new type of person of knowledge. They were neither 'free' like the Greek philosophers, nor educators like the university faculty. Towards the scholars in the monasteries these new court intellectuals directed particularly vicious criticism: the scholastics' commitment to the life of the mind – their *vita contemplativa* – was but a barren self-indulgence in words, they charged. Proper for the artist, the writer or the philosopher was the *vita activa* – a life about-town, fully engaged in politics and commerce. 'Humanists' is what they called themselves: students of and participants in human life in this world, rather than contemplating the next. Aggrandizing themselves and their princes, they titled their cultural project 'Renaissance': the alleged rebirth of ancient Hellenized Rome, reducing the thousand years since its fall to mere 'Middle Ages', the time in between, separating old and new glory. Petrarch (Francesco Petrarca, 1304–1374), perhaps the first of the humanists and the one who shaped their interests in the Greek and Roman classics, their disdain for the "Dark Ages" (a term he's credited with coining) and their innovative attitude towards their vernacular languages – Tuscan for him – worked for the most glorious prince of all, the pope.

Rather than theology, cosmology and natural philosophy, the humanists concentrated their interests in earthly and politically effective disciplines like history and rhetoric. Philology, in particular, received a new and exciting role (which we discuss more in Chapter 6) as a new translation project was underway. The humanists were fascinated by the masterpieces of their adopted classical heritage, but were contemptuous of the translated versions produced by the efforts of 200 years earlier (discussed in Chapter 4): Aristotle was a hero of a glorious past, while Scholastic Aristotelianism they found barren and distorted. They therefore set out to recover the original texts, equipped with the great wealth of their brave new world and enabled by the newly opened routes to the East. Monks of the Eastern Church, fleeing Muslim pressure, brought with them manuscripts – light, portable and now highly coveted by the deep-pocketed princes, seeking all genres of marvels and rarities to adorn their court. A heroic allure befitting of the age came to adorn the project: in 1406 one Jacopo Angelo saved himself and a manuscript of Ptolemy's *Geography* from a sinking ship; in 1417, the only surviving manuscript of Lucretius was brought from "a remote monastery" by

Poggio Bracciolini. The imploding Byzantine Empire, before finally falling to the Muslim Ottomans in 1453, provided not only manuscripts but the necessary linguistic skills as well: in 1396, the desperate Byzantine emperor Manuel Paleologus arrived in Italy in an attempt to recruit emergency assistance. He clearly overestimated the solidarity of his Christian brethren, and returning empty-handed, he also left behind an intellectual from his entourage: Manuel Chrysoloras (c. 1355–1415) remained in Florence, and began teaching Greek language and literature to Florentine courtiers and humanists.

The Meeting of Scholar and Artisan

The court was not the only site where the great new riches were generated and dispensed: the marketplace was another. Economic prosperity; erosion of social and political barriers; a new cultural celebration of earthly goods, challenging the traditional Christian ascetic values – all came together in the thriving urban marketplace, creating clear opportunities for both artisan and scholar to climb the walks of life. Everything promising to produce wealth or glory could find a public willing to pay: a new brilliant text – poetic or scientific; a new technological invention – practical or spectacular; a new tool – material or intellectual. The marketplace provided the scholar and the artisan with an opportunity to meet and the incentive to communicate: the scholar's abstract theories could help the artisan re-examine and improve traditional means and habits of doing and making; the artisan's practical, empirical skills allowed the scholar to question and subvert old theories. The market rewarded practicality and innovation. Even mathematics, the most ephemeral of scholarly disciplines, became useful, helping the merchant calculate interest rates and investment returns and the painter design visual harmonies. The ancient dichotomy between *knowing-that* and *knowing-how* was eroding, and in the trading zones between *episteme* and *techne* new knowledge was being created.

Filippo Brunelleschi (1377–1466) is an excellent representative of this stage in the construction of the cathedral that is science. Born to a Florentine notary and city official and related through his mother to the extremely rich Spini and Aldobrandini families, he was still trained as a goldsmith – the practical vocation befitting his artistic talents and apparently no longer a disgrace even for a wealthy and educated family. His theoretical education was originally limited to his father's teaching, until later in life a friendship with the mathematician, merchant and physician Paolo dal Pozzo Toscanelli (1397–1482) spurred his interests in mathematics.

In 1413, Brunelleschi demonstrated the power of the combination of practical and theoretical skills and talents – when the ambient culture allowed them to be exercised together – with an invention that would reshape European art: mathematical perspective. He didn't leave an account of the invention (a little more about his secrecy later) but a description of the way it was introduced in Florence's Piazza del Duomo (by his friend, the scholar Antonio Manetti, 1423–1497) provides good hints. The spectators were to stand in the dark entrance of the *Duomo* – the cathedral of Santa Maria del Fiore – and look across the piazza at the *Battistero*. In one hand, they held Brunelleschi's painting of the baptistery, facing away from them, and through a hole in the painting peeped at its reflection in a mirror they held in the other hand (Figure 5.1). Lowering the mirror and raising it again they were treated to a visual marvel: one couldn't tell the real building from the reflected painting (especially since the mirror, painted silver around the edges, reflected also the sky and the moving clouds). Medieval painters understood that in order to represent depth on a flat canvas, lines which in the real world are parallel should be drawn as converging, but the exact geometry of this convergence eluded them (note the raised figures and the slipping plates in Figure 5.2). Brunelleschi's theatrical display hints at the way he may have discovered the angles of convergence to create a convincing sense of depth: by tracing them on a mirror. Brunelleschi was putting to use the ancient mathematical science of optics and the new developments in the art of producing mirrors (befitting the age of vanity) to produce a

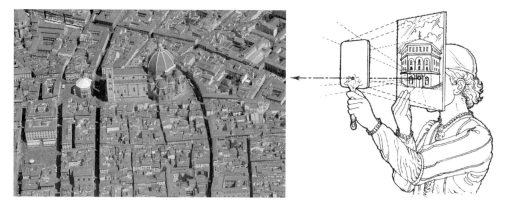

Figure 5.1 On the left: The *Piazza del Duomo* in Florence with the *Battistero* (baptistery) on the left and the *Santa Maria del Fiore* cathedral on the right, with Brunelleschi's octagonal dome on top of it. On the right: Jim Anderson's depiction of Brunelleschi's perspective spectacle. The observer stands inside the heavily shaded cathedral entrance (note the shadows in the photograph on the left), facing the baptistery. She (apparently both ladies and gentlemen were invited to the demonstration) holds Brunelleschi's painting of the baptistery with its painted side facing away, towards the baptistery, looking at it through a hole in the canvas. In the other hand, she holds a mirror facing her. When she lowers the mirror, she sees the baptistery; when she lifts it, she sees in it a reflection of the painting, and a reflection *in* the painting of the sky and clouds. According to Manetti, one could hardly tell the difference between the real baptistery and its reflected representation.

Figure 5.2 Medieval attempts at perspective: a fourteenth-century illumination of a manuscript of the *Chronicles* (of the Hundred Years' War) by Jean Froissart (c. 1337–c. 1405), depicting a banquet with the Duke of Lancaster and the King of Portugal (British Library Ms. Royal 14EIV, f.244 v.). Note that the illuminator attempted to portray depth and had an idea of how to achieve it: the lines of tile are converging and the more remote figures are higher up. Without the geometrical rules of perspective, however, he cannot create an illusion of three-dimensionality: the table appears tilted as if the plates and food are about to slip off it.

new spectacular technique, befitting *palazzo* and *piazza*: the ability to create a powerful illusion (Figure 5.3).

The Duomo provided the literal and metaphorical foundations for another demonstration of Brunelleschi's virtuosic ability to merge material and theoretical knowledge, or *knowing-how* and *knowing-that*. Five years after the invention of linear perspective, he submitted a bid to build a dome to the cathedral. The required size of the envisioned dome – bigger than Rome's Pantheon – turned its construction into such a challenge that the cathedral had remained dome-less since its construction a century earlier. The method of building a large dome at the time called for 'centering': filling the space beneath the dome with heavy scaffoldings to carry its weight until the keystone was put in place and the structure became self-supporting (see Chapter 1). For this massive size the method seemed structurally dangerous and prohibitively expensive, and Brunelleschi won the tender by promising to dispense with it altogether and meet the challenge at a fraction of the expected price.

Figure 5.3 An example of perfect linear perspective and the powerful illusion of depth it enabled the artist to create: Piero della Francesca's *Ideal City* (c. 1470). All the lines that would be parallel in reality – the street, the paving, the stories of the buildings on both sides – converge towards one point on the imagined horizon behind the circular building in the middle. The effect is an apparent three-dimensionality on the two-dimensional canvas.

Brunelleschi built two domes. The beautiful octagon we're accustomed to seeing (Figure 5.1, left) is actually a light roof, carried by a heavier vault into which Brunelleschi's ingenuity was invested. On the artisanal side, it was constructed with innovative 'fishbone' brick-laying techniques, which enabled him to wrap the courses around the curve, creating stability before the whole structure was complete. A combination of craftsmanship and erudition allowed engineering creativity: inspired by a new translation of Vitruvius (see Chapter 1), Brunelleschi designed and built cranes and hoists of unprecedented efficiency (Figure 5.4). Bringing in new mathematical skills, he apparently had in his own dwelling a scale model of his dome, on which he could experiment with his novelties. Most crucially and quite mysteriously, Brunelleschi's vault was not hemispheric but parabolic. Perhaps by calculations (though quite unlikely), or perhaps by experimenting on his model or in some other way, Brunelleschi realized that other, more complex curves provided more stable arches than the simple and perfect circle.

As with the invention of perspective, we're bound to use the words 'apparently' and 'seemingly' because Brunelleschi – like the artisan of the guild and unlike the scholar of the monastery or the university – kept his knowledge secret. Whether it was theoretical and abstract or practical and material, it was trade knowledge, a commodity for which one had to pay. Brunelleschi was an ambitious and scrupulous man: when the Florentine senate, wary that his promises were too good to be true, forced him to partner with Lorenzo Ghiberti (1378–1455), his erstwhile rival for this and previous projects, he accepted the condition so as not to lose the tender, but then feigned illness, returning only when receiving sole responsibility and 90 percent of the pay. When his Florentine masons demanded a raise, he replaced them with Pisan ones, taking the locals back only after a substantial pay cut. These are not mere anecdotes,

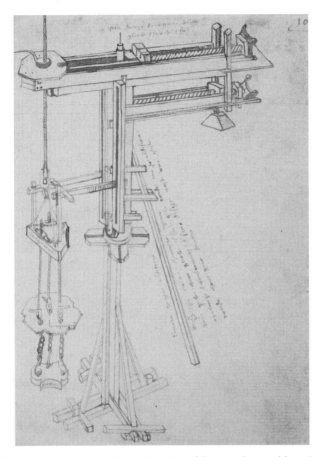

Figure 5.4 An example of Brunelleschi's combination of *knowing-how* and *knowing-that*: a revolving crane built for the work on the Duomo. This hoisting machine employs an assembly of the Archimedean 'simple machines,' discussed in Chapter 9, that would be crucial for Galileo: winch, pulley, infinite screw. It's difficult to know which part of this assembly Brunelleschi learned from traditional masons and which part he developed following theoretical principles, but as a working machine it was novel and exciting enough for this diagram to be drawn by Bonaccorso Ghiberti, the nephew of Brunelleschi's competitor/collaborator Lorenzo.

but characteristics of the new age and its new knowledge: Brunelleschi's virtuosity was his livelihood, and he traded it for profit and glory.

The Movable Press and Its Cultural Impact

Knowledge turned from spiritual elevation to material goods, technological innovation became prized and texts – a coveted commodity. So it is hardly surprising that there were those who sought to benefit from them.

Manuscripts were very expensive. The material they were written on – parchment (processed hide, also called vellum) – was so costly that scribes often tried to erase and write over manuscripts, creating *palimpsests* on which priceless classics are sometimes found, hidden under insignificant scribblings. More importantly: the work invested in copying manuscripts was enormous, and required rare learnedness. The ability to produce texts quickly, cheaply, regularly and in significant quantities loomed as a lucrative prospect. The feat was achieved around 1450 – some fifteen years after Brunelleschi completed his dome – by another goldsmith, a German, Johannes Gutenberg of Mainz (c. 1400–1468), with the invention of the movable press.

The Invention

Like his Florentine brother-in-trade, Gutenberg was an entrepreneur and a risk-taker, whose career saw rises and falls. He even shared Brunelleschi's interest in mirrors: one of the purposes of the press project was to relieve him of the debts incurred in a previous failed project of mirrors devised to capture the holy light of relics. Since, like Brunelleschi, he kept his knowledge secret, we know little about the process of developing his technology that appears to have taken him a whole decade. The main stages of this technology changed little for centuries:

A character – a letter or a punctuation mark – is carved onto a *punch*, which is then used to create an indentation in the *matrix*. The matrix is put into a *mold* and the mold into a wooden *block*, into which molten metal is poured to make a *type* (Figure 5.5). A collection of types of all the required characters is the *font*, which is arranged in a rectangular case with many compartments. The *typesetter* sets the types in lines – the letters positioned to form a mirror-image to the manuscript – and the lines are arranged into *galleys* held by a wooden frame. Ink is spread onto the galleys, a paper placed and pressed, and a page is printed (Figure 5.6).

Like all complex inventions, Gutenberg's was mostly an assemblage of known technologies: the press was a converted olive press (used to obtain oil); ink and paper were in common use (although Gutenberg also improved his ink); and printing from a mirror-image was a well-known technique, used originally in China for decorating silk, and employed in the Muslim world and Europe since the Middle Ages. Befitting a goldsmith, Gutenberg's main material innovation was an alloy: the metal composite which could be cast into durable characters that preserved their sharp relief upon cooling. This material invention embeds the crucial innovation of the movable press, which is a conceptual innovation: *standardization*. Standardization is what

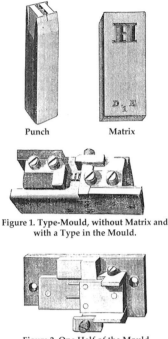

Punch   Matrix

Figure 1. Type-Mould, without Matrix and
with a Type in the Mould.

Figure 2. One Half of the Mould.

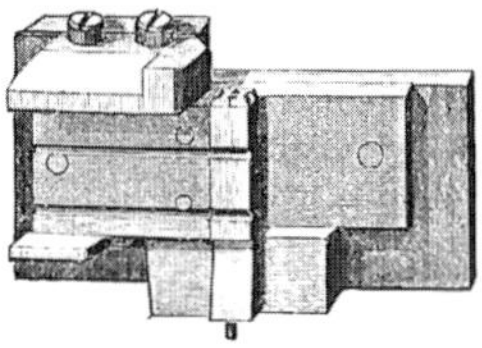

Figure 3. The Other Half of the Mould.

Figure 5.5 The punch, matrix and mold of the movable press. The punch and matrix changed little over the centuries. The mold here is a nineteenth-century version, easier to use than the original ones but still fundamentally the same. The core of Gutenberg's invention – the ability to produce varied texts by a standard process – persisted even as his technology gave way first to offset lithography late in the nineteenth century and then to digital printing. It is nicely captured by Theodore de Vinne, the author of the 1876 *The Invention of Printing* (New York: Francis and Hart) from which the images are taken: "The depth of the stamped letter, and its distance from the side, must be absolutely uniform in all the matrices required for a font or a complete assortment of letter [in order] to secure a uniform height to all the types, and to facilitate the frequent changes of matrix on the mould … For every character or letter really required in a full working assortment of types, the type founder cuts a separate punch and fits up a separate matrix; but for all the characters and letters which are made to be used together, there is but one mould. Types are of no use … if they cannot be arranged and handled with facility, and printed in lines that are truly parallel … The uniformity of the body … is as essential as the variety of face" (pp. 55–56).

Figure 5.6 The basic stages of the Guttenberg printing press, which changed little until the introduction of rotary and offset printing in the mid-nineteenth century. Here depicted in the illustration 'The Printer's Workshop' by Jost Amman to Hartmann Schopper's *Panoplia* (*Book of Trades*) from 1568 (Frankfurt am Main: Sigmund Feierabend). In the background two typesetters sit in front of type drawers and set the pages using a composition stick. In the foreground, the younger man on our right – the apprentice or novice – applies ink to the galleys and then lays a blank sheet of paper from the pile in front of him. The inked frame with the paper on top is then moved under the press and pressured by the screw. The older man – the master – on our left checks the freshly printed page and puts it on the pile in front of him.

made Gutenberg's press profitable: all types were of the same size, ensuring that the same matrix and mold, as well as the metal of an over-used type, could be used again and again with minimal adjustment. Standardizing the sizes of pages meant the frames could also be reused. Gutenberg's innovation was conceptual, but could not be achieved without his unique material skills: he was not alone among his contemporaries in attempting to standardize printing in similar ways; his way succeeded because his alloy and ink were simply better.

Standardized printing dropped the price of books to a fraction of their previous cost, and their availability rose exponentially. The cultural experience may perhaps be compared to ours with the advent of the internet, but it was significantly more dramatic: people who can now access the internet could have afforded books beforehand; but until the movable press, unless you were a scholar by vocation, you were unlikely to ever encounter a manuscript. A book, on the other hand, was a common artifact – it was not found in every household, but usually within the village, at least in the hands of the priest. And for scholars the change was even more dramatic: no longer was the master the only one holding the text, hence wielding final authority over its content and proper interpretation – every student could avail himself of one (Figure 4.8, left, nicely captures the transition). That also meant that censorship became much more difficult; for example, when Galileo was put under house arrest in Florence and the Inquisition prohibited the publication of his *Dialogue Concerning the Two Chief World Systems*, he had the book published in Strasburg. A marvelous imaginary community became possible: the 'republic of letters,' uniting scholars from all over Europe, and soon thereafter – all over the world. These scholars no longer needed to travel like the friars of yore to study a manuscript in a library: they could buy and collect books, establish their own library and correspond with other scholars who could own similar libraries and the same books.

Imitation and Inspiration

The *same* books: standardizing reproduction standardized the text. The manuscript used to be a unique artifact: copying required every copier to comprehend and interpret the text in order to recognize the idiosyncrasies of his predecessor's script and correct earlier mistakes of copying or misunderstanding, and every copier introduced his own. Each copy was personal and final, but each version was also open to reinterpretation and change by the next scribe to copy it. Once the typesetter replaced the scribe, copying ceased to be a scholarly re-creation of a text and became a technical task, oblivious to content. "Imitation was detached from inspiration, copying from composing" (Eisenstein, *The Printing Press as an Agent of Change*, p. 126): the 'author' became sharply distinguished from the copier, and the concepts of copyrights and plagiarism could be invented.

Standardization allowed accuracy and authority. The book was public and unchanging: the galleys were proofed and corrected before printing, and all

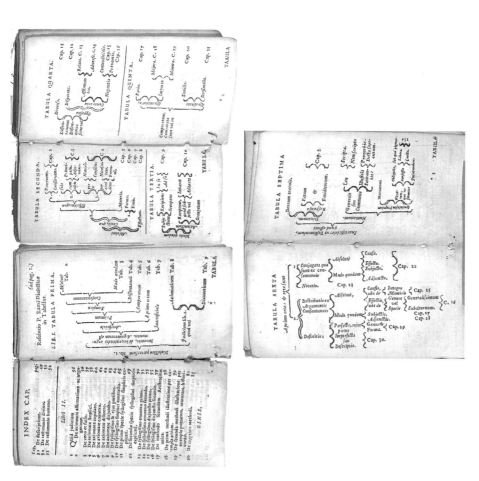

Figure 5.7 The hierarchical organization of logic according to Petrus Ramus, introduced as the table of contents for his 1594 *Dialectica libri duo*. *Logica* is divided into *inventio* (the formulation of arguments) and *iudicio* (their judgment); the former is then divided into 'artificial' and 'testimonial' and so on. None of the terms or the relations between them are particularly novel, but this way of presenting and thinking of them is new, and dependent on the standardized reproduction enabled by the movable press: such a complex image would not have been trusted to be copied accurately by hand.

copies printed from them – all books of the same edition – were identical. Standardization also allowed neutral indexing and cross-referencing. It enabled the law of the land or the authoritative interpretation of the scriptures to be unified and centrally enforced: it could be formulated in one place – the capital or a holy city – printed in another place, and distributed unchanged throughout the kingdom or among all believers, many travel-days away, overriding any local lore and custom.

Standard pagination meant indexes and cross-referencing. Proofreading allowed tables, calendars and diagrams – all of which were very precarious in a manuscript since even the most minor error could render them meaningless. Incorporating older techniques of printing – woodcuts and engravings – allowed images to be added, which had an impact far beyond decoration. Logic, in particular, which since Aristotle was a linguistic art, verbally analyzing modes of thinking and argumentation, was reconceived as a visual means of disciplining reason. A leader among the new generation of humanists pursuing this visual turn was Petrus Ramus (Pierre de la Ramée), who was born in 1515 and stabbed to death during the Bartholomew Day massacre on August 26, 1572. "All the things that Aristotle has said," wrote Ramus irreverently, "are inconsistent because they are poorly systematized and can be called to mind only by the use of arbitrary mnemonic devices" (quoted by Ong, *Ramus and the Decay of Dialogue*, pp. 46–47). For Ramus and scholars for whom the press was an integral part of learned life, logic needed to be organized, practical, creative and bring to mind new relations between things and concepts – all of which could be achieved visually, as in Figure 5.7. Logic could and should serve as an accurate map between concepts, and geographical maps could and should accurately represent the world: the standardized, proof-read press eliminated the inevitable inaccuracies in a copied manuscript that used to limit maps to a symbolic role (Figure 5.8).

Global Knowledge

Maps would become the emblem of the knowledge befitting the budding new age. Standardized, visual and impersonal, they allowed the skilled user to find his way towards and around previously unknown places.

Navigating the Open Seas

Navigation in foreign terrain had not been an important type of European knowledge. Voyages were almost exclusively on well-traveled routes and navigation was based on local know-how. Captains and pilots were

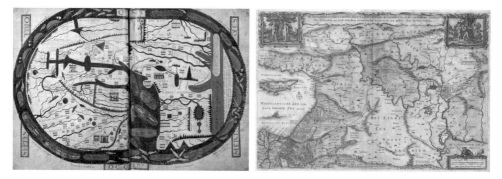

Figure 5.8 The changing idea of visual representation arising from the printing press, as exemplified by two *mappae mundi* (maps of the world). On the left, a so-called a T-O map (after its shape) from Isadore of Seville's 623 *Etymologiae* (see Chapter 4). Although this image is taken from its first printed version (1472), the text was written well before standardized reproduction could be imagined, so the map is drawn without any pretense of geographical details – all these would have been lost when copied by hand. Isadore obviously didn't think the world 'looked' like that – he was representing a religious idea of its arrangement, with Jerusalem at the center and the paradise at the outmost East – "east of Eden," as told in *Genesis* – and at the top (note 'Oriens' at the top and 'Occidens' at the bottom). The 1657 map on the right, by the Dutch engraver and cartographer Nicolaes Visscher I (1618–1679), is titled "Of the history of Paradise and the Lands of Kanaan." As its name demonstrates, the map is just as mythical, but it is drawn with a clear idea of visually capturing real geographic features, like mountains, rivers, seas and islands (note that Cyprus is quite accurate, though not to scale). This is possible because once engraved on a copper plate of the standard size, the image could be proofed and then pressed continuously and uniformly.

acquainted with the sea they were traveling through much like guides on land: they recognized currents and winds and features of the shores that they tried to keep within sight. This was knowledge acquired by apprenticeship and mastered through practice. But in the second half of the fifteenth century – the two generations following the invention of the movable press – Europeans departed the comfort of their land-bound seas and took to the oceans.

They could not hope for such immediate, personal acquaintance with such vastness. The islanders of the South Pacific, great nomads of the open sea, could navigate by cloud patterns and bird trajectories, but even they were island-hopping. Between 1405 and 1433, vast Chinese fleets were dispatched to explore the shores of South East Asia, India, the Arab Peninsula and East Africa. But even they, under the command of the eunuch-admiral Zheng He, traveled along well-known trade routes. Neither the Polynesian *Wa'a Kaulua* (the double-hulled canoe) nor the Chinese *Chuán* (the large sail ship) attempted what the European *Caravel* did: to sail from home to a destination on the other side of the globe, not knowing for sure if it was really there, how far away it was and what would lie along the way.

That they would even try such feats almost defies explanation. The prospects of returning from these voyages were so low that one wouldn't expect curiosity, honor or even greed to outweigh the fear. Yet they did try, and the reasons do seem to have been fortune and glory.

The thriving trade through Mongol-ruled Asia that we mentioned above insatiably whet Europe's appetite for goods from the East – silk and porcelain from China; spices from India; exotic fruits from Malaya and medicines from Sumatra. Their own new-found prosperity made the Italians, and then their neighbors and competitors north and west of the Alps, somewhat more capable of paying for these luxuries with wool and leather. Yet lucrative as it was, the new global trade left the European merchants dissatisfied. The main beneficiaries were their Muslim counterparts, particularly Arabs who ruled the trade routes through Asia and the Indian Ocean and brought these coveted commodities to the port towns of the Eastern Mediterranean – Alexandria, Tripoli, Constantinople – from where Genovese and Venetian ships would take them to markets throughout Europe. Bypassing the Muslim mediation – and for the Iberians and the French also the Italian stronghold on the Mediterranean – promised enormous profits but entailed tremendous risks. It also required the conviction that the mariners' own risk wasn't reckless, but that the unknown routes could ultimately be made known.

Discoveries

By the end of the fifteenth century, Europeans were on their way to mastering this new knowledge. First were the Portuguese, closely followed by the Spaniards. In 1456, Diogo Gomes made it to the Cape Verde archipelago near the west coast of Africa, and in 1460, Pedro de Sintra reached Sierra Leone. Within the next decade, João de Santarém and others made it past the equator to the islands of the Gulf of Guinea and to the coasts of today's Ghana, achieving the intended commercial goal: trading gold directly with the local Arabs and Berbers. A crucial breakthrough in the continuous encroachment around the coasts of Africa came in 1488, when Bartolomeu Dias sailed around the southern tip of Africa and into the Indian Ocean: a marine route was opened to the breathtaking riches of the spice trade. In 1498, Vasco da Gamma completed the task by making it to Calicut. Meanwhile, Christopher Columbus convinced the Spanish Court to allow him to circumvent the now-Portuguese stronghold on the Asian trade by sailing west. In 1492, he discovered, instead of a route to China, a new continent. The name it came to be known by steals some of Columbus' deserved glory but tells of the context of these discoveries. The name *America* originates with the Florentine Amerigo Vespucci, whose letter to his employer, published in 1502 and translated and distributed soon all over Europe, announced that it was indeed a *Mundus Nouvus* (for the Europeans) – as was the letter's title – on which Columbus had landed, rather than the eastern shores of Asia. The employer was none other than Lorenzo de' Medici, whose bank was financing many of the expeditions under Vespucci's close watch, sometimes as an active participant. Like capital, knowledge was transgressing boundaries: Columbus the Italian had offered his services to the Portuguese king before settling for the Spanish; his compatriot Giovanni da Verrazzano explored North America's Atlantic Coast under the French flag, from South Carolina to Newfoundland; the Englishman Henry Hudson sailed up the Hudson River on behalf of the Dutch.

Columbus didn't make it to China, but his cousin Rafael Perestrello did, landing in Guangzhou in 1516, inaugurating trade with the mainland. Japan opened to European trade – in particular in firearms, which the Japanese were quick to reverse-engineer, produce and sell to the Koreans – in 1543. Soon Japan was directly connected to the New World: in 1565, Andrés de Urdaneta deciphered the regular directions of the Pacific trade winds: eastward north of the equator; westward south of it – similar, if on a much larger scale, to the Atlantic. This opened up a direct route between Japan and the Philippines, which had just been violently claimed by King Philip II of Spain. Struggling to wrest some of the riches of the eastern trade from the Iberians, who ruled the South Seas, the Dutch, French and English turned north. They explored the shores of (now) California, Virginia and Canada searching for the Northern Passage to China and India. They could not find it, nor fortunes like those of South and Central America, but beyond unknowingly preparing the immense migration movement to follow, they were honing their marine skills: in sailing, navigation and piracy. In 1580, Francis Drake, who excelled in all three, led a captured Portuguese ship – the *Santa Maria*, renamed *Mary*, then renamed *Golden Hind* – in circumnavigating the globe.

The first to attempt this venture did not survive it. Back in 1513, Vasco Núñez de Balboa used much violence to cross the Isthmus of Panama and reach the shores of the Pacific, and in 1519, Ferdinand Magellan, crossing the straits now bearing his name, made it from the Atlantic to the Pacific by sea on his way back home. Only one of the five vessels that had left on the expedition survived to complete the circumnavigation and return to Spain, and only thirty-five men of the original 240, under the only remaining captain: the Basque Juan Sebastián Elcano (Magellan himself was killed in the Philippines). It was not a valid trade route – the fortunes that Drake would bring back from his circumnavigation fifty years after this first one were robbed from the Spaniards, rather than earned by commerce. But East and West were connected.

Global Commerce

Yet more than through these heroic – or reckless – voyages, the globe was now interconnected by the very booty for which Drake raided the Spanish colony in Panama: silver. Since the Chinese and Japanese merchants were far less impressed with European goods than the Europeans were with theirs, European purchasing power was limited. But East Asians needed no introduction to silver, which had served them as common currency for many centuries, officially and unofficially. Like gold, silver was useless beyond its symbolic and aesthetic value, and it was not as rare and precious, hence not as dangerous to handle. It was extremely coveted in all cultures familiar with it – even if these cultures were completely oblivious to the existence of one another. Around the Mediterranean, silver has always instigated and bankrolled wars: in the fifth century BCE the Athenians financed their war against the Persian invasion with silver from their Laurium mines; in the third to second centuries BCE Rome and Carthage waged wars over the silver mines in the Iberian Peninsula.

The Spanish conquerors of the New World thus knew what they were looking for, and in 1531, a decade after Mexico's bloody conquest by Hernán Cortés, three Spanish miners found it in the remotest regions of his occupation, near Taxco. Fifty years of prospecting and plundering later, the quest for the precious metal culminated in the materialization of the legendary *Sierra de la Plata – Silver Mountains –* in Potosí (today's Bolivia). On the backs of local forced laborers and African slaves, European merchants now had the wherewithal to buy porcelain from China, silk from Japan and spices from India.

What emerged was not only a two-way exchange of goods for silver, but a global network in which commodities, currencies and especially knowledge – technologies and skills – were competing and changing hands. We already mentioned the weapons that the Japanese bought from the Portuguese, taught themselves to produce and sold to the Koreans. Their evolving marine-military capacities – coupled with a willingness to use the extreme violence that made them effective – allowed the Portuguese to monopolize the trade between the Indian Ocean and Europe. Mariners of the South China Sea and the Bay of Bengal were not familiar with the ancient Mediterranean custom of violently enforced territoriality of marine trade routes, and soon the Portuguese could extend their monopoly to the inter-Asian trade, taking textiles from India to Indonesia, spices from Indonesia to China, and silk and porcelain between China and Japan. They – and the

Dutch and English who aggressively wrested away their monopoly in the early seventeenth century – were no longer selling goods but marine skills and technology: a ship built in Rotterdam could sail to the Indian Ocean where it would spend its entire life operated by a local crew under a Dutch captain.

Other kinds of knowledge were traveling in other directions. The porcelain trade with Europe was so lucrative that its manufacturers in Jingdezheng, more than 800 kilometers from the Hong Kong port from which it would be shipped, were reshaping their vases to suit what they learned or speculated to be European taste. The secrets of their trade were also moving. Japanese porcelain makers couldn't, and seemingly had, previously felt a need to emulate the quality of high-end Chinese porcelain. But by the middle of the sixteenth century, eager to tap into the European wealth, they were buying kilns and expertise from their competitors in Korea (on whose soil they were soon to wage war against China, using Spanish weapons). Europeans had been trying to mimic this marvelous art since the first Qingbai vessel was presented to Louis the Great of Hungary in 1338 by a Chinese embassy on its way to the pope. But with little success: Chinese clay – Kaolin – was necessary, and even more importantly: Chinese *know-how*.

Theoretical knowledge proved to be susceptible to the challenges and opportunities of the new global networks. European scholars, especially if they shared the humanists' sensitivity to the changing world around them, could not escape the realization that much of what they knew about the world was simply wrong. Augustine had convincing arguments why there should not be humans on the other side of the globe: there was no reason to believe that it wasn't covered in water and the scriptures didn't report any of Adam's offspring traveling south. Yet, convincing as his arguments were, he was wrong: there were humans in the Southern Hemisphere. Careful calculations based on the Biblical story concluded that Noah's flood supposedly occurred around 2350 BCE, but the Chinese had much earlier records – reliable records – of kingdom and worship. The Chinese in particular, but also the great cultures of Asia and the Americas, demonstrated that human dignity and morality could thrive without the knowledge of God. Europeans' knowledge of the non-human world proved as wanting: the Indian Ocean wasn't land-locked, as Ptolemy had said. Aristotle recorded had some 500 species of animals in his writings and Dioscorides' *De Materia Medica* – the definitive first-century encyclopedia – mentioned about 600 species of plants (see Chapter 8). But, by 1594, the botanical garden of the Faculty of Medicine of Leiden University listed 1,060 species in its inventory and many more were coming: from Mexico and India; Indonesia and the West Indies; Japan and Malabar; tropical Africa and China.

Knowledge for the New Age

For all participants in and observers of this global network of exchange, and in particular for Europeans – the most active and aggressive among them – new knowledge was required. More crucial than trying to fill the great lacunas with content was the realization that both *knowing-that* and *know-how* had to change their form. Knowledge was being created and exchanged by traveling and for traveling, so it had to take a shape that would allow it to travel.

Knowledge of this new form needed, to begin with, to be *factual and accurate*, especially since it often came from afar and therefore could not be immediately verified. Crucial decisions had to be based on trust: a merchant whose silk turned out as good as promised would be able to sell his next boatload before it saw a shore. In contrast, the Medici Bank would not extend *credit* again to an explorer whose previous reports had not proven *credible*.

This new knowledge needed to be *standardized and translatable*. Spanish and Chinese traders in Guangzhou or Portuguese and Indian merchants in Malacca needed to converse on the quality and purity of porcelain, silver and turmeric; an Indonesian crew and a Dutch captain in the Bay of Bengal needed to communicate expertise and experience gained on opposite sides of the globe and relate to one another information about winds, sails and currents.

Life depended on this knowledge being *precise*. Quantities of water and food needed to be precisely calculated; large enough to suffice until the next landing – yet not so large as to dangerously slow the ship. Marine maps needed to be precise enough that the pilot – quite likely unacquainted with the seascape, as we discussed – could steer away from unobservable hazards and into a safe but unfamiliar shore. In the middle of the ocean, the only clues for navigation were the position of the heavenly bodies, so for the first time in their history star maps and ephemerides (position tables) *had* to be precise. Around the equator, an error of one degree – completely acceptable for the academic astronomer – would result in a deviation of more than 100 kilometers off the planned path. The horrible fate of a ship missing an island where provisions were to be loaded is captured well in the legend of the 'Flying Dutchman' – the ghost ship floating aimlessly on the ocean.

Mathematics, the most abstract of disciplines, became the most practical tool for the banker on land and the sailor at sea. The ancients had an aesthetic fascination with the precision of mathematics and this very precision meant to them that it belonged strictly to the realm of the ideal (see Chapters 2 and 3). Now this precision was of immediate, vital, worldly value. The precise calculation of things like interest rates, risks and relative values was as indispensable to the financial survival of the entrepreneur in the complex, perilous and extremely lucrative new venture of global marine commerce as the precise calculation of heavenly positions was for the literal survival of the mariner who carried his goods.

This practical value turned mathematics into a coveted trade. Already since the fourteenth century in Italy, merchants and craftsmen had been sending their sons to study mathematics in local 'Abacus Schools' (sons, because girls were almost universally deprived of formal schooling). There, boys learned how to solve complex arithmetical problems that they might come upon in the course of their work: converging rates; compounding interest; finding unknown quantities from known ones. Their primary study material was the 1227 edition of the *Liber Abaci* (*Book of Calculation*) by Leonardo Pisanus (Fibonacci), which introduced the Indo-Arabic numeral and decimal system and quasi-algebraic strategies for using it.

There was now place and resources for new teaching material and new charismatic mathematician-teachers. Niccolò Fontana Tartaglia (1499–1557) was one, and his life and career represent the times similarly to the way Brunelleschi's represented the precious century. His father, a courier who delivered mail in the regions of the Republic of Venice to which their town Brescia belonged, was murdered by robbers when Niccolò was about 6. When he was 12, a soldier of the invading French army slashed his jaw and palate, leaving him a stutterer (hence the nickname 'Tartaglia'). His family could no longer afford tuition, but he taught himself Greek, Latin and mathematics so impressively that he came under the patronage of a wealthy nobleman, and by his late teens was already an abacus teacher in Verona, gaining enough reputation to take his teaching career to Venice. Tutoring provided him with an income, and mathematical spectacles provided the fame that brought the students and the patrons. The spectacles were comprised of public mathematical contests, in which participants competed to solve mathematical problems on a stage in the town square in front of a boisterous audience. The competitions also included solving difficult mathematical riddles – Tartaglia's solution to the cubic equation made him particularly famous (and involved him in a bitter plagiarism dispute with Girolamo Cardano, whose wealth and nobility left Tartaglia no chance to win). Less dramatic demonstrations of mathematical competence came by way of publication: a definitive translation of Euclid's *Elements* into Italian and a book boldly titled *Nova Scientia*. No less boldly, the book promised the Duke of Urbino, to whom it was dedicated, to teach his gunners how to properly aim their cannons with the help of mathematical analysis – a new arena for the scholar to meet the artisan.

Global Institutions of Knowledge

Trade Companies

Abacus schools were not the only new institutes of knowledge. Even more representative of the new age were the bodies established to maximize profits from the global trade. They didn't hesitate to use violence for that purpose, as we noted, but their main tool was creating, gathering, standardizing and monopolizing knowledge. The Iberian empires assigned the task to royal institutions: the Portuguese established the *Casa de India* in Lisbon in 1501 and the Spanish soon followed with the *Casa de Contratación* (House of Trade) in Seville in 1503. These *Casas* had control over customs, taxes, marine contracts and schedules, but most importantly: they were in charge of the *Padrón Real* (*Padrão Real* in Portuguese) – the Royal Register (Figure 5.9). This was a large, secret, constantly evolving world map, heavily annotated and supported by charts and records assembled by a small army of cartographers, navigators, record keepers and administrators, headed by the *pilot major* – a position created especially for Amerigo Vespucci in 1508. "No pilot shall use any other chart but only one which has been taken from the *Padrón Real*," commanded the royal appointment letter to Vespucci, "on pain of a fine of fifty dobles." Returning from the "lands of the Indies, discovered or to be discovered," the pilot would report, under oath of accuracy and secrecy, all "new lands, islands, bays, or harbors, or anything else that they make a note of them for the said *Padrón Real*" (Markham, *The Letters of Amerigo Vespucci*, p. 65).

The new knowledge was flowing in: from global periphery to center of epistemic authority and political power, where it was verified, standardized, formalized and dispatched again to the periphery to be applied and augmented. In the other trading super-powers, England and the Netherlands, this center was taken by the private joint-stock companies: the East India Companies (the Dutch and the English); the Muscovy Company; and the Virginia Company. These were private, rather than royal institutions, which organized themselves to allow their wealthy shareholders to maximize return on their investment and minimize their risk by joining their funds. But the royal charters they managed to obtain (the first from Queen Elizabeth to the English India Company on the last day of 1600) and their vast and superbly profitable monopolies – at its prime, the EIC held half (!) of world trade – made them almost sovereign entities. The companies took over territories, established their own laws, which they enforced with their own military power – yet like the Casas, they were, crucially, institutions of knowledge. They provided their outgoing captains with knowledge of their destinations and required

Figure 5.9 A presentation copy of the *Padrón Real* by the (Portuguese) master cartographer Diogo Ribeiro of the (Spanish) Casa de Contratación, offered to the pope in 1529 and kept in the Vatican Apostolic Library (no working copies of the map survived). The lines criss-crossing the oceans are 'portolan' lines, which designate sailing directions, primarily between ports.

that it be improved and expanded. And it wasn't just geographical knowledge and marine know-how that the companies were seeking and distributing, but any knowledge that could further their profits: new raw materials, from minerals to foodstuff; new crops, from spices to medicines; new products of art, from silk to porcelain. New knowledge of the people and cultures that the captains came across was particularly crucial: could they be subjugated? Could they be traded with? How? What were their languages, habits and customs, and could they be used to further trade or subjugation?

It's telling to compare the casas and companies to the emblematic medieval institution of knowledge: the guild. Like the guild, they were established to maximize profit by controlling and monopolizing knowledge. But it was a different kind of knowledge. The guild's was local and material; literally embodied in the master – and the only way to control it was to regulate the master's behavior: his relations to his brethren, apprentices and customers. The companies' knowledge, on the other hand, was global and abstract. It was inscribed into maps and registers, dispatched afar, and circulated. Its value was in its flow, which is what the companies sought to control.

The Jesuits

For the Jesuit Mission, global commerce was an opportunity to gather souls rather than pecuniary profit, and the lofty ambition and great challenges and dangers associated with it called for a careful formalization of the global flow of knowledge. The *Societas Jesu* was established in 1540 by Ignatius of Loyola (1491–1556), a Spanish-Basque soldier-turned-priest, in the midst of a great schism in the Church (about which in Chapter 6). Befitting its founder's personality and experience, the Society of Jesus was conceived as a fast-moving 'army' of preachers directly under the pope's personal command, yet it soon became apparent that the best weapon of the time was the new knowledge, education the best strategy. In 1548, the first Jesuit college opened in Messina, Sicily – a high school not only for future Jesuits but for all boys; the idea of educating girls was considered, but never materialized. By Ignatius' death there were thirty-three such colleges; by 1600 – 236; and by 1615, 372 all around Europe. In 1627, the Society of Jesus was educating 40,000 students in France alone. Its schools were so popular not only because they were free – sponsored by a patron or the town – but because their curriculum was innovative and up-to-date, including rhetoric, classics, theater and, crucially, mathematics.

For those who joined the Society and especially those identified as 'scholars', suitable for the mission, education was intensified: they became novices for two full years and moved from their hometown college to one of the central institutions such as those in Coimbra or Évora (in Portugal) where they continued to study Greek philosophy, Christian theology and the most advanced sciences of the day. If they made it to the *Collegium Romanum* – the order's university in the center of Rome – they could study with some of the most innovative scholars of the time. The missionaries-to-be would usually spend some time as masters before being dispatched to one of the colonial colleges near their assigned mission – Rio de Janeiro if they were headed to Brazil; Goa for India; Macao for China and Japan – where they would partake in a standardized program acquiring the local languages and customs.

This erudition was important firstly for turning the missionaries into better people, as the Jesuits learned from the Humanists' Latin heroes, Cicero and Seneca. The theology was meant to make them more devout Christians – the college was a "Trojan horse filled with soldiers from heaven, which every year produces conquistadors of souls" ("Baltazar Tales," mid-seventeenth century, cited in Brockey, *Journey to the East*, p. 211) – and local knowledge was of obvious practical necessity to conquer these foreign souls. Upon arriving at their mission, these "conquistadors of souls" would turn from students into active pursuers of knowledge. The regular composition of carefully structured "edifying reports" was an integral and mandatory part of the mission. The reports traveled in the opposite direction of the missionaries: from their residences, through the provincial father to the Father General in Rome, where they were edited, digested and distributed to the colleges. The practical information they contained would serve to prepare the novices for their missions, and the more exciting new knowledge would help recruit and enthuse new ones, as well as establish the credibility of the Society as an institution of knowledge.

This epistolary project proved an unquestionable success. From Peru, Spanish Jesuits sent in the cure for malaria – the bark of the Cinchona or *The Jesuit's Bark*. Camellias were named after the Czech Jesuit Georg Joseph Kamel (1661–1706) in honor of the definitive accounts of the Philippine flora and fauna he sent, mostly through the *Royal Society of London for Improving Natural Knowledge* – whose reliance on this model of correspondence we'll discuss in Chapter 9. The China mission produced more abstract knowledge. In the early decades of the seventeenth century, the astronomer Johannes Kepler – whom we'll discuss in Chapter 7 – was corresponding with the German Jesuit Johan Schreck. Schreck sent him an account of the Chinese method for calculating solar eclipses and Kepler repaid with a copy of his Rudolphine Tables, carried by another Jesuit missionary – the Polish Michael Boym. In the last decades of the century, the great philosopher Gottfried Wilhelm Leibniz learned about the Chinese language from the Italian Jesuit Claudio Filippo Grimaldi.

China was the most coveted missionary prize, because of its vast populace and its illustrious culture – and for the China Mission, erudition indeed turned out to be both a necessary and efficient tool. The custom that earned them their infamy (and eventually led to their disbandment in 1773) – to conquer souls through political power – seemed particularly fit for China and its powerful centralized government. This meant gaining the confidence and respect of the political-bureaucratic elite – the mandarins – whose status as *literati* was sanctioned by the elaborate system of exams which they had to pass to achieve their posts.

To Matteo Ricci (1552–1610), who arrived at the Macao college in 1582 to bolster the budding China Mission, the Mandarins appeared very much like the Humanists back in Italy, his home country. By 1595, his Chinese was good enough to try and impress them with *Jiaoyun lunyan* – an elegant collection of classical aphorisms on friendship that he translated. The little book generated enough interest and connections that he was allowed to travel north, towards the court, shedding on the way the Buddhist clergy robes the missionaries were wearing in their southern China residences for a full mandarin garb. This move came to symbolize the Jesuit doctrines of 'accommodation,' namely: adopting and adapting to foreign customs, sometimes quite questionable from a Christian perspective, in the service of conversions. Jesuits' detractors, both in Europe and in China, found this approach outright repulsive; a sign of the order's duplicity, religious shallowness and power-hunger. For us, however, it signals something particularly significant about the new global age: a new understanding of the erudite as a trans-cultural class and of their knowledge as transcending all boundaries.

Classical erudition, however, was of limited value. Ricci could smoothen his journey with gifts of paintings and exotic goods, but when he finally made it to Beijing, he discovered that what most impressed his adopted Chinese colleagues were neither those material objects nor ancient wisdom – they were content with their own. Rather, it was the new knowledge of the world and the means of gathering and recording it that truly caught their attention: clocks; prisms; maps; globes. Ricci was well-equipped to satisfy their curiosity: in the Collegium Romanum he had been a student of Christopher Clavius (1538–1612), one of the greatest astronomers of the time. In 1602, at the request of the emperor and in collaboration with the convert-mandarin Li Zhizao (1565–1630), he produced the emblematic artifact of that era of global exchange: a *mappa mundi* – a map of the world (similar to Figures 5.8, right, and 5.9; the original is sadly lost). It was a marvelous piece: printed on a folded screen of six panels, more than 3.6 meters wide and 1.5 meters high, and it was also a true object of knowledge. Befitting its cross-cultural import, it found its way, translated and annotated, to Japan, Korea and even Manchuria. In 1607, Ricci cemented the prestige gained by the great map with a series of mathematical treatises, capped by a translation of Euclid's *Elements* (Jihe yuanben), produced with Li and another converted scholar – Xu Guangqi (1562–1633). These demonstrations of skill anchored the Jesuits' presence in Beijing for the better part of a century – as specifically *foreign* experts. In the decades after Ricci's death they were invited to the Chinese Astro-calendric Bureau, where their astronomical expertise was employed in resolving the disarray in the imperial calendar.

Conclusion

Here are the two cultural elements of astronomy we discussed in Chapter 3: experts producing calendars and centralized government employing them for astrological and political-ritual purposes. That the experts and the government could meet across oceans was not because of some inherent universality: astronomy, like cathedrals, doesn't travel on its own (the Jesuits also brought the cathedral with them, particularly to South America). The reason why the Jesuits had the astronomical skills to impress the Chinese court was that a similar failing of the calendar was taking place in Europe (we'll discuss that in Chapter 7), for similar reasons, and in a similar context: the Church's bureaucratic-centralized interests. In fact, Clavius' involvement in the calendar reform is what gave him the credentials to institute mathematics at the core of Jesuit education, not least because of its claim to certainty and universality. This produced missionary-mathematicians like Ricci, and his even better skilled successors at the Chinese Astro-calendric Bureau, the German Johann Adam Schall (1592–1666) and the Dutch Ferdinand Verbiest (1623–1688). Moreover, their astronomical skills were of use to the imperial government because the astronomy practiced in the Astro-calendric Bureau was a product of an earlier era of cross-continental exchange: the Mongol Empire of the thirteenth to fourteenth centuries. Kublai Khan, Genghis' grandson, who declared himself the first emperor of the Yuan Dynasty in 1271, had imported Muslim scholars as both officials and experts, and they brought with them the Hellenistic-based science we discussed in Chapter 4. The most famous of these scholars was the Persian Jamal al-Din al-Bukhārī, who arrived in China equipped not only with the skills of theoretical Ptolemaic astronomy, but also the appropriate instruments. These instruments were modeled on the ones employed by Kublai's brother Hūlāgū and his court astronomer Nasir al-Tūsī at the Marāgha observatory (see Chapter 4). They were therefore Ptolemaic-style instruments that the Jesuit astronomers could easily recognize, even if by the time of their arrival, they had fallen into disrepair. Science is not inherently universal, but Hellenistic astronomy was globalized twice: it was imported once to Europe and once to China, with Muslim mediation in both cases.

For the history of science, however, early modern globalization of commerce, religion and war was particularly crucial: it ushered in a new mathematical age. Mathematics traveled well and was useful in crossing cultural barriers. Not only the mathematical skills had become valuable – much of what was known could be reduced to numbers and figures. For the Jesuits, it was the number of converts that counted (the numbers in China, despite the great investment and prestige, were disappointing). For the banker, goods were reduced to their monetary value; whatever their use or beauty may have been, they needed to fetch enough to allow for the next investment. Money itself was, in its very essence, nothing but an exchange rate: the value of silk could be measured in silver, and silver's value – in porcelain. For the mariner in the middle of the ocean, the sky above was no longer the heavens but a mathematical grid against which to plot his purely mathematical itinerary. There were no places on the open sea, just distances.

Discussion Questions

1. Does the discussion of the struggles, compromises and synthesis that characterize the encounter of monotheism with Hellenistic science suggest to you any new thoughts about the relations between science and religion in general?

2. Is the account of the encounter between the scholar and the artisan and the social and cultural conditions supporting it convincing? Does it carry any epistemological lessons?

3. The invention of the movable press and its cultural impact begs a comparison to the rise of twenty-first-century communication technologies. Is the comparison helpful? What insights does it suggest? Historically? Philosophically?

4. Does the idea of 'global knowledge' agree with the arguments about the locality of knowledge in general and scientific knowledge in particular made in Chapter 1? If not, which one appears to you more convincing and why? If the two concepts do share a middle ground, where is it to be found?

5. Does it make sense to think of the 'global corporations' of early modern times as trading primarily in knowledge? Does it make new sense of the relations between science and globality in the current day and age?

Suggested Readings

Primary Texts

Aquinas, Thomas, "Q. 14, Art. 9: Can Faith Deal with Things Which Are Known as Scientific Conclusions?" in *Truth (Quaestiones Disputatae de Veritate)*, Robert E. Schmidt (trans.) (Cambridge: Hacket, 1995 [1954]), Vol. 2, pp. 247–252.

Tartaglia, Niccolo, "Letter of Dedication" from *Nova Scientia*, in Stillman Drake and I. E. Drabkin (trans.), *Mechanics in Sixteenth-Century Italy* (Madison: University of Wisconsin Press, 1969), pp. 63–69.

Moryson, Fynes, *An Itinerary, Containing His Ten Yeeres Travell* (Glasgow: James MacLehose & Sons, 1907 [1617]), Vol. 1, "to the Reader," xix–xxi and ch. V, pp. 112–117; Vol. 2, ch. VI, pp. 122 ff.

Secondary Sources

On science and monotheistic thought:

Funkenstein, Amos, *Theology and the Scientific Imagination* (Princeton University Press, 1986).

On the meeting of the artisan and the scholar:

Smith, Pamela H., *The Body of the Artisan: Art and Experience in the Scientific Revolution* (University of Chicago Press, 2004).

Zilsel, Edgar, *The Social Origins of Modern Science*, D. Raven, E. Krohn and R. S. Cohen (eds.) (Dordrecht: Kluwer, 2000).

On the cultural import of the printing press:

Eisenstein, Elizabeth L., *The Printing Press as an Agent of Change: Communication and Cultural Change in Early Modern Europe* (Cambridge University Press, 1979).

Ong, Walter J., *Ramus, Method, and the Decay of Dialogue: From the Art of Discourse to the Art of Reason* (Cambridge, MA: Harvard University Press, 1983).

On navigation and mathematical knowledge:

Alexander, Amir, *Geometrical Landscapes: The Voyages of Discovery and the Transformation of Mathematical Practice* (Stanford University Press, 2002).

Markham, Clements R., *The Letters of Amerigo Vespucci and Other Documents Illustrative of his Career* (Farnham: Ashgate, 2010 [1894]).

On Jesuit science:

Dear, Peter, *Discipline and Experience: The Mathematical Way in the Scientific Revolution* (University of Chicago Press, 1995).

On Jesuits in China:

Brockey, Liam, *Journey to the East: The Jesuit mission to China, 1579–1724* (Cambridge, MA: Belknap Press of Harvard University Press, 2007).

On global trade and knowledge:

Brook, Timothy, *Vermeer's Hat: The Seventeenth Century and the Dawn of the Global Age* (New York: Bloomsbury Press, 2008).

Cook, Harold John, *Matters of Exchange: Commerce, Medicine, and Science in the Dutch Golden Age* (New Haven, CT: Yale University Press, 2007).

On the Chinese adoption and adaptation of European knowledge:

Elman, Benjamin A., *On Their Own Terms: Science in China, 1550–1900* (Cambridge, MA: Harvard University Press, 2005).

6 | Magic

Spectator vs. Participant Knowledge

Here is one thing that the masons building the cathedral and the scholars occupying its chapels would have agreed on: that the stones out of which it is constructed are heavy, and that there is nothing one can do about it. We can, of course, accommodate this heaviness. The scholars may try to understand why stones are heavy and what heaviness is; the masons may try to tackle it "with winch and pulley." But both would agree that some things in the world are light and some are heavy, and no knowledge can change that.

But what if there were such knowledge? What if we could make stones light? What if we could "defy gravity" and have "hewn rock" hoist itself up "into heaven" (recall Ormond's poem in Chapter 1)? Or be "hoisted" with the help of some creature much stronger than us? The aspiration to knowledge of this kind – immediate, useful, free of the limitations set by the regularities of nature – is what we call magic.

We discussed earlier why it's a mistake to use 'science' as a term of praise, and why and how we should think about it as a historical, contingent phenomenon, bounded in time and place. Similarly, it is a mistake to use 'magic' as a pejorative, synonymous with 'superstition,' or connoting some strange and mysterious occupation. We need to understand magic on its own terms. This does not mean we should collect recipes for concoctions or formulae for incantations, but that we need to understand who were the people engaged in magic and what *they* (and their contemporaries) thought they were doing. We need to understand how they related to one another and to their neighbors; and most importantly for our interests in the history of science: what they knew about their world and how they tried to affect it. Historians and anthropologists sometimes try to define 'magic' as a universal category, common to all cultures and distinguished from science and religion in some fundamental way. We, however, will study magic as we've been studying science – as a *particular tradition* of knowledge that developed around the Mediterranean from Antiquity until early modern times. Whether this tradition, with its claims and aspirations to knowledge, resembles other traditions in other times and places, and whether such comparisons are valuable, are difficult questions which need not divert us here. For us, the important comparison is between this magical tradition and mainstream philosophy of nature, the main origin of what we call science. These two sets of beliefs, practices and institutions (or, in the case of magic, the conspicuous lack of the latter) had complex and intense relations. Sometimes they defined themselves against each other, sometimes in alliance; sometimes the one borrowed from the other, sometimes vehemently rejected it. Magic is worth studying for its own sake, but it is also crucial to understanding the history of science, both in its similarity and its differences.

The Magical Tradition(s)

We've started by presenting a category – magic – and characterizing it in contrast to other types of knowledge vested in the cathedral of science. We cannot avoid such categories and characterizations. History, after all, is not – and cannot be – an attempt to relive the past. It is a story about the past, a story that *we* tell, to explain *to ourselves* major features of *our* culture – the culture of science. But these *are*, crucially, *our* categories; they are neither obvious nor necessary, nor would the people whose work they describe necessarily abide by them. This is true about 'science,' and it is even truer about 'magic.' Indeed, the word 'magic' itself has a distinguishable history, which gave it a meaning that did not always agree with our characterization above. It comes from *magi*, the title of the Persian Zoroastrian priests, about whom the Greeks knew from at least the sixth century BCE. To a large degree, what the Greeks knew, or thought they knew, about the practices of these priests came to comprise what was customarily included under magic: not just the production of mysterious wonders, but also fortune telling and necromancy (summoning the dead), as well as alchemy and astrology, which, as we'll see, can only questionably be described as 'magical.'

From its very beginning, then, the use of the 'magic' connoted esoteric, foreign knowledge, relating to mysterious religion. It gradually extended to denote the alleged skills of the Egyptian priests, finally replacing the term *goēs* (γόης), which the Greeks originally used for their own wizards. Greek and Roman thinkers were quite ambivalent on whether they should believe the fantastic reports about the wonders produced by these wizards and sorcerers. The great naturalist Pliny the Elder (see Chapter 4), for example, hardly remarks on the magical properties of plants, yet is very interested in those of animals – sometimes writing about them with disgust, sometimes with ridicule, sometimes providing a recipe, apparently convinced of its power. He's contemptuous of the *magi* as charlatans, but admits that "[t]here is no one who is not afraid of spells and incantations" (Pliny, *Natural History*, 10.xxxvii.15, quoted from Kieckhefer, *Magic in the Middle Ages*, p. 24). Galen (c. 130–c. 200) shared this contempt, but also the conviction that herbs should be gathered with the left hand, before sunrise. Politicians seem to have shared intellectuals' ambivalence: a century earlier, Emperor Augustus (63 BCE–14 CE) had 2,000 magical scrolls burned; he apparently wanted to take no chances.

When the term 'magic' was picked up by Christian thinkers, the connotation of dangerous heresy was added to its esoteric flavor: whatever the magi did, they did with the help of their gods. Since these could not be real gods, they were obviously demons. Yet the ambivalence did not disappear: monotheistic thinkers more often denounced magic as heretical than rejected it as impossible. The Jewish scriptures do not, of course, use the Greek term, but are full of sorcery, wizardry and necromancy. Aaron competes with the Egyptian priests in turning staves into crocodiles; the prophet Elijah competes with the priests of the Baal in summoning fire to the altar; "the witch of Endor" convenes Samuel from the dead to talk to Saul. The Torah (the Pentateuch) does not embrace or even condone magic – it prohibits it on penalty of death – but it has no doubts over its efficacy. Saul desperately searches for the witch after having banished all sorcerers from his kingdom, and he and his stock are condemned to demise for this last transgression – by Samuel himself, who did indeed rise: the prohibition of magic went hand-in-hand in with the belief in its efficacy. In Christianity, this ambivalence is particularly striking, because divine magic, in the form of miracles, is at the heart of its mythology. Much of the Gospel is comprised of stories about Christ's miracles, and such miracles are (together with martyrdom) a distinguishing feature of saints. Moreover, the Eucharist – the central Christian ritual – is a magical act: the priest, using an ancient incantation, defies the laws of nature by turning one substance – bread (and wine) – into another – flesh (and blood). From the monotheistic perspective, God created the world and its laws, so it is in His power to circumvent them. He can make the Sun stop at Gibeon and the Moon in the Valley of Aijalon and will do so if approached by the right person in the right way.

The crucial distinction during much of antiquity and the Middle Ages, then, was not between magic on the one hand and more earthly knowledge – intellectual and practical – on the other, but between divine and demonic magic. The difference between the two kinds lay in the origins of power, and one could usually tell by its effects: evil magical deeds were demonic, and benevolent ones were miracles. Yet both kinds were *supernatural*; both the witch and the saint made things behave against their nature. But wondrous deeds could also be accomplished within the realm of nature. Nature was full of secrets (on which we'll discuss more later), and manipulating them was magical, because it produced irregular and unexpected effects. Since these effects were still within the realm of nature, this was natural magic. This allowed for another essential distinction, which was formulated by William of Auvergne in the thirteenth century: between natural and demonic magic. This distinction became very useful for sixteenth- and seventeenth-century natural philosophers and physicians who were seeking to adopt the knowledge and intellectual tools developed within the magical tradition without the risk of heresy. We will return to these thinkers towards the end of the chapter.

Tense Relations

There is one final distinction we need to consider: between practical and learned magic. This is a complex and dynamic distinction, whose boundaries are both important and shifting. One may think of the village wise-woman as the archetype of a practitioner: someone who knows the curative (and poisonous) properties of local flora and fauna and is also called in to help in childbirth or set a broken bone. This woman was surely present in antiquity and throughout the Middle Ages (see Figure 6.1), and was still to be found in rural communities: someone like Matteuccia Francisci who was healing villagers in Todi, near Perugia, in the first half of the fifteenth century (Kieckhefer, *Magic in the Middle Ages*, pp. 59–60); or "Barbe, wife of Jean Mallebarbe," who dispensed curative (and allegedly poisonous) powders in Charmes, Lorraine, in the late sixteenth century (Briggs, *Witches & Neighbors*, p. 57); or Walpurga Hausmännin, a midwife practicing in Dillingen in Southern Germany at about the same time (Stephens, *Demon Lovers*, p. 1). Like the mason, she – almost always 'she' – possesses a great deal of know-how, gained either by direct inquiry of her immediate environment or by apprenticeship. Like the mason, she is most likely semi-literate at best, so she leaves little trace of this knowledge, and has little access to the sophisticated considerations of learned magicians. We can't even tell if she thinks of herself as a 'witch' until – perhaps because an untimely death has frightened her neighbors and raised their suspicion, or perhaps because of a neighborly grudge – she finds herself in front of the learned magician: the inquisitor. This may be someone like Heinrich Krämer (1430–1505), who in the last thirty years of his life prosecuted between 48 and 200 witches, according to his various testimonies.

It may be strange to think of the inquisitor, whose business is to eradicate magic, as a magician himself, but he's definitely an essential contributor to the magical tradition because he (the inquisitor is always 'he') spent his career carefully defining magic and gathering knowledge about it. It's the inquisitor who made the nuanced but fateful distinction between acceptable *saga* and the demonic, evil-doing *malefica* – an ominous legal term for a witch. And it's a legal text – the tenth-century *Canon Episcopi*,

Figure 6.1 A female practical physician from a fifteenth-century manuscript of John of Arderne's medical treatise, acquired by Sir Hans Sloane in 1694 for 5 shillings and now in his collection at the British library. The woman applies heated cups and uses a pointed instrument, seemingly mounted on a spring, to open a fistula (bottom left). It is easy to imagine why such practices and such instruments, and the women employing them, would be both revered and suspected.

incorporated into the canon law in the twelfth century – that learnedly enumerates for us all the deeds witches were believed to do; most importantly, riding domestic animals through the air.[1]

Almost all we know of witches like Barbe, Matteuccia and Walpurga comes from the inquisitors' protocols of their trials. We can learn from the protocols that they knew and understood little of the stories about conferences with demons which were so important for Krämer and his peers – they had to guess what stories these men were expecting to hear from them so the torture would cease and they would be allowed to die. But we can also see, from witnesses' testimonies and her own, that many of the witch's practices conform to the inquisitor's (and our) expectations: she is secretive, speaks in the name of ancient knowledge, employs incantations and refers to the symbolic relations between stars, plants, animals and body parts. These are the practices that the learned magician ponders and gives meaning to. It is the same type of complex relations between practical know-how and learned knowing-that that we discussed in Chapter 1, which the secretive nature of magic makes even more intricate.

---

[1] Broomsticks were added by later imagination, apparently inspired by ways of applying hallucinogens through rubbing an anointed staff. The original imagery of riding a broom has its brush pointing forward – the aerodynamic version is a modern (and modernist) bias.

Most learned magicians are further removed from the practical magician than the inquisitor: they are philosophers or theologians who read and write about magic, and usually find it repugnant. Such, for example, is Jacob Sprenger (1437–1495), Krämer's colleague, who was, like him, a Dominican friar and theologian, but preferred contemplating witches as a professor at the University of Cologne to prosecuting them. Yet this was not always the case: Nicholas of Poland, a thirteenth-century Dominican monk, is a rare example defying the rule. Academically trained and well versed in mainstream medicine, he wrote rebellious pamphlets against Hippocrates and Galen and became well-known as a healer and influential as a preacher and mentor, treating many – from the poor to the Duke of Sieradz (in Lesser Poland) – with amulets made of frogs and snakes. Isaac Luria, the greatest theoretician of the Kabbalah (see below), is another example: he was also a practical magician – healing, talking to the dead, and producing great wonders.

Indeed, one may think about the history of magic in Christianity through the perspective of the relations between the practitioner and the scholar. In the case of magic, all political, educational and religious institutions support the latter. The ambivalence towards magic, shared in Late Antiquity by even the most careful thinkers such as Augustine, gave way in the Middle Ages to a much stricter view. On the one hand, scholars like Aquinas put much effort into supporting the belief in demons by giving naturalized accounts of them and their interaction with humans. On the other hand, the *Canon Episcopi* didn't prohibit the *exercise* of magic, as older texts did, but the very *belief* in magical practices. The author (or authors) of this text did believe in the witches' encounters with demons, but the stories the witches told about these encounters and their consequences had to be delusions forced on them by the devil – to believe otherwise was pagan heresy.

This attitude would change again in the fifteenth century. It was perhaps a reaction to the rise of Aristotelianism, which we discussed in Chapter 4. This causal, naturalizing way of looking at the world suggested that the miracles at the heart of Christian mythology could perhaps be also explained naturally, threatening belief. Christian theologians resigned themselves to the idea that miracles could no longer be expected, so some sought evidence for their possibility in *maleficia* – acts of demonic magic. Since demons were fallen angels, the magical deeds they effected through collaborating with witches were devilish reversals of miracles. The most famous text to argue and explain this line of thought is the *Malleus Maleficarum* – 'The Witches' Hammer' – of 1486. Written by our acquaintances Krämer and Sprenger, the *Malleus* narrates in great, sometimes pornographic, details the encounters between witches and demons. Walter Stephens in *Demon Lovers* (see

Suggested Readings) explains that there is a strong epistemological assumption behind these stories, which makes this text interesting to the history of science. It is an empirical assumption: the witches need to testify to outlandish occurrences – the supernatural acts of demons. Being so strange, these stories have to be based on the most reliable type of sensual knowledge, and that would be knowledge by touch, the epitome of which is knowledge of the flesh. The *Malleus'* misogynistic account of women's sexuality is thus an argument for the possibility of this most repugnant sexual intercourse, and hence of their reliability as witnesses about the demon and his *maleficia*.

It is in texts like the *Malleus'* that we find the theoretical basis for the most horrific chapter of the history of European magic – the witch hunt. The great witch trials began in the middle of the fifteenth century, spread through Europe during the sixteenth century and finally subsided by the middle of the seventeenth century – the period between Galileo and Newton. Some 90,000 people (by conservative estimates) – 80 percent of them women – were prosecuted, and about half of them executed. Most of this important story is out of our scope, but one thing is worth noting as far as the history of science goes. Demons, we observed, were the explanation that Christians of Late Antiquity and of the Middle Ages provided for magical powers. For the Renaissance theoreticians of witchcraft, it was the other way around: magical powers testified to the existence of demons.

Magical Cosmogonies

Yet not all scholars were hostile to magic. Indeed, as we'll see towards the end of the chapter, even as the witch hunt was raging, the relations between learned and practical magicians began to change again and magical ideas and practices came to play a crucial role in the New Science of the seventeenth century.

To those scholars sympathetic to magic, however, one insight was particularly challenging. To believe in magic is to believe that humans – at least some of them – can interfere directly with the very make-up of the world: to make something happen that defies the regular order of things. This is an idea that the scholar finds very hard to admit. Whether pagan or monotheistic, almost all authorities agree on one fundamental assumption: we know this world as spectators. We dwell in it, so we can learn its ways and gain a measure of control within it, but we can never change its elements or the rules governing them. The Biblical cosmogony (creation story) makes a point of stressing this: God created the world in a way humans cannot comprehend, let alone mimic – from nothing at all and in a word. He created humans on the sixth and last day and put them in it when it was already complete. He allows them (within limits) to study this world – in Genesis 2:20, "man gave names to all cattle, and to the fowl of the air, and to every beast of the field" – but not to change it. When, in Babel (which is the same place as Babylon), they tried to challenge the hierarchy between human and divine capacities – when they tried to build "a tower, with its top in heaven" (*Genesis* 11:4) – God deprived them of their main source of power and the apparent reason for their arrogance – their common language. Most traditional Jewish interpreters of the scriptures, from Antiquity through the Middle Ages and into early modern times, adopted a strict version of this idea: we can only know Creation as it was presented to us; we can neither know why and how God made the world, nor make any of His creations behave other than He commanded them to. Divine knowledge is not only beyond human grasp – even aspiring to it is a sin. Aristotle, of course, does not provide a cosmogony – his cosmos was never created – nor does he worry about sinful knowledge. But since he allows for no supernatural realm, it is even harder to imagine how, in his cosmos, order can be diverted. Aristotle's cosmos is all there is and the cosmos just *is* its order and regularity, so just as it makes no sense to ask what is beyond the cosmos or what was before it, it makes no sense to try and circumvent its order.

There were, however, other sources, more esoteric but still popular, which provided cosmogonies that gave an explanation for the idea that humans can interfere with the laws of nature. They supported this idea of human magical powers by presenting creation as a more human-like accomplishment, either in addition to or by giving humans an active role in it. Some of these sources we have encountered in previous chapters, Plato's *Timaeus* being the most influential among them. In this dialogue, Plato narrates a cosmogony that allows much more leeway for human intervention than the Biblical one. Rather than creating all from nothing, the 'artificer' god – the 'demiurge' – takes the elements (the same four Greek elements), and, like a human artisan, arranges them into patterns to make the material substances. Both the substances and the patterns are there for him to work with, and Plato even feels comfortable in allowing us a glimpse into the demiurge's blueprint, which comprises those orderly sequences of numbers and figures that so impressed the Pythagoreans (see Chapter 2). No less important is that humans are not mere spectators in the *Timaeus*: their personal development reiterates the making of the world, and their soul reflects its orders and harmonies. We have come across this idea – that the world has a soul, and that the human soul is immediately related to it – in another text we have mentioned: Plotinus' *Enneads*, the basic text of so-called *Neo-Platonism*.[2] Unlike the strict submission of humans to the order of nature in Aristotle and the strict dichotomy between Creator and Creation in *Genesis*, the 'Great Chain of Being' gives humans a special role in creation: it is the human soul through which the ideal forms in the *Nous* – the cosmic intellect – are embodied in otherwise formless matter. This means that without humans, the cosmos, with all its varied creatures and species, would not have existed, so it does make some sense that they may also interfere with its make-up as magic requires.

Kabbalah

It was not only in pagan texts that magical thought could find grounding and legitimation. A short Jewish text in Hebrew – *Sefer Yetzirah* (*Book of Creation*) – demonstrates how the monotheistic tradition could also support such thought. Neither the author of this book nor its exact origins were (or are) known, but this just heightened its venerated status. It was probably composed around the turn of the Christian era or in the following few centuries, but as was common with similar texts, it was thought to be truly ancient – coming perhaps from Abraham himself. Creation, in *Sefer Yetzirah*, is a linguistic act. God creates the foundations of the world by 'carving,' 'inscribing' and 'composing' into matter the twenty-two letters of the Hebrew alphabet, which he first endows with particular powers (Figure 6.2). He then strikes a covenant with Abraham, and, by giving him the Hebrew language, reveals to him the secret of the world.

According to *Sefer Yetzirah*, then, the world was inscribed in a particular human language – Hebrew – and the code was revealed to humans in ancient times. This is a crucial amalgamation of magical ideas, to which we'll return: the veneration of antiquity; the belief in the power of words; and the essential secrecy of this knowledge. It also reflects and justifies a particular set of magical practices: *Gematria*, or the search for prophecies and divine insights in the Torah through calculating and analyzing the numerical values assigned to letters (א=1; ב=2; י=10; ק=100, etc.). Jewish scholars of Late Antiquity were familiar with (and usually suspicious of) magical ideas, in both Gnostic and Neo-Platonist versions, as well as those in *Sefer Yetzirah*. But these ideas only took a real hold on their imagination

---

[2] 'Neo-Platonism' is not a term that Plotinus or his disciples would have used – from their point of view, they were simply continuing the great tradition initiated by Plato. There are many good reasons as to why we should stick to the categories used by the heroes of our story, but since this is a very common term in the literature, especially in relation to magical thought, we will not be pedantic in this case.

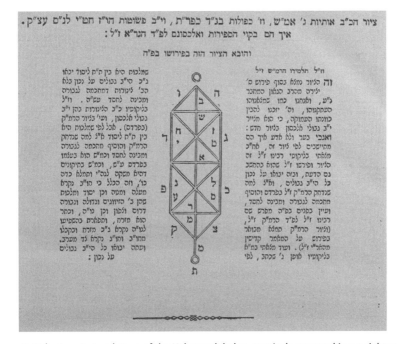

Figure 6.2 The twenty-two letters of the Hebrew alphabet creatively arranged in an eighteenth-century interpretation of Sefer Yetzira. The interpreter – Elijah ben Solomon Zalman, known as the Gaon (genius) of Vilna – was trying to capture stanzas like the following: "Twenty-two Foundation Letters:/ He placed them in a circle/ like a wall with 231 Gates./ The Circle oscillates back and forth./ A sign for this is:/ There is nothing in good higher than Delight (Oneg – ענג)/ There is nothing evil lower than Plague (Nega – נגע)." The title of the diagram is: "Image of the twenty-two letters … and how they are in the lines of the Sefirot and their obliques."

with the appearance of *Sefer HaZohar* (*Book of Splendor*) in the thirteenth century. *Sefer HaZohar* was probably composed around the time of its 'discovery' by Rabbi Moshe di-Leon (Moses de León, c. 1240–1305) in Spain, perhaps by di-Leon himself or his milieu. Yet, written in Aramaic, it professed to be an ancient work from the period after the destruction of the Jewish temple in Jerusalem by the Romans in 70 CE. It was adopted by the sages of Zefat (Safed) in Palestine (or as they called it – the Land of Israel) in the sixteenth century, and especially their leader Rabbi Isaac Luria (known as 'Ha'Ari Hakadosh'). In their interpretation it became the cornerstone of *Kabbalah*, the Jewish magical tradition, which has greatly excited thinkers with occult tendencies – Christian as well as Jewish – since early modern times.

The main part of *Sefer HaZohar* is a symbolic line-by-line reading of the Torah: an interpretation of the Torah as if it contains another message, hidden behind the literal meaning of the words. It is a mystical message in that it pertains to the place of humans in the world, and, especially in Luria's version, it comes in the form of a truly human-centered story of creation. In this

Kabbalist cosmogony (as in the Manichean), humans were not created last, but first: God begins His work by making The Primordial Man (*Adam Kadmon*), which comprises all forms – animate and inanimate – of the world to be created. Through the eyes, ears and mouth of the Primordial Man, the divine light emanated to create a world in the harmonious form of the human body. But creation in this story is traumatic and disastrous. To create the world, Luria tells, God has to first retract Himself from His infinity in order to free the space into which the world *could be* created; so it is already a God-forsaken world into which humans are put, abandoned even before it is created. Worse: creation itself was a catastrophic failure. It was an eruption of divine light, and God had created vessels – *Sefirot* (Figure 6.3)[3] – into which this light was to

Figure 6.3 A diagram of the ten Sefirot, composed of the first letter of each Sefira, from Moshe Cordovero's *Pardes Rimonim* (*Orchard of Pomegranates*). The name of the book is taken from a line in the Bible's *Song of Songs* (or *Song of Solomon*), an erotic poem which the Kabbalist tradition reads as an allegory for the relations between God and his people: "your branches are an orchard of pomegranates, with choicest of fruits" (4.13). The outermost is כ (KH-K), which is the first letter of כתר (*Keter* – crown), because "the first to nobility is big and surrounds all"; the innermost is מ (M), the first letter of מלכות (*Malkhut* – kingdom), and so on. From the interpretation of the order, Cordovero summarizes, "we learn about the nobilities, which are like the spheres turning in the one within the other like onion skins, and for this reason are called heavens [or firmaments]" (Gate 6, ch. 4). The original text was written in 1548 in Safed, but the image is from a 1592 version printed and published in Europe – another example of the impact of the movable press discussed in Chapter 5. The inscription above the diagram reads: "the crown (כתר) is home to all and origins of all."

---

[3] It is not clear if the Hebrew term comes from the Greek 'sphaira' (σφαίρα – sphere), relating to the Hellenic cosmology, or from the Hebrew 'sappir' (ספיר – sapphire), which connotes radiance.

flow. But the vessels could not withstand the immense potency of the emanating light and were crushed, scattering shards and sparkles into the primeval abyss. The shards are the material world in which we dwell, and it is upon us, humans, to search for and gather the divine sparkles strewn among them.

Hermetica

The most popular magical cosmogony in Christian Europe came in a text, or rather a group of texts, all in Greek, whose origin is still shrouded in mystery – the *Hermetic Corpus*:

> Mind, the father of all, who is life and light, gave birth to a man like himself whom he loved as his own child. The man was most fair: he had the father's image; and god, who was really in love with his own form, bestowed on him all his craftworks. And after the man had observed what the craftsman had created with the father's help, he also wished to make some craftwork, and the father agreed to this. Entering the craftsman's sphere, where he was to have all authority, the man observed his brother's craftworks; the governors loved the man, and each gave a share of his own order.
>
> *Corpus Hermeticum* I.12–13 in: Copenhaver, *Hermetica*, 3

This story is told by *Poimandres* (sometimes transcribed as Poemander, sometimes Pimander), a divine "mind of sovereignty," to the alleged author of these texts: Hermes Trismegistus – the thrice-great Hermes. The title *Hermetic Corpus* commonly refers to the main group of *Hermetica*: seventeen dialogues between Hermes and his disciples. Another important text usually attached to them is the *Perfect Dialogue*, known in Latin as the *Asclepius*, after the Greek god of medicine, who leads the dialog. Yet another – the *Emerald Table* (*Tabula Smargadina*, allegedly found on such a table in Hermes' grave) – is in a sense the most influential of the corpus, as it contains the foundations of the metaphorical language the alchemists would use for centuries to come. One book of Arabic origins should also be mentioned here, because, although not exactly of the same corpus, it came into European hands at a similar time and occupied the same intellectual niche. This is the *Ghayat Al-Hakim* (غاية الحكيم – *Purpose of the Wise*), known in Latin as *Picatrix*, which combines distinctly material recipes of astral magic with pronouncements on elevated magical principles (Figure 6.4). Many more related texts were known in Antiquity, of which only fragments survive; the largest collection is found in a fifth-century *Anthology*.

In this quote, taken from the first of the Hermetic dialogues, one finds perhaps the best expression of the main themes of all magical creation stories and a model of how stories make sense of and justify the aspirations of the magician. In such cosmogonies, creation is not a one-time divine

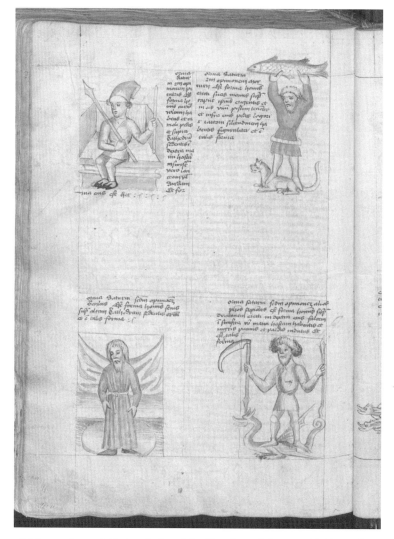

Figure 6.4 A page from a Latin translation of the *Picatrix*, made from a mid-thirteenth-century Spanish translation from the Arabic, prepared by an order of King Alphonso X of Castile. Four images of Saturn, with explanations of the significance of its placement in the various signs, culminating in the bottom right with the one Saturn is most associated with: death (the sickle) and rebirth (the serpent). The manuscript is now at the Biblioteka Jagiellonska of the Jagiellonian University in Krakow.

utterance, incomprehensible and unchangeable. It is an ongoing "craft-work," in which "man" – the primordial, androgynous representation of all humans – is explicitly invited to take part. The magician is not necessarily, therefore, a delusional heretic, trying to take an impossible bypass around unchangeable general laws of nature; he is rather a "loved" participant in the process of shaping these laws.

Magical Epistemology

Antiquity and Secrecy

The figure of Hermes Trismegistus captures another crucial aspect of magical thought. The *Hermetic Corpus* was likely written in the second century, which is the age of the oldest manuscript we have at hand. This manuscript was only discovered in 1945 (in Nag Hammadi in Upper Egypt), so it could not have been known to early-modern readers of the *Hermetica*. Yet Hermes, its alleged author, is said to have been an Egyptian priest, a contemporary of Moses. This allowed his Christian devotees to find in the *Corpus* prophesies of the coming of Christ, supposedly pre-dating the Old Testament, giving credence to their belief in both Christ and Hermes. But this emphasis on antiquity was not unique to the Christian magicians. The magical text, as we also saw in the Kabbalistic examples, draws its authority from its (often feigned) antiquity. This is not to say that the ideas in the *Hermetica* or *Sefer Hazohar* are *not* ancient: modern scholarship has shown that many of them had indeed originated in Egypt and Mesopotamia many centuries before these two texts (and the others we mentioned) were written. What is important is that magicians never proclaim novelty or demand credit for innovation: the more antique their knowledge, the more trustworthy it's supposed to be, and the more potent.

The story of the re-emergence and consequent decline of the *Hermetic Corpus* in Renaissance Europe illustrates this point well. In 1462, Cosimo de' Medici obtained a fourteenth-century manuscript of the first fourteen dialogs of the *Corpus*. Byzantium was collapsing under Ottoman pressure and many of its cultural treasures were becoming available, at the proper price, to the supremely wealthy and marvel-hungry princes of Italy and their Humanist clients. But in the eyes of Cosimo this clearly was no run-of-the-mill marvel. Marcilio Ficino (1433–1499), the best scholar of his entourage and probably the most skilled scholar of his generation, was at the time busy with no less than a new translation of Plato. Cosimo instructed him to abandon it and fully dedicate himself to preparing an edited Latin version of these newly discovered magical texts. Within less than a century, Ficino's 1471 *Corpus Hermeticum* went through two dozen editions and was translated into French, Dutch, Spanish and Italian – an imposing best-seller. Then, in 1614, the Hermes craze ended abruptly. A French scholar by the name of Isaac Casaubon (1559–1614) demonstrated, using new philological techniques, that the *Hermetic Corpus* could not have been that ancient: he found terms, phrases and puns which referred to ideas and events that happened much later than Hermes' alleged time. And although magical ideas continued to thrive, Hermes and the texts ascribed to him sunk into the realm of antiquarians and uninformed acolytes.

Why would such a thing have happened? If ideas are important and convincing, why should it matter if they were authored a millennium later? The first words of Ficino's preface demonstrate why:

> At the time when Moses was born flourished Atlas the astrologer, brother of the natural philosopher Prometheus and maternal grandfather of the elder Mercurius, whose grandson was Mercurius Trismegistus [Mercury is the Latin name for Hermes] ... They called him Trismegistus or thrice-greatest because he was the greatest philosopher and the greatest priest and the greatest king ... Among philosophers he first turned from physical and mathematical topics to contemplation of things divine, and he was the first to discuss with great wisdom the majesty of God, the order of demons and the transformations of souls. Thus, he was called the first author of theology, and Orpheus followed him, taking second place in the ancient theology.
>
> Copenhaver, *Hermetica*, p. xlviii

For Ficino, Hermes was more important than even Plato because he was *more ancient*. Casaubon was proud of his novelty – of his new philological skills and of his ability to produce new knowledge with them. This pride is at the heart of the natural philosophical tradition: the great veneration of the authority of Aristotle notwithstanding, the *discovery* or *invention* of something not known before has always been its main pride and purpose. Figure 6.5 provides a nice illustration: it is the first page of a sixteenth-century edition of a thirteenth-century text on optics by the Polish friar Erazmus Ciołek Witelo (c. 1230–c. 1300). Although he follows the work of the great Muslim optician Ibn al-Haytham (Alhazen – see Chapter 4) very closely, Witelo makes a point to celebrate his own novelties, and the very first words on this page read: "you have in this work, lucid reader, a great number of geometrical elements, which you would not find in Euclid ..." For the magician, in contrast, there is no new knowledge – only a secret truth revealed once, and corrupted by time and fading memory, so only direct access to that secret, as from a truly ancient text, is worth careful attention.

The idea that magical knowledge is ancient and declining – *Scientia Prisca* as it was sometimes called – is closely related to the capacities the magician assumes to possess. Observing nature as it is, we may discover the ways in which it behaves *generally* and *regularly*: we can see that stones fall, so we infer that they are heavy. But the magician expects to make nature behave in an *inordinate* way at a *specific* instance: he intends to make a stone rise. So by its very essence, magical knowledge pertains to what is

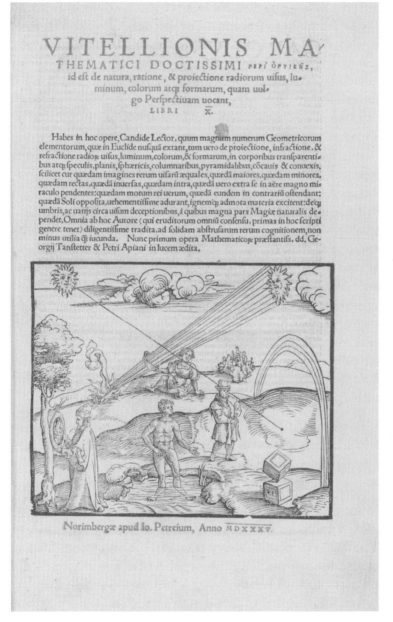

Figure 6.5 The frontispiece of the 1535 printed edition of Witelo's *Perspectiva* (or *Opticae libri decem*) from the 1270s. The text above the image promises explanations, practicality and innovation: "you have in this work, lucid reader, a great number of geometrical elements, which you would not find in Euclid … concerning the projection, infraction, and refraction of rays … of light in transparent bodies and in mirrors plane, spherical, columned, pyramidal … on which the visual deceptions of Natural Magic mostly depend" (see Figure 6.8).

hidden; what nature herself holds as a secret – magic is knowledge of the *occult*. This is why humans cannot discover this knowledge on their own. It's knowledge of a secret, and as such, it must have been divulged – in a mythical past, presumably by a divinity, since it is an intimate knowledge of the workings of creation. From the moment of its entrusting to humans, magical knowledge could only deteriorate: by the passage of time, by its handling through the generations and by fading memory.

The idea that nature holds some secrets from us may seem very esoteric, but it does have roots in Aristotle and Galen, the empirically bent forefathers of natural philosophy and medicine. Some properties of natural substances, Aristotle explains, arise from the fundamental qualities of matter; from the combination of the properties of the four elements. We can sense those properties. But others we can't: the essence of a substance is the particular combination of matter and form defining it as an acorn, a horse or a person, and some of its essential properties just belong to a substance by virtue of what it is. They cannot be reduced to some other, simpler properties: horses beget horses because they are horses, and the loadstone attracts iron because it has a magnetic virtue. Nothing more can be known about these properties – they cause change in matter, but are not themselves caused by anything else. This is the sense in which the great seventeenth-century Aristotelian scholar Kenelm Digby (1603–1665) still calls them "qualities occult, specificall, or incomprehensible":

> The loadestone and Electricall bodies are produced for miraculous, and not understandable things; and in which, it must be acknowledged, that they worke by hidden qualities, that mans witt cannot reach unto.
>
> Kenelm Digby, *Two Treatises* (Paris: Gilles Blaizot, 1644), Preface

Clearly, secrecy for the magician is not only an abstract philosophical consideration but also an important practice and a mode of being and self-understanding. Practically, secrecy is crucial for the safety of the magical practitioner, because magic is at best subversive and at worst – as when it offers an alternative creation story – heretical. Secrecy is also self-imposed restriction: magical knowledge, because it's so potent and particular, is dangerous, and should not be divulged to anyone. Moreover: we saw that because magic defies the observable regularities of nature, the magician is not a discoverer or inventor but a custodian of ancient knowledge. For the same reasons, this almost-divine knowledge is well beyond human capacities and the magician cannot be in real control of it; he is only a medium of its transmission and application. As a medium, the magician needs to be prepared, not unlike the primordial vessels of Kabbalah; he needs to be in the right elevated state to receive the great secret that can be delivered through him. Magical knowledge is therefore occult also in the sense that it cannot be studied by everyone, only by the select few who have ascended to such a state; it takes initiation.

Circumventing Reason: The Strange Role of Language

Magic demands initiation because, unlike natural philosophy, it cannot be studied by observation. But, again unlike natural philosophy, it can't be studied from books either: words are made to capture the regular and the shared; magical knowledge is inordinate and personal.

This should seem paradoxical: isn't magic all about words? Doesn't a witch like Walpurga enact her power by uttering incantations? Doesn't a wizard like Nicholas empower his talismans by writing magical verses on them? Isn't the main point of *Sefer Yetzirah* that the world comprises a linguistic code? The atomists' responses to Parmenides and Ptolemy's sub-version of Aristotle exemplify how inconsistencies, tensions and compro-mises are fruitful to the original thinkers (and telling to the historian) in all branches of thought. Magical thought is of course no exception: the magician is both superbly powerful and completely powerless; the greatest magician is both a heathen priest and a prophet of Christ; magical deeds are signs both of God's omnipotence and of the existence of rebellious demons; and so forth. But the paradoxical import of language for the magi-cian is especially interesting. Here is an example, from a fifteenth-cen-tury household book, of how magical words can be both meaningless and all-powerful:

> Amara Tonta Tyra post hos firabis ficaliri Elypolis starras poly polyque lique linarras buccabor uel barton vel Titram celi massis Metumbor o priczoni Jordan Ciriacus Valentinus.
>
> Kieckhefer, *Magic in the Middle Ages*, p. 4

This formulation, meant to drive away demonic possession, is a mixture of confused Latin, quasi-Greek and complete gibberish. "Rex, pax, nax in Cristo filio suo," from the same manuscript, is a more modest version: only the word 'nax' is meaningless. Clearly, these words are not supposed to fulfill their powerful function by representing things in the world the way ordinary words do; magical knowledge, unlike natural philosophy, aspires to affect the world, rather than represent or describe it. Famous magical words – like *abraxas* or *abracadabra* – may have meant something once, in Greek, Aramaic or Hebrew, but they no longer do, and their power stems from the order and rhythm of their consonants and the presumed antiq-uity of their source. The enchantment with the alliteration of consonants is particularly apparent in magical inscriptions like this one, from an Egyptian papyrus from the first century BCE:

ablanathanablanamacharamaracharamarach

ablanathanablanamacharamaracharamara

ablanathanablanamacharamaracharamara

ablanathanablanamacharamaracharamara

Kieckhefer, *Magic in the Middle Ages*, p. 20

The belief in the direct power of language is not unique to magic: both the magical cosmogony of *Sefer Yetzirah* and the common one in *Genesis* involve the use of language. But whereas in *Genesis* God speaks clearly, in words and sentences, in *Sefer Yetzirah* He only deals with letters. Fittingly, the magician looks for a meaning hidden *behind* the literal text of the scriptures – not in what the words signify, but in patterns of their arrangement.

The paradox, then, is that magic ascribes immense power to language while depriving it of its usual force – to convey meaning. Perhaps, however, it should not be looked at as a paradox, but as another aspect of the idea that magic aims at the extraordinary, and therefore can't be constrained by our common paths to knowledge. Observation and books, we said, don't help. Reason, from this perspective, is even worse: it is actively in the way of magical knowledge. By forcing everything into general concepts and categories – indeed, by using language – it prevents us from perceiving the deep truth in particular things and from utilizing unique, occult properties in order to achieve wondrous effects. The great early modern physician and magician, Jan Baptist van Helmont (1579–1644), had a particularly dim view of reason:

> Reason ... not onely feign perswasions, for the deceiving and flattery of it self ... but also ... plainly yield it self for a Parasite, and to the servitude of the desires ... Reason being left with us, came to us, as it were, a brand from a tormentor, for a remembrance of Calamities, and of our fall. And that the knowledge of good and evil, attained by eating of the Apple, was Reason its very self, which is so greatly adored by mortal men.

Jan Baptist van Helmont, "The Hunting, or Searching Out of Sciences" in *Van Helmont's Works Containing His Most Excellent Philosophy* ... J. C. (trans.) (London: Lodowick Hoyd, 1664), pp. 16–17

For van Helmont, as for many witches over the centuries (whose practices he aims to emulate and legitimize), this meant trying to circumvent reason – to arrive at a mental state in which perception was not hampered by reason's judgment. This could be achieved through fasting, dancing, avoiding sleep, alcohol or drugs – like the famous unguent produced from toads and smeared on the witch's skin. It could also be achieved by reciting, repetitively, words and formulas – since 'enchantment' comes from 'chanting,' this repetitive intonation is a fundamental magical practice.

Magical Cosmologies

The Symbolic World

Language, in the magical tradition, is not an aid and a tool of reason, but a means to avoid it, and magical words are powerful not for the meaning they carry, but for their direct, vocal effect. But while magic takes meaning away from language, it ascribes it to everything else in the world. For the natural philosopher, we saw, the substances in the world are related to one another in causal and deterministic ways. For the magician, their relations are symbolic – they have meaning.

Things in the magical cosmos resemble, reflect and express one another, and these similarities are potent. Mars is red like an angry face, so it causes war. Mercury is a name for both a planet and a metal, so they must be somehow related – and indeed, they are both quick (hence the metal's other name: quicksilver). Mercury is also the name of a god – who is fleet-footed and therefore the messenger god – and his priest by the same name (or in Greek – Hermes) is the messenger of the great magical secrets. Wine mixed with water resembles blood, so it can be used by the necromancer (and indeed, by a priest conducting the Eucharist); and the tongue of a frog, put on a sleeping woman's heart, will untie her tongue. Some symbolic relations are sympathetic: a walnut is folded like the brain, so it can cure headaches; draconium, whose leaves look like dragons, cures snakebites; the eye of the keen-sighted vulture, wrapped in wolf skin and hung around the neck, cures eye ailments. Some relations are antipathetic: the antipathy between the sheep and the wolf is such that a drum made from wolf skin will mute one made out of sheepskin; and because the shrew-mouse is afraid of the wheel rut, dirt from the rut will cure its bite.

The most important and general symbolic relation for the magician is that between the human body – the 'microcosmos' – and the world as a whole – the macrocosmos (or 'megacosmos'). The terms seem to have been coined in an influential twelfth-century manuscript, *Cosmographia*, by Bernard Silvester – another example of those who dabbled in both practical and learned magic – and we also find a version of this idea in the Kabbalistic concept of *Adam Kadmon* (primordial man). John Dee (1527–1608/9), who succeeded in combining life as an extraordinary practical mathematician – Elizabeth's most cherished and busiest expert on land surveying and navigation – with an active magical career ranging from astrology to communicating with angels, provided a particularly poetic version of this grand analogy:

> The entire universe is like a lyre tuned by some excellent artificer, whose strings are separate species of the universal whole. Anyone who knew how to touch these and make them vibrate would draw forth marvelous harmonies. In himself, man is wholly analogous to the universal lyre.
>
> John Dee, Aphorism, XI; cited by Szőnyi, in Hanegraaff,
Dictionary of Gnosis & Western Esotericism, p. 304

Dee's fascination with the mathematical harmonies of the cosmos places him simultaneously within the Platonic tradition and within the new mathematical natural philosophy of his time (we'll return to this convergence below). It is in Plato's *Timaeus* that one finds the first canonical version of the microcosm-macrocosm analogy, when Plato asks, "in the likeness of what animal did the Creator make the world?" and answers that "the Deity, intending to make this world like the fairest and most perfect of intelligible beings, framed one visible animal comprehending within itself all other animals of a kindred nature." He goes on to detail how this "intelligent creature" which is "by nature fairest and best" is made:

> First, then, the gods, imitating the spherical shape of the universe, enclosed the two divine courses in a spherical body, that ... we now term the head, being the most divine part of us and the lord of all that is in us: to this the gods, when they put together the body, gave all the other members to be servants, considering that it partook of every sort of motion. In order then that it might not tumble about among the high and deep places of the earth ... they provided the body to be its vehicle and means of locomotion; which consequently had length and was furnished with four limbs extended and flexible ... Such was the origin of legs and hands, which for this reason were attached to every man; and the gods, deeming the front part of man to be more honourable and more fit to command than the hinder part, made us to move mostly in a forward direction. Wherefore man must have his front part unlike and distinguished from the rest of his body.
>
> Plato, Timaeus, XVI

The Organic World

What one can notice in the *Timaeus* is that the microcosm-macrocosm analogy was tied, from early on, to another fundamental assumption of magical thought: that the world is an organism – an "animal." Not only is "[t]he world a living creature truly endowed with soul and intelligence by the providence of God," explains Plato, but it grows and changes like a living creature: "the elements severally grow up, and appear, and decay":

> ... water, by condensation ... becomes stone and earth; and ... when melted and dispersed, passes into vapour and air. Air, again, when inflamed, becomes fire; and again fire, when condensed and extinguished, passes once more into the form of air; and once more, air, when collected and condensed, produces cloud and mist; and from these, when still more compressed, comes flowing water, and from water comes earth and stones once more; and thus generation appears to be transmitted from one to the other in a circle.
>
> Plato, *Timaeus*, XVIII

The magical world is thus not the causal system that science inherited from Aristotelian natural philosophy; this is why this world is opaque and resists both reason and the senses. But magical nature is not completely obscure and capricious – even if the powers that magic aspires to harness and manipulate are occult. It has a structure that can be deciphered: an organic network of signs and representations whose symbolic heart is the human body. These assumptions gave learned and practical magicians common grounds: they were embedded into the talismans, amulets and incantations of the practitioner, and analyzed and elaborated upon by the scholar. They are also the core beliefs of the most academic of magical pursuits: alchemy and astrology.

Scientific Magic

It is far from clear that alchemy and astrology belong in this chapter, because they lack many of the fundamental properties by which we distinguished magic. They are hardly secretive: astrology was taught in universities well into the seventeenth century (primarily in relation to medicine), and alchemy, although never established as part of the curriculum and quite guarded about its substances and operations, was a respectable subject matter of learned texts and encyclopedias throughout the Middle Ages. Although alchemists and astrologists shared their veneration of antiquity with all mediaeval scholars, they were much less averse to innovation than the magicians whom we discussed above. Alchemy was an empirical discipline, whose practitioners were proud of new means and new effects, and astrology went through a series of reforms intended to improve its empirical accuracy and philosophical foundations. Neither had much use for demons or difficult cosmogonies – indeed, they fit well within the Aristotelian framework and the fundamental source for astrology, the *Tetrabiblos* (simply 'Four Books'), was authored by the great father of medieval astronomy – Claudius

Ptolemy. But no discussion of magic can be complete without considering these two disciplines, because within them are crystalized, in theory and practice, the two fundamental beliefs that distinguished the magical tradition from mainstream, institutionalized, natural philosophy: the belief in the symbolic make-up of the cosmos and the belief in the vital, hierarchical order of its elements. Things in the world reflected, expressed and signified one another; they grew and transformed into one another, evolving from base to noble; they related to one another passionately and intently and influenced one another in ways reminiscent of human language. The world was full of meanings and intentions that the alchemist and the astrologer deciphered and manipulated with great skill and erudition.

Alchemy

Alchemy took its name from the Arabic *al-kimiya*, a term whose origin could have been the Coptic word for pouring and mixing or the ancient Egyptian word for black earth. Both meanings are based on good reasoning: the alchemist attempted to untangle the weave of meanings within the realm of matter; to unlock matter's powers of transformation and assist and hasten the process by which substances ennobled themselves. The goal was itself noble, even heroic: to uncover the all-transforming Philosopher's Stone and the all-curing Panacea. The process made good Aristotelian sense: all things in the world strive to actualize their potential – acorns to oaks and children to adults – and so did matter. Base metals were formed inside the earth, from dry earthy and moist watery evaporations, and evolved into the most perfect of metals – gold. This evolution, explained Aristotle (as discussed in Chapter 2), was driven and directed by the form – matter itself was passive – so the alchemist trod on solid theoretical grounds in trying to strip the base metal down to primal matter – *materia prima* – and grafting onto it new forms. This process – "The Great Work" (*Opus Magnum*) – began with *calcination* by slow heating; continued with *solution* in sharp liquids; *putrefaction* in warm compost; and *reduction* in "philosopher's milk" (perhaps lime water of some concentration). This produced the *materia prima*, which after *sublimation* in spiritual substances, *coagulation* and *fermentation* with yeast of gold, was elevated into superior matter (*materia ultima*) or Philosopher's Stone – *lapis philosophorum*: "this stone which isn't a stone, this precious thing which has no value, this polymorphous thing which has no form, this unknown thing which is known to all" (cited in Hanegraaff, *Dictionary*, p. 25), as it's described by Zosimos of Panopolis, whose fourth-century fragments represent the earliest alchemical texts in existence. The Philosopher's Stone could then be multiplied and spread onto the base metal to produce gold.

Many of the processes and materials employed in this process are shrouded in mystery, including the substance with which The Great Work (and "The Lesser Work" – producing silver) begins, and the language the alchemists used is often purposefully coded: "red rose" for an elixir; "eagle" for evaporation; and "raven" for black. But there are many hints, such as the colors matter takes, after which the stages are called: *nigerdo* (black); *albedo* (white); *rubedo* (red) and so forth. Some of the remaining alchemical notebooks from periods far apart, like those of the Persian Rhazes (Abu Bakr al-Razi, 865–925) or the Bermuda-born, Harvard-educated George Starkey (1628–1665), are orderly enough to be deciphered, and with an impressive combination of chemical and textual skills historians have recently replicated some of those stages and effects (although not the production of gold, regrettably), like the *Philosophical Tree* in Figure 6.6, made of mercury and "a seed of gold." The modern replication may tempt us to think of

Figure 6.6 The 'Philosophical Tree' produced by Lawrence Principe in his laboratory. Following directions deciphered from cryptic instructions given by George Starkey, Principe sealed a pasty mixture of gold and a specially prepared mercury in a long-necked flask (the "philosopher's egg") and heated it. After several days, the enclosed mass that initially occupied less than a fifth of the flask suddenly 'grew' upwards into a tree-like structure that filled the flask. For Starkey and his fellow alchemists, this striking phenomenon proved that they could cause the 'seed of gold' to vegetate – a significant step towards producing the transmutatory Philosopher's Stone.

alchemical knowledge as nothing more than practical chemistry couched in strange terms, but this would be a mistake. We may translate 'calcination' into 'metallic oxidation,' for example, and from this translation we do gain some understanding of how we can comprehend and perform the process. But by the same token we lose some of our understanding of the way the alchemist comprehended and performed it. The organic connotations of 'putrefaction,' for example, are not empty: the alchemist sees all matter – metal, wood or human flesh – as decomposing and decaying in similar ways. Similarly, the moral connotations of 'sublimation' are crucial for the alchemist: he expects to be elevated by the process no less than the material he works on, and knows that the success of the material transformation depends on his own capacity for spiritual transformation. Here is how this close affinity between the alchemist and his subject matter is expressed by a very cool-headed and forward-looking thinker – Francis Bacon:

> But certain it is, whether it be believed or no, that as the most excellent of metals, gold, is of all other the most pliant and most enduring to be wrought; so of all living and breathing substances, the perfectest (Man) is the most susceptible of help, improvement, impression, and alteration. And not only in his body, but in his mind and spirit. And there again not only in his appetite and affection, but in his power of wit and reason.
>
> Francis Bacon, "A Discourse Touching Helps for the Intellectual Powers," *The Works of Lord Bacon* (London: Henry Bohn, 1854), Vol. II, p. 46

These reflections are not just after-the-fact musings: they guide alchemical practice and empirical exploration. For example, the alchemist manipulates matter with potent organic substances – snake's poison, lion's urine, wolf's bane, blood and sperm – because self-perfection is natural to the base metal, and if it does not happen – if lead remained lead, and didn't evolve into gold on its own – it means the metal is ailing, and needs to be healed. And he trusts this understanding because it does produce many marvelous effects, though perhaps not yet the most desired ones – gold and the panacea.

Astrology

If alchemy studied and practiced the ennoblement of base matter (mostly in monasteries), astrology (often in the court) deliberated and analyzed the influence of the noble – the heavenly bodies – on the base material world. What this influence consisted of remained an open question – one would find one answer in Plato, another in Aristotle and yet another in Christianity, and from the thirteenth century this astral influence was commonly understood with light as an analogy. But for the practicing astrologer these questions made little difference: influences, for him, had symbolic underpinning and weren't reducible to physical causes. And even if there were causes underlying the heavenly phenomena, he had little use for them: his working material was the positions and geometrical configurations of the stars and the planets. Indeed, it was the astronomer who provided the astrologer with orderly tables of these phenomena, but the two (who may just be the same person, wearing different hats at different times of the day) perceived them in very different ways. Where the astronomer saw featureless dots of light whose changing positions needed to be reduced to impassionate, simple motions, the astrologer saw a splendid theater, whose scenes he aspired to comprehend in their full complexity. The main characters in this theater were the planets, which – far from being indistinguishable and unknowable light spots – had rich and distinct personalities. The Sun was the ruler, wise and trustworthy; the Moon his mistress – luxurious and wandering. Mars was the warrior – angry and violent, brave and obstinate; Venus was beautiful and desirable. They controlled human life accordingly: the Sun ruled the head and indicated courage and leadership; the Moon governed marriage and travel; Mars caused wars, murder and plunder and was responsible for the genitals; Venus brought erotic love and sweet music. There is an internal coherence to this cosmic astro-psychology, and some of these personality traits make good astronomical sense: the Sun is in the middle of the heaven and his light rules the cycle of life; Mercury is the fastest moving planet (its period around us – or the Sun – is the shortest), so fits being the messenger, hence the patron of letters and logic. But most other features had no inherent reason – they were just so: for the astrologer, the heavens were as intrinsically complex as human life and times.

Indeed, like human life, the complexity didn't come only from the characters and gender of these individual personalities, but from the relations between them. The Sun, Saturn and Jupiter belong to the diurnal sect, and are in tense relations with the nocturnal sect comprising the Moon, Venus and Mars (Mercury is a hermaphrodite, and belongs to both). But this does not mean that the relations within the sect were necessarily warm: Venus and Mars are nemeses, perhaps because the former (like Jupiter) is 'benefic', or benevolent, and the latter (like Saturn) is 'malefic', or an evildoer. Relating astrology to alchemy and through it to medicine, planets also had their counterparts in metals: the perfection and color of the Sun related it to gold and the Moon to silver; the swiftness of Mercury is captured by quicksilver and the malevolence of Saturn by the baseness of lead. Mars was red, hence angry and violent, so it naturally related to iron, also red and the metal of swords.

These attributes and attitudes of the planets seem to mostly originate in myth and imagination, but to fully grasp the complexity of the heavens' influence the astrologer needs the astronomer's tables: the 'ephemerides.' This is because the planets' powers and relations depend on their position – in the heavens and relative to each other. The *ecliptic* – the Sun's path around the Earth – and the 10° band about it, in which the planets travel, are turned by the astrologer, as noted in Chapter 3, into the *Zodiac*: divided into twelve 'signs' of 30°, each 'ruled' by a constellation. Each planet rules a sign (or two) and is 'exalted' – that is, its power is heightened – in another: the Sun in Aries; Venus in Pisces. Each is also 'humiliated' – its power diminished – in some signs: Mars in Capricorn, the Moon in Scorpio. Some signs belong to light, like Virgo; some to darkness, like Libra; some are male, like Aries; some female, like Taurus (the complexity does produce paradoxes. The planet's influence is enhanced by signs compatible with its properties and vice versa. The angles, or 'aspects,' between the heavenly bodies are also crucial: two planets, or a planet and a star, could be in *conjunction* – namely at the same angle from us; or in *opposition* – 180° from each other. Some of these aspects, like *trine* (120°) or *sextile* (60°), are favorable – they enhance the stars' influence. Others, like *squared* (90°), are unfavorable. Finally, their daily motion is also crucial: the celestial sphere rises and falls as a whole, and with it the signs of the Zodiac. The great circle through which the signs move through the day – they 'ascend' in the east and 'descend' in the west – is also divided into twelve segments. Known as 'places,' or 'houses', each one is relevant to a different part of one's life: the first, just under the horizon to the east – to life; the seventh, just above the horizon to the west – to marriage; and so forth. The angle between this circle and the horizon is of course different for every location on Earth (on the equator they're perpendicular), giving astrological analysis yet another level of complexity, relating to time and place of Earth.

The astrologer's main tool, then, is a map – what we call a 'horoscope' – of the planetary positions on the Zodiac and the Zodiac's position in the heavens, as viewed from a particular place, at a particular time. This time is either in the past – if it is a nativity for someone already born – or in the future, if the astrologer is asked to advise on taking some action, such as going to war or signing a treaty. The map, plotted from the astronomer's ephemerides, is finite and can be as accurate as allowed by the empirical data and the astrologer's skills. The interpretation of this map, however – what it signifies about the qualities of the person whose nativity it represents and what the future holds for them; what will be an auspicious time to assault or retreat – this interpretation remains open-ended. The countless intricacies in which the properties, places and groupings of the celestial bodies may relate to and transform one another produces a tapestry of inexhaustible complexity. One can keep probing the map deeper and find contradictory answers, or return to it after the fact and find reasons why the original forecast had failed. Aware and weary of this fact, honest astrologers like Johannes Kepler tried to veer away from predicting particular events, but to no avail: these predictions were, understandably, what their powerful patrons and wealthy customers were mostly willing to pay for.

Magic and the New Science

What happened to magic? Where did it go? Although Tarot readers, talisman writers and weekend magazine astrologers still abound, magic clearly no longer holds the same cultural role it used to: a viable claim to knowledge, a serious resource and even an alternative to mainstream natural philosophy.

At this stage of the book the reader will hopefully not fall for the simplistic answer that magic's fortress of superstition fell to the progressing rationality of science. Pointing out the magical interests of many of the heroes of the so-called 'Scientific Revolution' is equally futile: Johannes Kepler (1571–1630) and Galileo Galilei (1564–1585) made much of their livings as astrologers; Robert Boyle (1627–1691) and Isaac Newton (1643–1727) were active alchemists, to mention just the biggest names.

Of historical interest, rather, is what made early modern thinkers – in their time, for their reasons and motivations – lose their trust in the magical alternative, and what they retained of it. Or from a slightly different perspective: which of the magical beliefs and practices were incorporated into the New Science – which of the stones comprising the cathedral of science were carved by magic – and which were rejected. The answers are as complex as the questions.

Natural Magic
It is not difficult to see how attractive this wealth of empirical and practical knowledge was to curious intellectuals, even if they could not fully endorse the complex system of beliefs that came with magical practices, whether for religious or intellectual reasons. And indeed, even at times

when either those beliefs or those practices were condemned or perse-cuted (as we saw – persecuting the practices necessitated adopting at least some of the beliefs), many mainstream scholars found ways to engage with magic. Many of the texts concerning learned magic arrived in Europe with the great translation projects described in Chapter 4, and to many scholars magic represented an integral part of an ancient, mysteriously coherent corpus of wisdom. The challenge facing Christian natural phi-losophers in accommodating magic into their work resembled the chal-lenge that they faced with pagan science in general and the Aristotelian corpus in particular. This knowledge was exciting, but it was also based on clearly heretical assumptions: how could these assumptions be excused or bypassed? One common solution, as mentioned above, was to revert to the concept of 'natural magic.' Carefully avoiding, and usually making a point to admonish, all demonic influences, university-educated and often university-employed scholars gave themselves license to tread these diffi-cult grounds with relative impunity.

The great Franciscan scholar Roger Bacon (c. 1215–c. 1292) is a good example. A student and then a teacher in Oxford and Paris, he composed, by commission of the pope, a major educational reform, at the center of which lay astrology and alchemy. Bacon was as impressed with a book named *The Secret of Secrets* (*Secretum Secretorum*), which he believed was Aristotle's, as he was by the core Aristotelian texts, and was moved to write at length on the power of words to cause action. However, he was anything but a backward-looking believer in superstitions. In fact, he developed a combination of the empirical dimensions of practical magic and the con-cepts of similarity and mutual representation that are explored in learned magic. This enabled him to formulate great novelties, especially in optics. Theoretically, he turned Muslim optics into a theory of vision; practically, he is credited with the invention of the spectacles. For Bacon, the magi-cal ability to really *see* – into things, to see them as they *really* are – had far-reaching theological implications. But while the looking glasses were very popular among his scholarly peers (as Figure 6.7 nicely captures), his magical-theological "novelties" were anything but. Bacon was condemned in 1277 – apparently for providing astrological analysis of Christ's birth – and perhaps even spent some time imprisoned.

The Magical Renaissance

Natural Magic is thus a category that allowed scholars to meld magic with mainstream Aristotelian natural philosophy into *Scientia Prisca*, and it was used similarly 200 years later by that great emblem of the Renaissance

Figure 6.7 A great achievement of practical magic – Bacon's spectacles. The image is of Hugues de Provence (Hugh of Saint-Cher), part of Tommaso da Modena's 1342 fresco series of "Forty Illustrious Members of the Dominican Order" in the Chapter House of the Seminario attached to the Basilica San Nicolo in Treviso, Northern Italy. The other illustrious Dominicans apparently had better eyesight or less tolerance to natural magic, because one of them is holding a magnifying glass.

Man – the Italian scholar, rhetorician, free thinker and nobleman, Giovanni Pico della Mirandola (1463–1494). Like Bacon, Pico was greatly impressed by the influx of ancient knowledge; not just of Aristotle and Plato but also Jewish learning and especially Kabbalah, all of which he ventured to unify. Like Bacon, he was only partially protected by his strong admonition of demonic magic; his over-enthusiasm for the occult – which for him related not only to controversial theology (like Bacon) but also to radical political thought – landed him in house arrest. Rather than Bacon's medieval monastery and university, Pico's milieu was that of the Renaissance court – he was a member of Ficino's circle – and as befitted a humanist, he was more interested in language that in technical natural philosophy. Pico's responsibility for bringing Jewish learned magic into Christianity and his criticism of divinatory astrology are crucial landmarks of early modern magical thought – but there is a more important, though subtle, way in which his work served as a conduit for magical values into mainstream knowledge. With his magical fascination with antiquity, Ficino, Pico's mentor, turned to look for the origins of Plato's philosophy in ancient Babylon and Egypt, in the *Hermetic Corpus* in particular. Pico took this idea even further, finding in Kabbalah the foundations of Christian theology. To support these bold speculations, Pico taught himself not only 'proper' classical Latin and Greek, but also Hebrew and Aramaic, laying the foundations for empirical philology. These, paradoxically, were the skills that Isaac Casaubon would use half a century later to dismiss the antiquity of the *Hermetic Corpus*.

The main themes of Pico's magical work and life – the antiquity of magic; the authenticity and venerability of its sources, especially Jewish ones; the strict distinction between evil demonic magic and benevolent natural magic (which still could not protect against a brush with authorities) – were picked up again a generation later, by Cornelius Agrippa (1486–1535/6): a German itinerant scholar, academic, spy and feminist (his *Declamation on the Nobility and Pre-eminence of the Female Sex* wasn't finished, but was still published in 1529). Agrippa's *De occulta philosophia* became the main source for Renaissance and early modern thinkers interested in magic. It was learned magic: Agrippa didn't provide incantations or recipes, but an attempt at an elevated systematization of the pagan, Jewish and Christian concepts and narratives discussed above: a cosmogony with Man at its center; a metaphysics with hierarchy between matter and spirit; an epistemology of symbols and secrets. For Renaissance and early modern thinkers, this amalgamation of learned magic provided an alternative to the highly formalized discourse of medieval Christianized Aristotelianism, with its strict rationalism on the one hand and strict limitations of human knowledge on the other.

Knowledge Is Power

The most influential adaptation of this alternative came from the English courtier, lawyer, politician and philosopher Francis Bacon (1561–1626). Being the sophisticated practical man that he was, Bacon was unimpressed by magic's claims to antiquity and had little respect for Agrippa's eclectic scholarship, but he wholeheartedly adopted the main principle that always distinguished magical thought from mainstream natural philosophy, the idea of knowledge with which we started this chapter; in his formulation: "knowledge is power" (*scientia potestas est* – power here in the sense of potency; capacity to do). Bacon gave a compelling illustration of what he had in mind in an unfinished piece of science fiction – *New Atlantis*. A ship is wrecked on the shores of an unknown island, ruled by an order of philosophers tellingly titled 'Salomon's House,' and the awed travelers are taken to observe its achievement:

> We have also perspective-houses, where we make demonstrations of all lights and radiations; and of all colours; and out of things uncoloured and transparent, we can represent unto you all several colours; not in rain-bows, as it is in gems and prisms, but themselves single ...
>
> We have also sound-houses, where we practice and demonstrate all sounds and their generation. We have harmonies which you have not, of quarter-sounds, and lesser slides of sounds ...
>
> We have also perfume-houses; wherewith we join also practices of taste. We multiply smells, which may seem strange. We imitate smells, making all smells to breathe out of other mixtures than those that give them. We make divers imitations of taste likewise, so that they will deceive any man's taste ...
>
> We have also engine houses ...
>
> We have also a mathematical house, where are represented all instruments, as well of geometry and astronomy, exquisitely made ...
>
> Francis Bacon, *New Atlantis* (London: Tho. Newcomb, 1659), pp. 33–35

These great deeds are all natural; the members of Salomon's House don't call upon any demons or angels to assist them or try to bypass the order of nature. But the deeds *are* magical, in the sense discussed at the beginning of the chapter: they are not simply spectacular but aspire to manipulate nature at its very foundations.

Bacon was not alone among his contemporaries, the advocates and virtuosi of the New Science (whom we'll consider in detail in the following chapters), in adopting this magical confidence in the power of human knowledge to mimic and even compete with the divine. Galileo boasted such confidence unabashedly: "with regard to those few [things] which the human intellect does understand, I believe that its knowledge equals the divine in objective certainty" (Galileo Galilei, *Dialogue Concerning the Two Chief World Systems*, Stillman Drake (trans.) (Berkeley, CA: University of California Press, 1967), p. 103). This was the type of epistemological pride that the Bible powerfully rebuffed with the story of the Tower of Babel, and it may well have been the main reason for Galileo's difficulties with the Church. Yet Kepler, whose piety is unquestioned, puts it as strongly: "geometry is coeternal with God" (Johannes Kepler, *The Harmony of the World*, A.

J. Aiton *et al.* (trans and ann.) (Philadelphia, PA: American Philosophical Society, 1997 [1619]), p. 146).

Bacon's science is magical, and so is his world: it's full of many particular facts, not clearly related to one another. He tried to collect as many instances as possible in his *Sylva Sylvarum* (*Forest of Forests*, in ten parts, titled "Centuries") – whose name, again, reveals Bacon's clear and explicit indebtedness to the magical tradition. Some of the facts are quite inane, like: "It is observed by some, that all Herbs wax sweeter, both in smell and, if after they be grown up some reasonable time, they be cut, and so you take the latter Sprout" (Francis Bacon, *Sylva Sylvarum* (London: William Lee, 1670), Cent. V, p. 99). Some of them are a little more practical and esoteric:

> Take a *Glasse* with a *Belly* and a long *Nebb*, fill the *Belly* (in part) with *Water*: Take also another *Glasse*, whereinto put *Claret Wine* and *water* mingled, Reverse the first *Glasse*, with the *Belly* upwards, Stopping the *Nebb* with your Finger; then dipp the Mouth of it within the Second *Glasse*, and remove your Finger: Continue it in that posture for a time; And it will unmingle the *wine* from the *Water*: the *wine* ascending and setling in the topp of the upper *Glasse*, and the *water* descending and setling in the bottom of the lower *Glasse*. The passage is apparent to the Eye; for you will see the *wine*, as it were, in small veine, rising through the Water. For handsomnesse sake ... it were good you hang the upper Glass upon a Naile.
>
> Bacon, *Sylva Sylvarum*, Cent. I, pp. 3–4 (I left the spelling, capitals, italics and punctuation as they are in the original, but in the following, for ease of reading, I'll use the modern transcription.)

This combination of useful tidbits and domestic spectacles demonstrates that Bacon was assimilating into the New Science not only the learned magic of Agrippa's type, but also the practical natural magic of which the most famous book, at the time and since, was *Magia Naturalis*, by the Neapolitan polymath Giambattista della Porta (1535–1615; like many of the heroes of this chapter, della Porta's nobility didn't save him from an uncomfortable brush with the Inquisition). Della Porta's cheerful table of contents (Figure 6.8) is the best explication of this tradition: "Of the Causes of Wonderful Things; Of the Generation of Animals; ... Of the Wonder of the Load-Stone; ... Of Beautifying Women ..." Natural magic was a treasure trove of empirical facts that the harbingers of the New Science were eager to probe, and it also offered much by way of "experiments" to be emulated, but in the attitude towards these experiments one can discern a crucial difference between the two traditions. For della Porta, the experiments have to be spectacular and entertaining:

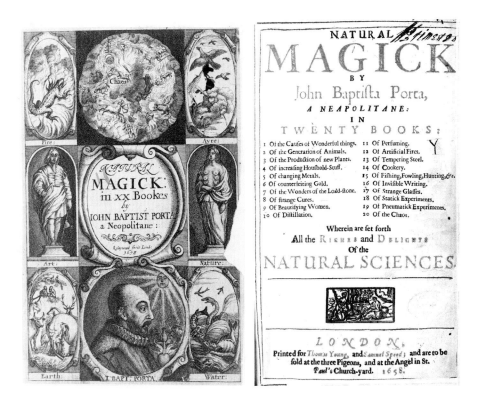

Figure 6.8 The frontispiece of *Natural Magick*, the best-selling 1658 English translation of della Porta's *Magia Naturalis*. Note the alchemical symbols on the left: the many-breasted *Nature*, the Salamander in *Fire* and so on; and the mixture of mundane and occult topics in the Table of Contents on the right: from "Of the Causes of Wonderful things," through the "Production of new Plants" and "Cookery" to "Invisible Writing," "Pneumatic Experiments" and "Chaos."

For what could be invented more ingeniously, then that certain experiments should follow the imaginary conceits of the mind, and the truth of Mathematical Demonstrations should be made good by Ocular experiments? What could seem more wonderful, then that by reciprocal strokes of reflexion, Images should appear outwardly, hanging in the air, and yet neither the visible object nor the Glass seen? That they may seem not to be the repercussions of the Glasses, but Spirits of vain Phantasms?

Giambattista della Porta, *Natural Magick* (London: John Wright, 1669), p. 355

This is not to say that della Porta doesn't take his experiments seriously, only that he does not expect them to tell him something new about the nature of things: they are performances; "demonstrations" of known "truths" and of his skills as a magician. For Bacon, on the other hand, the experiments are ways to examine the validity of the magical claims, which he takes with many grains of salt, and find order in the phenomena that they produce:

> I heard it affirmed by a Man, that was a great Dealer in Secrets, but he was but vain; That there was a Conspiracy (which himself hindred) to have killed Queen Mary ... by a Burning-Glass, when she walked in St. James Park ... But thus much (no doubt) is true; that if Burning Glasses could be brought to a great strength, (as they talk generally of Burning-Glasses, that are able to burn a Navy) the Percussion of the Air alone, by such a Burning Glass, would make no Noise; No more than is found in Coruscations, and Lightnings, without Thunders.
>
> Bacon, *Sylva Sylvarum*, Cent. I, p. 34

Bacon was very much setting the tone for the way in which the new natural philosophers would approach magic: adopt the practical knowledge while attempting to maintain a distance from all of its suspect baggage. A particularly important version of this approach was the assimilation of alchemical knowledge and magical remedies into medicine, especially by Philippus Aureolus Theophrastus Bombastus von Hohenheim, better known as Paracelsus (1493–1541), and Jan Baptist van Helmont whose words we saw above. We will discuss the place of magic in their medical work in detail in Chapter 8. For our discussion here it's more significant to stress how important it was for Paracelsus and van Helmont to naturalize these magical means and practices, and, for all the obvious reasons, detach themselves sharply from any association with demonic magic. In similar ways, Kepler could incorporate astrology's fascination with mathematical harmonies into his new astronomy, and Newton could find in astrology's 'influences' a way to conceptualize celestial forces in his new celestial mechanics, while both steered carefully clear from any problematic association with the supernatural.

Alchemy provides a particularly telling example of this way in which magic was adopted and adapted into the new ways of doing natural philosophy, especially from the second half of the sixteenth century. Even when they began calling themselves 'chymists' and then 'chemists,' practitioners kept the procedures and processes of the alchemists that we discussed above – calcination, sublimation, putrefaction, etc. – as well as the alchemists' equipment – mortars, crucibles, scales, cucurbits, alembics, retorts, etc. Nor did they relinquish the hope of producing gold. What the new chemists refused to accept was the way the alchemists explained to themselves what it was that they were doing:

The world hath been much abused by the Opinion of Making of Gold. The work itself I judge to be possible but the means (hitherto propounded) to effect it, are in Practice, full of Error and imposture and in the Theory, full of unfound Imaginations. For to say, that Nature hath an intention to make all Metals Gold; and that if she were delivered from Impediments, she would perform her own work; and that, if the Crudities, Impurities and Leprosies of Metal were cured, they would become Gold, and that a little quantity of the Medicine in the Work of Projection, will turn a Sea of the baser Metal into Gold by multiplying: All these are but dreams, and so are many other Grounds of *Alchymy*.

Bacon, Sylva Sylvarum, Cent. IV, p. 71

Bacon has every intention of keeping all the practical knowledge amassed by alchemists, and indeed of improving upon it – he dedicates many pages to "Means to enduce and accelerate Putrefaction" (Bacon, *Sylva Sylvarum*, Cent. IV, p. 73) and other alchemical procedures. But he has no place for the antiquity of knowledge, for the symbolic infrastructure of the cosmos, or for the personal elevation of the magician.

Conclusion

This, then, is what happened to magic as natural philosophy began shaping itself into what we recognize as our own science. The New Science adopted many magical practices and ways of thinking without the theoretical framework in which they thrived. Some of the rift between the natural philosophy and learned magic had clear intellectual reasons: with the decline of trust in Aristotle and Galen, the idea of 'occult properties' lost its philosophical legitimation. Indeed, the reference to such mysterious, irreducible qualities of substances became a symbol of ignorance masquerading behind fancy language – the 'virtus dormitiva' in *Le Malade Imaginaire* of the great French playwright of the time, Molière (Jean-Baptiste Poquelin, 1622–1673), is an excellent example of such ridicule: the crook doctor explains to the gullible patient that poppy will make him sleep because it has (in gibberish Latin) 'the power of sleepiness.' But there were other fundamental reasons for the rift.

Most crucial, perhaps, was that natural philosophy was no longer willing to entertain the belief that was at the core of all strands of the magical tradition – practical and theoretical; marginalized and mainstream: that the world is an organic, symbolic whole, full of hidden meanings. With it disappeared the related idea that knowledge is a secret code for deciphering these meanings, handed down from a divine and ancient origin, degenerating with time. The craving for antiquity was giving way to a great belief in novelty, and little place was left for the great secret inscribed into things and accessible only to a selected few. "We may ... well hope that many excellent and useful matters are yet treasured up in the bosom of nature ... still undiscovered," said Bacon, but these "undiscovered ... matters" were not really secrets.[4] They "will doubtless," he promised, "be brought to light in the course and lapse of years, as the others have been before them" (*Novum Organum* (London: Pickering, 1844), Aphorism 109, pp. 91–92). Without the secret there was also little point in the deeply personal and enigmatic practices of the magus and his initiates. The new ideal (though of course not always the reality) became that of open, public, cooperative knowledge.

But while rejecting the theoretical and philosophical assumptions underlying magic, the emerging New Science of the sixteenth and seventeenth centuries picked up and bequeathed magic's fundamental ambition: to know and manipulate nature at its most fundamental level, almost as if we were its maker.

---

[4] Some translations do use "secrets" here, but the original Latin does not: "esse in naturae sinu multa excellentis usus recondita."

Discussion Questions

1. Have we lost an important resource of knowledge with the relative decline of the place of magic in our lives? Is there indeed some ancient fountain of knowledge which requires a different approach than that of science?

2. Secrecy plays an important role in the modern culture of knowledge – from state secrets, through trade secrets, to scientists' data, withheld until publication. Is this the same secrecy as that of the magical tradition? What is similar? What is different?

3. Some astrology is still practiced today – there are still experts who can have one's nativity drawn and interpreted. Can it therefore be argued that the tradition has survived? If not – what has been irretrievably lost?

4. Is the replication of alchemical experiments an interesting undertaking? What can be learned from it? What cannot be recovered?

5. Is it useful to think of some of the inquiries and achievements of current science – genetic engineering, nano-technology, cloning – as 'magical'? In what sense is it enlightening? What does it obscure?

Suggested Readings

Primary Texts

Herms Trismegistus, *Poimander I* in Copenhaver, *Hermetica*, pp. 1–7.

Plato, *Timaeus* (http://classics.mit.edu/Plato/timaeus.html: from "All men, Socrates, who have any degree of right feeling" to "visible gods have an end").

Augustine, *On Christian Doctrine*, Book II 30–36; 45.

Giambatista della Porta, *Natural Magick*, Book 1, chapters I–III; Book 5, chapter I; Book 9, chapter I.

Secondary Sources

General history of magical thought and practice:

Briggs, Robin, *Witches & Neighbors: The Social and Cultural Context of European Witchcraft* (London: Penguin, 1998).

Copenhaver, Brian, *Magic in Western Culture: From Antiquity to the Enlightenment* (Cambridge University Press, 2015).

Eamon, William, *Science and the Secrets of Nature: Books of Secrets in Medieval and Early Modern Culture* (Princeton University Press, 1994).

Hanegraaff, Wouter J. (ed.), *Dictionary of Gnosis & Western Esotericism* (Leiden: Brill, 2006).

Kieckhefer, Richard, *Magic in the Middle Ages* (Cambridge University Press, 2000).

On the *Hermetic Corpus*:
Copenhaver, Brian, *Hermetica* (Cambridge University Press, 1992), Introduction.

On alchemy:
Newman, William R., *Promethean Ambitions: Alchemy and the Quest to Perfect Nature.* (University of Chicago Press, 2004).
Principe, Lawrence M., *The Secrets of Alchemy* (University of Chicago Press, 2015).

On astrology:
Beck, Roger, *A Brief History of Ancient Astrology* (Blackwell, 2007).

The magician:
Ginzburg, Carlo, *The Cheese and the Worms: the Cosmos of a Sixteenth-Century Miller*, 2nd edn. (Baltimore, MD: Johns Hopkins University Press, 2013).
Grafton, Anthony, *Cardano's Cosmos: The Worlds and Works of a Renaissance Astrologer* (Cambridge, MA: Harvard University Press, 2001).
Yourcenar, Marguerite, *The Abyss*, Grace Frick (trans.) (London: Weidenfeld & Nicolson, 1976).

Magic and science:
Bono, James J., *The Word of God and the Languages of Man: Interpreting Nature in Early Modern Science and Medicine* (Madison, WI: University of Wisconsin Press, 1995).
Webster, Charles, *From Paracelsus to Newton: Magic and the Making of Modern Science* (Cambridge University Press, 1982).
Yates, Frances A., *Giordano Bruno and the Hermetic Tradition* (University of Chicago Press, 1964).

On theories behind the witch hunt:
Stephens, Walter, *Demon Lovers: Witchcraft, Sex, and the Crisis of Belief* (University of Chicago Press, 2002).

7 | The Moving Earth

Introduction

The movable press was both a product and a harbinger of a new age of knowledge in Europe: commercial, expansive, urban, adventurous. It befitted a new world which was both much larger and much smaller: its horizons extended well beyond what could be imagined just decades earlier, but what lay beyond those horizons was now reachable and negotiable. Yet perhaps the most resounding impact of the press was close to home: it shook the foundations of *the* European institution of knowledge, the Catholic Church.

Press and Reformation

Here are some measures of the change brought about by the press: by the beginning of the sixteenth century, there were approximately 240 printing shops in Europe; producing a printed title was an estimated 300 times cheaper than the equivalent handwritten version; European libraries today hold more than twice as many books from the half-century after the invention of the press (known as *incunabula* – Figures 3.14 and 5.8 are a good example) than manuscripts from the whole of the Middle Ages and Antiquity.

And the most important title, the first and most popular book printed by Gutenberg in 1455 and his competitors immediately after, was the Bible. The movable press made a physical copy of the scriptures available to almost anyone, and one can hardly overstate the cultural importance of this material fact. When the fifteenth century rolled in, it became a religious maxim: if everyone *can* have a Bible, Martin Luther decreed, everyone *should* have one. Everyone should be able to read the Bible in their own language (Luther's own German translation of both Testaments and Apocrypha was published in 1534), take personal responsibility for understanding it, and, added Luther, be granted authority to interpret it.

Luther (1483–1546) was a German Augustinian friar who taught theology at the newly founded University of Wittenberg and entertained strong views on the corruption of the Church and its failure, as an earthly institution, to properly embody the divine lore it was supposed to represent. These were not new complaints. Many had expressed them over the centuries, and they would occasionally develop into real malcontent and even uprising. Usually, these revolts could be suppressed by violence or be redirected towards establishing another order devoted to poverty and the true teaching of Christ, but the press created a dramatically different option.

Luther (and many others) found one Church practice particularly offensive: the pope granting sinners reduction of their time of posthumous punishment in exchange for a donation. When the costs of building of St. Peter's Basilica in Rome pushed the Church to greatly expand the selling of these so-called *indulgences,* Luther's willingness to compromise was exhausted. On October 31, 1517, he formulated an angry and polemical *Disputation on the Power of Indulgences* which he sent to the archbishop, and – much more effectively – printed and posted as *Ninety-Five Theses* on the doors of the churches of Wittenberg. The printed *Theses* spread like wildfire, first in Germany and then, translated and edited, throughout the rest of Europe, causing a schism of a depth and magnitude the like of which the Church had never experienced before and has never recovered from since – the *Reformation.*

There are many important aspects to the Reformation – theological, political, cultural and ethical – but for us here, its epistemological significance is crucial. First, let us stress again that Luther's revolution was predicated on the movable press: all attempts to suppress his original theses and subsequent prolific writing by both the Church and secular authorities completely failed against the new power of printed publication. Within three years of the posting of the *Theses,* thirty-two of Luther's tracts were published in more than 500 editions, with some 3 million copies by the time of his death in 1546. These numbers do *not* include his translation of the Bible, which was to become an indispensable part of every Protestant household. Let's further stress the significance of Luther's Bible. The practical and theological personalization of the reading of the scriptures, which this book represented, was a direct assault on the foundation on which the Catholic Church established its authority: its expertise in interpreting God's word.

This authority, as discussed in Chapters 1 and 4, was not limited to questions of worship or abstract theology: "a Christian," as we saw Augustine claim, "is a person who thinks in believing and believes in thinking." The Church had worldly ambitions and responsibilities. Its claim to knowledge of the divine – since the time of Augustine and definitely since the establishment of its education system and the university – was closely related to its search for and ability to demonstrate knowledge of this world. Astonished by its inability to subdue the new heresy, the Church convened a conference of high officials and expert theologians – the extremely long *Council of Trent,* lasting from 1545 to 1563 – which resulted in many decrees and doctrines of faith uncompromisingly re-affirming orthodoxy. These comprised the theological core of what came to be called the *Counter-Reformation*, but more crucial to the development of science were some practical steps taken by the Church to re-establish its authority on earthly knowledge.

Counter-Reformation and the Calendar Reform

We already discussed (Chapter 5) one of the measures: the establishment of the Society of Jesus and its reform of education. The Jesuits, as we've pointed out, turned from the strict logicism of traditional Church education to disciplines aimed at moving and persuading: rhetoric, theater and dance. Even mathematics was taught (at least to the missionaries-to-be) for its persuasive power. The Jesuits were embracing the Humanist call for *vita activa* and the earthly disciplines required by it, but they were also reacting in a very telling way to the challenges of the Reformation. Luther and his disciples stressed belief based on the simplicity of heart and a personal understanding of the revealed word (reminiscent of the teaching of early Church Fathers like Tertullian we encountered in Chapter 1). The Protestant Church was therefore more interested in affecting the passions than persuading reason. The Jesuits' reform was thus doing more than just succumbing to the fads of Humanism; it was quietly acknowledging the powers of the new version of Christianity and adopting its means. The spectacular nature of the visual art sponsored and commissioned by the Catholic Church following the Council of Trent – which came to be known as 'Baroque Art' – represents a similar insight: the flock needed to be awed into belief rather than convinced by argument. The Catholic Church was losing confidence in its centuries-old synthesis of knowledge and belief – the one sketched by Augustine and formalized by Aquinas.

A little more than a century after Luther's *Theses* this synthesis would indeed collapse, unable to carry the burden of its own inconsistencies, exasperated by attempts at compromise and shaken by the pressures of a quickly changing science. The most crucial of these changes was, in its turn, closely related to another measure the Church adopted in order to face the challenge of the Reformation: the drive to reform the calendar.

For the Church to reclaim its centralizing, knowledge-based role – the very role that Luther had defied – it needed to demonstrate that it could still provide its believers with the wherewithal for worship. First and foremost was the need to regain control of *time*. An accurate, standardized measure of time was required, so the Holy Days could be planned confidently and well enough in advance to be celebrated with the proper pomp and circumstance befitting the Church's authority as God's representative on Earth. In other words: the Church needed an accurate calendar.

It no longer had one. The Julian calendar it had inherited, along with so much else, from Roman administrative law (the calendar took its name from Julius Caesar, who had enforced it) was fundamentally the Hellenistic calendar of the last century before the Christian Era. Built into it was a cumulative error of about a quarter of an hour a year – a result of what you may remember from Chapter 3 as Hipparchus' 1° a century 'precession of the equinoxes.' This is the difference between the year calculated as the period the Sun requires to return to a given position among the fixed stars versus the year calculated as the period it requires to return to the equinox. For Hipparchus, the precession was a very minute change that he was proud to account for with a very slow cyclical change. But the Church's interest in astronomy was practical and linear. It wasn't satisfied with the knowledge that in 26,000 years the heavenly pole would return to its same position above the terrestrial pole; it needed to know a specific number of years, representing the exact time that had passed from one particular event: the birth of Christ. An extra quarter hour a year amounted to a day in a century, and over a millennium and a half this had become a fortnight – so by the time they had to deal with the Reformation, Church authorities also had to worry if they were even celebrating the great rites within the appropriate month, let alone accurately.

The most important of these rites – Easter – was also the most difficult to calculate. The Gospels tell that Jesus rose from the dead on the third day after his crucifixion; which happened on the day of Passover (the last supper was the Jewish Seder), that is celebrated on the fifteenth day, or the full Moon, of the spring month; which occurred on a Friday. So Easter had to be celebrated on the first Sunday after the first Monday after the full Moon following the spring equinox – a complex triangulation of annual, monthly and weekly calculations.

Let's emphasize: it's not obvious that the problem had to be delegated to science – namely, to astronomy and its theoretically driven calculations. The equinox, and even more so the full Moon, could be determined locally, from straightforward observation, the way Passover and other holidays were determined in the Jewish tradition. But local initiative and authority were the very challenges set by the Reformation that the Church was trying to stave off. This is a fine example of the cultural import of science that we discussed in Chapter 3: astronomy is only of use in a complex, centralized political system. It thrives as a tool for a regime with claims to heavenly patronage; with customs of grand ceremonies to underscore this patronage; and with a need to coordinate these ceremonies centrally, over large distances. The Church was not unique in these demands: the Chinese court of the Late Ming and Early Qing Dynasties that we encountered in Chapter 5 shared them. The Church and the Chinese imperial court also shared, we saw, the reliance on Hellenistic-based astronomy and the calendar crisis that it was causing at that particular point in time.

So in its very first year – 1545 – the Council of Trent authorized the pope to effect a reform of the calendar, which concluded in 1582 with Pope Gregorius' institution of a new calendar, named after him, that is still in use. But the problem with the Julian calendar was clear for centuries – Bede, in the ninth century, had already mentioned Easter being three days later than the date officially set for it by the first ecumenical council, the Council of Nicaea, in 325. Moreover, the fact that the error now amounted to two whole weeks, meaning that Easter might be celebrated in the wrong month, created a sense of crisis shared by astronomers for decades. With the complete Greek original version of Ptolemy's *Almagest* finally making it to Europe in the fifteenth century – translated into Latin and then printed – astronomers were already putting forward bold suggestions very actively attempting to reform Ptolemaic astronomy in general and the calendar in particular. The boldest proposal came in 1514, in a short manuscript fittingly called *Commentariolus* (*Little Commentary*), which a Polish friar, Nicolaus Copernicus (1473–1543), circulated among astronomers and knowledgeable acquaintances. Many of the observational discrepancies in existing astronomy, Copernicus claimed, and even more so the unconvincing compromises built into Ptolemaic theory, would disappear if the Earth and the Sun would change places. Instead of resting at the center of the cosmos with all the planets and the sphere of the fixed stars moving around it, so Copernicus suggested, the Earth moves around the Sun, together with the other planets.

The Copernican Revolution

A close contemporary of Luther (who found Copernicus' theory bizarre), Copernicus was a product of these changing times; his education and livelihood straddling both a fading era and the new era that was just being ushered in. He was born in Northern Poland, then under Prussian rule, to a family of new merchant wealth and old Church connections. He first received traditional, Aristotelian education at the University of Krakow, followed by studies of Church law in Bologna, but in trying Medicine in Padua and receiving an LLD from Ferrara (in 1501) he actively took to the Humanist fads of the time. Yet Copernicus had no interest in *vita activa*: throughout his life he made his living as a canon in Frombork (then Frauenburg), a medieval *privilegium* (a personally conferred status) with few obligations, which he received thanks to the very Renaissance-style nepotism of his uncle, the Bishop of Warmia.

He was a skilled astronomer, versed in the newest theories and techniques, but never a professional one: his activities were supported by the patronage of his uncle, in whose Warmia house he set up his observatory.

Conservatism

Copernicus' famous 'hypothesis' is a product of this combination of novelty and conservatism. From a purely astronomical perspective, his hypothesis changed little: the Sun moved to the Earth's place as the reference point for all planetary motions, and the Earth correspondingly moved to the Sun's place, carrying the Moon with it. The Sun wasn't exactly at the center of each orbit, just as in Ptolemaic astronomy the Earth had not been. It was presumed to be at the center of the cosmos as a whole, but this assumption had no immediate implications for the astronomer. All other planets kept their order, and Copernicus fully embraced the fundamental requirements of Hellenistic astronomy – circular, concentric, uniform motions. He also maintained most of the theoretical means used to meet these requirements, established by Ptolemy and refined in the medieval tradition (though importantly not all, as we'll see): independent circular orbs, eccentricity, deferents and epicycles.

As Figure 7.1 shows, Copernicus' astronomy looks and feels well within the Ptolemaic tradition. And against common lore, neither he nor his contemporaries saw in the motion of the Earth a cause for religious worry – this

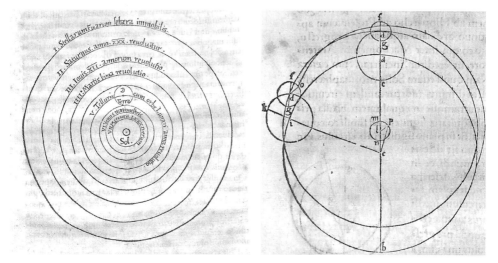

Figure 7.1 A couple of pages from Copernicus' *De Revolutionibus* of 1543. On the left: the planets, now with the Sun at the center and the Earth, together with the Moon, as the third planet. Compare to Figure 3.10 and note how this is, in a sense, just a straight change of place between the Sun and the Earth. On the right: one of the many diagrams in the book. Again: note how traditional it seems, similar to the diagram in Figure 3.14.

would take another century. In fact, when he eventually came to print and publish his hypothesis as a book, Copernicus actually dedicated it to the pope. More importantly, in the opening paragraphs of the *Commentariolus*, Copernicus does not display his system as a great innovation at all. Quite the opposite: it is a return to the "principle of regularity" of "our ancestors," which "the planetary theories of Ptolemy and most other astronomers, although consistent with the numerical data," abused. The compromises through which this empirical consistency was achieved, he claims, left a system that is "neither sufficiently absolute nor sufficiently pleasing to the mind" (Copernicus, *Commentariolus*, p. 57).

Revolutions

Copernicus' hypothesis was perhaps a minor transformation from a purely astronomical perspective – a mere exchange of places between the Sun and the Earth. But from physical and cosmological points of view it was revolutionary: it put the Earth in three different motions. In his book, Copernicus put the idea quite poetically:

> It is the earth ... from which the celestial ballet is beheld in its repeated performances before our eyes. Therefore, if any motion is ascribed to the earth, in all things outside it the same motion will appear, but in the opposite direction, as though they were moving past it.
>
> Copernicus, *On the Revolutions* II, Edward Rosen (trans.),
> (Baltimore, MD: Johns Hopkins University Press, 1992
> [1543]), p. 12

The heavenly bodies seem to rise and fall and move with the seasons, but all this "celestial ballet" is but an appearance:

> Such in particular is the daily rotation, since it seems to involve the entire universe except the earth and what is around it. However, if you grant that the heavens have no part in this motion but that the earth rotates from west to east, upon earnest consideration you will find that this is the actual situation concerning the apparent rising and setting of Sun, moon, stars and planets.
>
> Copernicus, *On the Revolutions* II, p. 12

The Sun and the stars don't move daily from east to west – it's the Earth that rotates around its poles from west to east. Similarly, it is not the Sun that moves yearly on the ecliptic from west to east – it's the Earth that moves from east to west, like any other planet. Finally, since the poles of the daily rotation always point in the same direction (the north pole at the North Star), the axis itself must rotate in the opposite direction to the annual motion and at the same velocity. This is because the axis of the Earth's daily rotation is oblique to the ecliptic (the 23½° that were the difference between the ecliptic and the celestial equator), and the physics available to Copernicus could only conceive of the Earth as if held solidly in its position in the orb of its annual revolution. It meant that the oblique pole must constantly change the point at which it's directed – unless you assume that the pole keeps rotating backwards, towards its original inclination. Think of a straw in a tilted can of drink that you hold while turning on your heels[1] – it'll change the point on the ceiling at which it is directed, unless you constantly twist your hand in the opposite direction to your rotation. In our physics this problem is solved, more simply, by the assumption of the conservation of angular momentum, which means that the axis of rotation maintains its orientation in space, always remaining parallel to itself, but for Copernicus the assumption of backward rotation also had a side advantage. Instead of assigning an independent (and very slow) motion to account for the precession of the equinoxes (see Chapter 3), he explained it as a slight difference between the angular velocity of the Earth around the Sun and the angular velocity of the axis around the pole (in the opposite direction).

Whether one finds these technical details exhilarating or excruciating, they should not obscure the enormous intellectual price Copernicus had to pay for putting the Earth in all of these motions. From the point of view of both science and common sense, these motions were hardly believable. As we saw in Chapter 3, this was not the first time that the motion of the Earth had been suggested – Aristarchus of Samos was its most famous ancient champion, already in the third century BCE. But the arguments that led to its rejection in antiquity had only become more powerful with time.

To begin with, there were excellent physical reasons why the Earth should be at the center of the cosmos. It was clearly at the center of the realm of elements, with heavy bodies moving naturally towards it and light ones moving away. If it keeps on moving, how are the earthly bodies supposed to know where to go to, and where from? Moreover: if the Earth isn't at the center of the cosmos, why don't these bodies move towards and away from the cosmos' real center?

To these scientific queries Copernicus had quite a good answer, well rooted in Aristotelian cosmology: the Earth is not only the soil under our feet. It's the whole realm of elements, and it all moves together. The elements move towards and away from this center, but in order to do so, the Earth doesn't also need to be at the center of the whole cosmos. After all, as we said, it is not at the center of any orbit in the Ptolemaic system (no more than the Sun is at the center of any of the orbits in Copernicus' own system).

---

[1] I owe this explanation to Keith Hutchison.

Ptolemy said that the Earth is at the common center of all orbits, and now the Sun is at that center, but the very concept of a common center has little astronomical significance, and Copernicus reminds his readers of this in the first of the Assumptions (*Petitiones*) opening the *Commentariolus*: "1. There is no one center to all the celestial circles or spheres." With no common center, he feels secure to add as his second Assumption: "2. The Center of the earth is not the center of the universe, but only of gravity and of the lunar sphere." The Sun can thus be "the center of the world (*mondus*)" in the same way that the Earth used to be the center of the Ptolemaic cosmos, as the "mid-point" of all orbits, and Copernicus adds this idea as his third Assumption.

An even more severe challenge is the main reason why the motion of the Earth was rejected when originally offered by Aristarchus (Chapter 3): because we observe no parallax among the fixed stars, the cosmos would have to be enormous for the Earth to be moving; much bigger than is reasonable. To the necessary enormity of the cosmos Copernicus' answer is simpler, and even more astonishing: it is simply so. This is his fourth Assumption: the world is indeed so huge that the motion of the Earth around the Sun doesn't change the angles of the stars in a detectable way. If required by science, he implies, what counts as reasonable will have to change.

But the most difficult objection to Copernicus' theory was completely straightforward: if we move, how come we don't feel it? If the Earth rotates from west to east, how come things not attached to it don't drift in the opposite direction? Why don't we feel a constant and extremely fast easterly wind? Why don't clouds and birds disappear to the west? Why don't cannon balls shot to the west, against our rotation, travel much further than those shot to the east? Why don't stones dropped from towers fall to the west of them? These questions didn't require much natural philosophy or cosmology to understand, but they could also be formulated in a sophisticated, technical way: nature is change, and change is motion. If the motion of the Earth is wholly imperceptible, if it creates *no change* in our world, in what sense is it motion? Moreover: the motion that Copernicus' hypothesis claimed was very fast, and it wasn't difficult to calculate just how fast. The circumference of the terrestrial globe, you'll remember, had already been measured by the ancient astronomers, and to traverse such a distance in 24 hours would mean that Europe had to travel at about 1,500 kilometers per hour. In a world in which the fastest motion that anyone could have experienced was a horse galloping at around 30 kilometers per hour, the idea that we may be traveling so fast without an inkling of our doing so was absurd.

Copernicus answered this challenge with the second Assumption we quoted above: that Aristotelian claim that the Earth is not only the soil under our feet, but the whole realm of elements under the Moon, and this terrestrial globe moves together. There is no wind and the clouds don't drift because the air also rotates around the Earth's axis.

To us, who have learned the physics that was developed in the seventeenth century (we will discuss this project in the coming chapters), this solution may sound very reasonable. We are used to the idea that when whole physical systems move together in uniform velocity, this motion cannot be noticed from within the system. But it's crucial to stress that Copernicus did *not* have such physics – it was in fact developed in response to the challenges of his moving Earth. Moreover, suggesting that the elements move together begged the question: why did they move at all? Again, the only physics Copernicus knew did not allow for such motion. Aristotle explained why within the terrestrial realm the four elements moved towards or away from the center while the heavens, made of the fifth element, moved around this center. But what could make all the elements move together around the center? What was the cause of this motion? Was it natural or forced?

As he did with his answer to the question of the lack of parallax, Copernicus demanded of his readers to submit their common sense to whatever was implied by reasons and arguments internal to one science – astronomy – even though these considerations also militated against the rest of science; they departed from natural philosophy and cosmology. Recalling that the ruling Aristotelian natural philosophy put a particular stress on good reason and good sense, the difficulties with the idea that the Earth moves seemed overwhelming.

Motivations

Copernicus was himself well aware of the unreasonableness of his hypothesis. It took him almost thirty years to turn his *Commentariolus* into a complete book – *On the Revolutions* – whose final proofs he reviewed on his death bed, never to see it actually in print. In the very first paragraph of the book's dedication to the pope, he says specifically:

> To be sure, there is general agreement among the authorities that the earth is at rest in the middle of the universe. They hold the contrary view to be inconceivable or downright silly.

> Copernicus, *On the Revolutions* II, p. 11

What, then, was so attractive in this "downright silly" (*ridiculum*) hypothesis that made him cling to it and accept all its "inconceivable" implications? The reasons for Copernicus' preference couldn't have been empirical. As we pointed out, he remained intensely loyal to the assumption of circular motions and incorporated the fundamental Ptolemaic tools of eccentricity and epicycles, and under these conditions the heliostatic and geostatic systems were strictly equivalent. We should insist on 'static' here rather than 'centric' in order to stress exactly this point of equivalence of the Copernican and Ptolemaic systems: both applied eccentricity, and the Sun was not at the center of the Copernican orbits anymore than the Earth was at the center of Ptolemy's orbits. Copernicus was well versed in the geometrical transformation between the systems, which was worked out by an astronomer of the previous generation – Johannes Müller von Königsberg (1436–1476), known as Regiomontanus. The commonly held idea that Copernicus' system was simpler than the Ptolemaic also needs a careful modification. Copernicus left in place not only eccentricity, but also the epicycles which Ibn al-Haytham found so deplorable (Chapter 4), using as many of them as Ptolemy did. The heliostatic system did allow, for example, for a simpler lunar theory, and in the context of the search for Easter's accurate date it was not a negligible achievement. In other respects, however, like the cumbersome explanation for the constant direction of the Earth's pole, it was actually more complex than the geostatic system.

Copernicus himself tells his readers on the first page of the *Commentariolus* what it was that made traditional astronomy "neither sufficiently absolute nor sufficiently pleasing to the mind": the equant point. There's no doubt that the equant was the most embarrassing of all the compromises built into Ptolemaic astronomy: if the planet moved in uniform angular velocity around some other point, it clearly *did not* move in circular, concentric uniform motion around either the Earth or the geometrical center of its orbit. And Copernicus indeed made good on his word and provided an equant-free model, but his model had little to do with the question of the Earth's motion. It was a double-epicycle model – basically a version of the Tusi couple discussed in Chapter 4 – and it worked equally well with a reference to an immobile Earth as it did to the Sun. Whether he was familiar with the Muslim work or developed the double-epicycle technique on his own is a subject of much debate among historians of astronomy but of little consequence to us here, exactly because it has nothing to do with the motion of the Earth.

So it wasn't one particular theoretical issue either that motivated Copernicus' dramatic revision of astronomy. However, the concentration on the equant does tells us something. It was crucial for Copernicus to compromise as little as absolutely necessary on the fundamental ancient criteria of order: concentric circular motion of uniform velocity. He didn't see himself as demolishing traditional astronomy, but as returning it to its former glory. None of the theoretical advantages that he gained from the motion of the Earth was, in itself, important enough to justify the dramatic change they necessitated. But together they provided what Copernicus called "symmetry": a coherence which replaced the troubling arbitrariness of the Ptolemaic system. This coherence suggested that the mathematical-geometrical theories of the astronomer didn't just 'save the phenomena' through clever trickery, but reflected real order in nature. It was a paradoxical order, because, as we saw, it contradicted the order ascribed by physics and cosmology, but for Copernicus, a mathematician – and recall that astronomy was a branch of mathematics – this paradox was apparently not too daunting, and perhaps even exciting. In fact, the quote we started with suggests that this physics impressed him little: why assign motion to the enormous "entire universe" if the "rising and setting of Sun, moon, stars and planets" could be explained by the much more modest rotation of "the earth and what is around it"?

Planet	Copernicus value (AU)	Modern value (AU)
Mercury	0.38	0.387
Venus	0.72	0.723
Earth	1.00	1.000
Mars	1.52	1.52
Jupiter	5.22	5.20
Saturn	9.17	9.54

First of all, revolving around the Sun made sense of the sequential order of the planets. Traditional astronomy simply assumed that the length of the planet's period indicates its distance – the longer the period, the further is the planet. Rotating around the Sun, there were now two groups of planets: Mercury and Venus are the 'inferior planets' – between the Earth and the Sun; Mars, Jupiter and Saturn the 'superior planets' – further away from the Sun than the Earth. Some facts about their periods and distances, which were previously just contingent facts, followed directly from this new order. For example, Mercury's angle to the Sun is always the smallest and its period the shortest; now these two facts followed from the same reason: it is the closest planet to the Sun, not to the Earth. Even more interestingly, the fact that neither Mercury nor Venus are ever in opposition to the Sun – the greatest 'elongation' (angle to the Sun) of Mercury is 28° and of Venus 47° – reflects that same order: since we observe these planets from outside their orbit, their maximum elongation is the angle by which we perceive the radius of their orbit (Figure 7.2, left and center). It is not that the Ptolemaic system could not account for either fact – all it took was calibrating the velocity and size of the corresponding deferents and epicycles

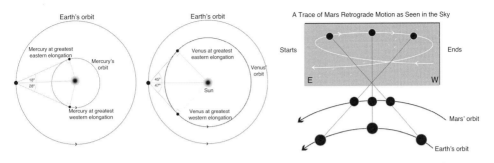

Figure 7.2 The explanatory innovations of the Copernican Hypothesis. Right: Copernicus' explanation to retrograde motion: when the Earth moves past a planet (here Mars) it appears as if this planet is moving backwards. Left and center: Copernicus' explanation for the fact that the angles we observe between Mercury and Venus and the Sun are limited (maximum 28° for Mercury and 47° for Venus). Had we been observing them from the center, it should have been possible for those planets to be 'in opposition' – namely, 180° away from the Sun. The Ptolemaic system has to forcibly synchronize the motions of these planets and the Sun, but if the Earth is moving around the Sun then this is no longer required. Since our orbit is further from the Sun than the orbits of Venus and Mercury, it means that we always observe them from 'outside': the angle we observe between either of them and the Sun is never larger than the angle by which we perceive the radius of their orbit. From this follows a way to calculate the radii of the interior planets in astronomical units (AU, the radius of the Earth's orbit or its mean distance from the Sun) according to the Copernican Hypothesis. When we observe the planet (take Venus) at its 'maximum elongation' – the largest angle from the Sun (in this case 47°) – the angle between the Sun and the Earth observed from the planet has to be 90°. The largest distance between Venus and the Sun is thus AUcos47°=0.72AU. The radii of the exterior planets are marginally harder to calculate. Here are the planets' radii according to Copernicus' calculations, compared with the modern values.

and synchronizing those of Venus and Mercury with the Sun's. The point was that in Copernicus' system all these facts had one real cause: the true order of the planets.

Even more interesting was how the motion of the Earth accounted for the phenomenon which most engaged Hellenistic astronomy: retrograde motion. Again: it was not that traditional astronomy couldn't explain this phenomenon; the epicycle did so quite satisfactorily. But Copernicus' hypothesis 'saved' this apparently disorderly motion in the most elegant way: by showing that it's merely apparent. The planets always move in the same direction; they only *appear* to slow down and reverse when the Earth bypasses them in its own motion, the way the shore seems to be moving backwards from a departing ship (Figure 7.2, right). Moreover, this explanation creates another "symmetry" where coincidence used to rule: the retrograde intervals of the planets nearer the Earth appear larger and more frequently than those of the planets further away (regardless of their distance from the Sun); Venus more than Mercury; Mars more than Jupiter; Jupiter more than Saturn. In the Ptolemaic system, each was an independent fact, accounted for by an independent set of epicycles. With the Earth set in motion, all these facts follow from one simple principle: the closer a planet is to the Earth, the more often the two planets (the Earth now a planet) pass each other by.

After Copernicus

Andreas Osiander and the Timid Interpretation

Copernicus didn't see a religious scandal brewing in his work – he was more worried that his astronomer peers would find his hypothesis "ridiculous" – but the following decades proved him wrong on both counts. The reception of his hypothesis among astronomers was quite favorable. His *Commentariolus*, in spite of its limited distribution as a manuscript, became well known enough that a young Lutheran admirer, Georg Joachim Rheticus (1514–1574), left his duties as professor of mathematics in Wittenberg to come and study with him in 1538. In 1540, Rheticus published an excitedly supportive *Narratio Prima* – 'first exposition' – of Copernicus' hypothesis, paving the way for the 1543 publication of *De Revolutionibus*. The title of the book is quite telling: for Copernicus and his contemporaries, 'revolutions' meant the continuous, circular, cyclical motion of the planets. When it came to mean the exact opposite: a drastic change from one situation – social, political or intellectual – to another, it was quickly applied to the change brought about by Copernicus' hypothesis. But the drama of this change would take some time to develop. One reason may have been that the person charged with bringing the book to print upon Copernicus' death and Rheticus' departure – another Lutheran scholar by the name of Andreas Osiander (1498–1552) – added an unsigned preface that could easily be read as Copernicus':

> There have already been widespread reports about the novel hypotheses of this work, which declares that the earth moves whereas the sun is at rest in the center of the universe. Hence certain scholars, I have no doubt, are deeply offended and believe that the liberal arts, which were established long ago on a sound basis, should not be thrown into confusion. But if these men are willing to examine the matter closely, they will find that the author of this work has done nothing blameworthy. For it is the duty of an astronomer to compose the history of the celestial motions through careful and expert study. Then he must conceive and devise the causes of these motions ... Since he cannot in any way attain to the true causes, he will adopt whatever suppositions enable the motions to be computed correctly from the principles of geometry for the future as well as for

the past. The present author has performed both these duties excellently. For these hypotheses need not be true nor even probable ... if they provide a calculus consistent with the observations, that alone is enough.

Osiander, Preface to Copernicus, *On the Revolutions*, p. xvi

Osiander's preface does little justice to the great ambition and self-confidence embedded in Copernicus' project and expressed in his dedication to the pope, but is worth quoting at length for a number of reasons. First, because the question of whether such epistemological humility befits scientists, or is reasonable to expect of them, still engages philosophers. Secondly, because it reminds us that this complex distinction between what is to be accepted within philosophical discussion and what is to be 'truly' believed was an old technique that allowed the Church to ignore the contradictions between naturalistic philosophy and monotheistic belief and retain its position as the leading European institution of knowledge. Osiander, though protestant at the time, was a Catholic convert for whom this difficult intellectual maneuver made perfect sense when facing such an unsettling challenge to "the sound basis of the liberal arts." Lastly, this strange constellation demonstrates well the upheaval of the times: a Protestant priest using a medieval Catholic ploy to protect an astronomical doctrine created by a Catholic friar in the context of the Counter-Reformation; moreover, a doctrine Luther was apprehensive about.

Giordano Bruno and the Radical Interpretation

What Osiander seems to have feared – a true religious scandal – took some forty years to materialize – but it finally did. Claims with deep cosmological implications, such as those advanced by Copernicus, are bound to have religious implications. The scandal came in the hands of a known rascal, Giordano Bruno (1548–1600), who reasoned as follows: since Copernicus' theory demands that the world be so enormous, we may as well consider it infinite. And an infinite world is no longer a *cosmos*; it is a *universe*: it has neither center nor periphery, nor any privileged points. It can, and therefore should (because God is omnipotent), contain infinitely many suns and, around them, infinitely many planets. Most crucially: this implies infinitely many earths, and on them, presumably, infinitely many human races. It is easy to see how different this image of the universe – and the place of humans in it – is to the traditional one, represented so emblematically in Figure 1.12. In that convincing and reassuring marriage of Aristotelian cosmology and Christian cosmogony, deeply rooted for many centuries, humans were nestled at the center of the cosmos, with God's loving eye resting on them. In its stead Bruno was suggesting an infinite universe scattered with infinite humans – how could one God attend to them all? How could He reveal himself to them? And how could He do it in the flesh?

In 1593, Bruno found himself in the hands of the Inquisition, and in 1600 he was burned at the stake for heresy in Campo di Fiori in Rome. The exact nature of the allegations against him remains uncertain, as the protocols were lost, in sloppiness not characteristic of the Inquisition – but there were many good reasons for the authorities to consider him an unrelenting heretic. Born in Nola, in southern Italy (he subsequently called himself "the Nolan" in many of his writings), he became a Dominican friar in Naples in 1575, and already in 1576 was accused of Arian heresy (a doctrine distinguishing between Jesus and God), which pushed him to flee to the more tolerant north and commence a life of wandering. In 1579 Bruno was in Geneva, where he converted to Calvinism, only to be excommunicated later on charges of disrespect. He moved on to France, where he published works on the nature of memory and on Neo-Platonic philosophy, as well as anti-Church satires. He made his way to England and wrote extensively and brilliantly enough that by 1583 he became very popular in the court and in Oxford, but when he turned his satire against his new benefactors he was forced to return to the continent and to Catholicism. He then made the mistake of returning to Italy in 1591, where a Venetian patron, disappointed with him as a tutor, denounced him to the Inquisition. It's hardly likely, then, that Bruno was burned strictly for his support of Copernicanism, but the pattern is clear: in a climate of religious volatility, with the radical ambitions of astronomers and natural philosophers, and with enthusiasts adopting new scientific ideas to support deviant theological doctrines, the delicate balance on which Church-sponsored science depended was quickly eroding.

Tycho Brahe and the New Empirical Astronomy

For astronomers in the second half of the sixteenth century, the hypothesis of the recently departed Copernicus brought enthusiasm of a different sort. As long as the Earth was assumed static and at the mean center of all orbits, the changing positions of the planets against the backdrop of the fixed stars meant only that: changing angular positions. Even the notion that they were moving was in a sense an interpolation: all that could be observed was that a planet was in one place one day and in another the next. From an off-center vantage point, these changing angles could be translated into real motions, over real distances – not just theoretically, as we saw in Copernicus' reasoning above, but empirically. Figure 7.2, center and left, demonstrate how, with simple trigonometric calculation, Copernicus could infer the real distances of the planets from the Sun: the radii of their orbits, expressed in Astronomical Units (the caption explains the procedure).

After millennia in which they were concerned almost exclusively with *theoria* – with constructing geometrical models – European astronomers found new excitement in observations, an excitement Johannes Kepler (1571–1630) would poetically express a couple of generations later:

> ... if the earth, our home, did not measure out its annual circuit in the midst of the other spheres, changing place for place, position for position, human reasoning would never struggle to the absolutely true distances of the planets, and to the other things which depend on them, and would never establish astronomy.
>
> Johannes Kepler, *The Harmony of the World*, A. J. Aiton *et al.* (trans. and ann.) (Philadelphia, PA: American Philosophical Society, 1997 [1619]), p. 496

The person most responsible for the empirical surge was Kepler's short-term employer, Tycho Brahe (1546–1601). Tycho was a Danish nobleman who had received his education in German, Protestant universities (where he famously lost his nose in a duel, to be replaced with a metal one). Returning to Denmark in 1567, he secured the support of King Friedrich for a project the like of which Europe had never seen: a very large, purpose-built astronomical observatory, which he had constructed on the Island of Hven, in the straits between Denmark and Sweden.

We can borrow the term 'Big Science' to describe Tycho's *Uranienborg*, or 'The Castle of Urania' (astronomy's muse – Figure 7.3, left). It was financed by the court and employed some 100 people at any given time, from servants, through instrument-builders, apprentices, calculators and expert astronomers, to Tycho himself. Most crucially: it featured large, expensive, purpose-built instruments (note how much bigger they are than the astronomers operating them in Figure 7.3, center). The largest was a building-sized mural quadrant, reminiscent enough of Marāgha observatory (see Chapters 4 and 5), that historians have speculated about there being a direct relation between the structures. These instruments were the same angle-measuring, naked-eye instruments described in Chapter 3, if expertly built and more accurate. What made them unique was their sheer size, which allowed minute division, and hence a much higher resolution than their traditional, hand-held versions (Figure 3.7), enabling Tycho to make observations many orders of magnitude more accurate than those of his predecessors (Figure 7.3, right).

This accuracy emboldened Tycho to make some audacious empirical claims. Already in 1572 – before moving into Uranienborg – Tycho observed an extremely radiant object that had suddenly appeared in the sky, dazzling enough to be seen even in daytime. Confident in the new accuracy of his instruments, Tycho concluded that the fact that the parallax

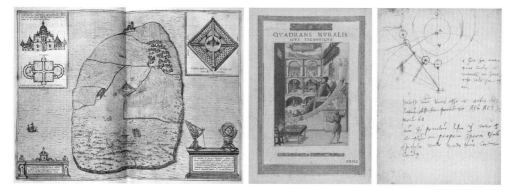

Figure 7.3 Tycho's observatory, his instruments and his observations. On the left: a map of the island of Hven with *Uranienborg* at its center. The large proportion of the island taken by the observatory is telling of the court power supporting it: the land was confiscated from the locals and they were forced to assist in its construction and maintenance, much to their resentment. At the center: the working of the Observatory according to Tycho's *Astronomiae instauratae mechanica* (1598). Observers on the roof, dwarfed by the huge instruments; 'computers' on the ground floor; and alchemical workshop in the basement. Tycho himself acts as the manager and master-observer, siting by his giant mural quadrant. On the right: the products of the observations – an entry from Tycho's notebooks from the period 1587–1596.

of the new object was smaller than the Moon's meant that it was indeed further away than the Moon – outside the terrestrial sphere, in the heavens. It was a *Stella Nova*, as the title of his publication asserts – a New Star. In 1577, he added a similar claim about a comet: this fleeting, ephemeral object was also above the Moon, outside the realm of elements and in the heavenly sphere.

Tellingly, Tycho was not exactly a Copernican. He appreciated the empirical advantages of having the Sun, rather than the Earth, as the reference point of the planets' orbits, but the idea that the Earth moved seemed to him just too outlandish. Instead he offered a compromise: the planets orbited the Sun – that was all that was necessary for the new calculations of motions and distances – but the Sun orbited the Earth. It may seem cumbersome to us, but for the best part of the century to follow, this became the preferred option for European scholars, as beautifully demonstrated by the frontispiece of the 1651 *Almagestum Novum* of the Italian Jesuit Giovanni Battista Riccioli (1598–1671; Figure 7.4). In fact, through Jesuit teaching, Tycho's system also came to represent European astronomy in China throughout the seventeenth century.

Figure 7.4 The frontispiece of Giovanni Battista Riccioli's *Almagestum Novum* (Bologna, 1651). Riccioli's version of Tycho's system outweighs the Copernican by the balance of theory – represented by Urania, the Muse of Astronomy. Riccioli's system is slightly different to Tycho's in that the comet's orbit, still outside the atmosphere, is very eccentric, so it passes on the other side of the Sun. Observational evidence is represented by the 100-eyed Argus, who holds in his hand a telescope – by then a fully respectable instrument. The Ptolemaic traditional, geostatic model is completely cast to the floor, but Urania assures the motionlessness of the Earth (which Tycho's system preserved) by reciting Psalm 104:5: "thou didst fix the Earth on its foundation that it never can be shaken."

But Tycho's theoretical caution could not mask how unsettling his empirical findings were to cosmology and to common sense. The distinction between heaven and Earth was the most basic for common sense, science and religion. It was shared by pagan and monotheistic thinkers; it allowed Christians to adopt Hellenistic science; and made good sense to Muslim and Chinese scholars alike: changes belonged under the Moon, while above it, all was eternal and unchanging. This was also why the heavenly bodies moved in simple and perfect motions – concentric circular orbits with uniform angular velocity. As we saw, even Copernicus strictly maintained this distinction: the idea that the sphere under the Moon, the terrestrial sphere comprising the realm of the elements, was sharply distinct from the celestial sphere and moved together as a whole, was his explanation as to why we don't feel this motion. Tycho's findings – the celestial position of the new star and the comet – meant that the heavens were *changing*.

Kepler and the Physicalization of the Heavens

The Marvelous Order of the Copernican Heavens

Whether Copernicus and Tycho were willing to admit it, or even fully grasped it, the literally world-changing implications of their work didn't need a reckless religious radical like Bruno to notice. If the Earth were a planet, then the planets must be like Earth, subject to the same vicissitudes; if the heavens were changing like things on Earth, they needed to be studied in the same way we study changing things on Earth: by discovering their real motions and assigning these motions real causes. This forceful idea underlies the quote from Kepler above, and it would drive him throughout his long and productive life: the causes would have to be mathematical, as is appropriate for the perfect orderliness of God's creation that he firmly believed in, and as was suitable for his own expertise as a mathematician. But these causes also needed to be physical. This, for Kepler, was the most important lesson from the theoretical speculations of Copernicus and the empirical work of Tycho: astronomy needed to become what he termed *physica coelestis* – a physics of the heavens.

When he first introduced himself to Tycho in 1597, Kepler had already demonstrated his way of inferring real physical properties of the heavenly bodies from the mathematical data provided by astronomical observations. In the *Mysterium Cosmographicum* he'd published a year earlier, aged only 25, he'd calculated that the distances between the planets (from Mercury to Saturn)

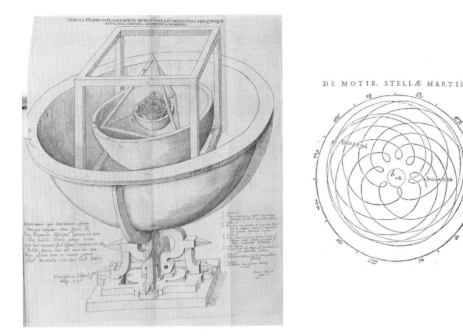

Figure 7.5 Kepler's evolving ways of realizing the idea of mathematical physics of the heavens. On the left: his diagram of the nested perfect solids from his *Mysterium Cosmographicum* of 1596 (probably in his own hand). The physical distances between the planets stand in the ratios of the five Platonic solids, from the dodecahedron between Saturn and Jupiter to the cube between Venus and Mercury. On the right: the "accurate depiction of the motions of the star Mars" according to his 1609 *Astronomia Nova*. This is the trajectory that Mars *would have* followed between 1580 and 1596 *if* either Ptolemy's theory or Tycho's was correct, and the reader is expected to see that such an orbit simply makes no *physical* sense. Whereas the youthful diagram resembles images from the magical tradition, the mature one is the first of its kind to appear in an astronomical text.

stood in relation to each other like the five perfect solids (Figure 7.5, left): it was as if one could place a cube between the orb of Saturn and that of Jupiter; a triangular pyramid between Jupiter and Mars; and so on.

Kepler was very proud of this discovery throughout his life. It was a certain mathematical truth, proven already by Plato in his *Timaeus*, that there are exactly five perfect solids; five kinds of three-dimensional bodies whose angles, sides and surfaces are all equal: the triangular, four-faced pyramid (tetrahedron), the six-faced cube (the hexahedron), the octahedron (eight-), the dodecahedron (twelve-) and the icosahedron (twenty-faced). Kepler's claim was purely about mathematical proportions: one of the implications of Tycho's claim that the comet was above the Moon was that the heavens were empty – if there were things like heavenly spheres or real material orbs, the comet would have to crash through them. But this suited Kepler just fine, because the ability to demonstrate these proportions meant two very exciting things. First, that one could indeed infer physical facts – real distances – from mathematical data. The second was that, as he saw it, he had proven Copernicus' hypothesis with mathematical certainty: since there were only five Platonic solids, hence five spaces between orbits, there could only be exactly six planets. Traditional astronomy had seven planets (count them!); in Copernicus, since the Earth and the Moon counted as one (and the Sun was no longer a planet), there were only six.

It's easy to see in Kepler's fantasy of cosmic mathematical harmony remnants of mathematical-magical thinking, of the sort we discussed in Chapter 6 – like Platonism or Neo-Platonism – which was gaining much popularity during the Renaissance. Kepler was keenly aware of the magical resonance of his ideas and bitterly denied it, and his letter to Tycho demonstrates that he had a point. Unlike the mathematical magicians among whom he didn't want to be counted and whose claims remained suggestive and speculative, he was offering a hypothesis with clear empirical content. In the letter he was asking for the specific, up-to-date, empirical figures for the planets' eccentricities. He was searching for those because some leeway was required when mathematical idealizations were applied to the world of matter. For the hypothesis of the *Mysterium* to work, he needed to know how thick the shells circumscribing the actual orbits should be (see Figure 7.5, right, again), so they could contain the eccentric orbits between their innermost part on the one side and outermost on the other. For the rest of his life, Kepler would argue that the *Mysterium* remained the foundation of all his work, even though within a decade he would no longer believe in concentric circular orbits. What he may have really referred to, therefore, is this invaluable contribution to the cathedral of science: the combination of reckless mathematical speculation and meticulous empirical investigation.

Kepler's Life and Times

Writing to Tycho, perhaps the most venerable astronomer of his time, was no trivial matter for Kepler. He was a high-school mathematics teacher in Graz, now Austria, of very humble origins. He had lost his mercenary father in his childhood, and his mother Katharina, a daughter of an inn-keeper, was one of those wise-women whom we discussed in Chapter 6, and whom Kepler inadvertently implicated in a witch trial. Kepler took a student exercise in Copernican astronomy – imagining and calculating how the heavenly motions would appear from the Sun – and dressed it as a story about a trip to the Moon, in which the narrator is delivered there by a demon summoned by his witch mother. To Kepler's dismay, the man-uscript of this piece of science fiction, which he would end up publishing as *The Dream* (*Somnium*), was used as evidence for Katharina's witchery. He knew her as a loving mother, keen on his education: she took him to watch the 1577 comet and a 1580 lunar eclipse, and sent him through the Protestants' version of a Church education – grammar school, seminary and the University of Tübingen. It was there that he became a devoted student of Michael Maestlin (1550–1631), one of the very first mainstream astronomers to teach Copernicus' system as the centerpiece of astronomy.

Tycho wasn't taken with Kepler's speculations, but was impressed enough by his mathematical skills to try and recruit him as a 'calculator.' This wouldn't happen in Hven: Tycho soon lost his favor with the Danish court and was forced to abandon his island. He accepted the role of Imperial Astronomer to the Holy Roman Emperor Rudolph II in Prague, with a promised new observatory to be built in Benatky castle, about 40 kilometers north of the city, and Kepler was happy to join his astronomical entourage there. He arrived in Benatky in early 1600, only to find himself, within six months and just removed from his obscure provincial post, taking Tycho's position: a beer-drinking bout with the emperor resulted in Tycho's sudden and painful death from kidney failure.

The position was not as illustrious as it may sound – Rudolf surrounded himself with alchemists, necromancers and magicians, exactly the intellectual milieu Kepler had always tried to distance himself from. He was mostly engaged as a court astrologer, and although he had no qualms about the discipline and was happy to draw nativities (including his and his dear ones'), he was deeply unhappy with his obligation to provide astrological prognostications and the political intrigues that this forced him into. By 1611, Rudolf's colorful court brought about his demise, as Rudolf was forced off the throne by his brother Mathias, meaning that Kepler had to leave Prague. The last two decades of his life were sad: his financial and intellectual standing deteriorating, he moved back to the German-speaking lands – first to Linz, then Ulm, then Regensburg, and when his applications to university posts declined, he took increasingly lower positions as a provincial mathematician. In 1613, his wife Barbara and son Friedrich died of small pox and from 1615 to 1621 his means and energy were exhausted by his mother's trial and imprisonment (although she was ultimately saved). In 1618, his daughter Susanna, his favorite, died as well. Throughout these sad wanderings he continued to work; when Susanna died, he tells, he could no longer work on his ambitious *Harmonies of the World*, and instead moved to work on the *Rudolphine Tables*, published in 1627. He died in poverty in Regensburg, in 1630, and his self-authored epitaph, which survived even though his grave didn't, is telling of his mood: *Mensus eram coelos, nunc Terrae metior umbras. Mens coelestis erat, corporis umbra jacet* ("I used to measure the Heavens, now I measure the shadows of Earth. The mind belonged to Heaven, the body's shadow lies here.").

The New Physical Optics

Yet in the relatively blissful decade he'd had in Prague, Kepler reshaped the mathematical sciences. Even Aristotle, with all his suspicion of mathematics (see Chapter 2), agreed that there were some entities in the world – things like planets, musical consonants and visual rays – that were by their very nature mathematical. The mathematical disciplines of the quadrivium, together with optics and mechanics, were therefore appropriate for dealing with these entities, and in the Middle Ages they received the title 'middle sciences' or 'mixed sciences,' because they studied mathematically things within the world of matter. But because the world of matter was complex and changing while mathematics was simple and perfect, it was clear to medieval mathematicians that they could only handle idealizations, and – unlike natural philosophers – offer no causes. The subject matter of optics were lines of vision and angles of reflection and refraction; of mechanics – diagrams representing dimensionless bodies; of music – the proportions of an abstract monochord.

This was a limitation Kepler was no longer willing to accept. The disciplines of his expertise – optics and astronomy – were to offer real causes for real physical entities, and to do so without relinquishing any of their mathematical prowess. The *Mysterium* hypothesis was his first attempt at this project. When, in 1600, he found himself entrusted with Tycho's observations (for the right of which he had to struggle with Tycho's family) and the task to develop a theory on Mars, he set about doing it in his own physical way. In 1603, he took a break from the astronomical project and turned to reshape optics.

Traditional optics was the mathematical theory of vision. It studied visual rays: straight lines, which could only change direction: refracted by changing media or reflected by polished surfaces. Whether these visual rays were physical entities or just mathematical representations of the process of vision, and what this process consisted of, was much debated. Aristotle taught that the mind is literally *in-formed*: it receives the form of the object, separated from its matter. The Platonists argued that the object multiplies itself as *species* through the medium (mainly air). The Atomists claimed that the object constantly releases *simulacra* – films of atoms retaining their order (and constantly replaced by atoms from the medium) – which flowed through the medium. But there was no debate that vision is a direct, cognitive relation between the object and the mind, through the eye. Light, in all of these theories, had an important, but secondary role: it was a state and a property of the medium; it was the transparency of substances like air or water, which allowed the visual rays to pass through. What optics dealt with was these rays, representing that authentic relation between object and mind.

Kepler abolished this assumption. Nothing *of the object*, he claimed, comes to and through the eye. The subject matter of his optics was no longer vision but *light*: flowing from the Sun (or any other luminous source), bouncing off objects and falling on screens, causally producing images. If the screen happens to be the retina of the eye, we could call it 'vision,' although how the mind deciphers those images, he admitted, became a complete mystery. The eye was a flesh-and-blood optical instrument – like a camera obscura (which he was using to observe the Sun), it had, in front of the retina-screen, a hole (the pupil) and a lens (the 'crystalline humor'). The images on the retina were therefore not reliable missives from the objects, but two-dimensional, inverted, fuzzy collections of stains of light shaped like the pupil. If this turned vision into a mystery, so be it: what was important to Kepler was that if the eye was but an instrument, then instruments were no worse than the eye. That meant that there was nothing particularly suspect in Tycho's great project of instrumental observation.

This was very important to Kepler: the far-reaching cosmological implications of Tycho's claims that the comet and the new star were in the heavens didn't escape his contemporaries, and their validity was fiercely questioned. The heavens, his rivals claimed, were too far and too different from the terrestrial realm for us to be able to observe their working, and the use of instruments in the observations made them doubly suspect: if proper vision is a direct relation of object through the eye, then any mediation is a distortion. But if the eye is just one kind of instrument, Kepler reasoned – flesh and blood, but instrument nonetheless – then what is observed through the instruments is as reliable as what is observed through the eye. If vision *is* mediation – a physical motion of light rather than cognitive connection between object and reason – then there isn't any fundamental difficulty about observing the planets. With Kepler's new optics, what could be claimed about the heavenly bodies could be as causal and physical as what could be claimed about the things we see around us.

The New Physical Astronomy

Making optics *physical* meant introducing a physical agent – light – that physically moved and physically produced images on the retina, although light did remain, for Kepler, a peculiar agent: both physical and mathematical, terrestrial and heavenly. Changing light from a property of the medium (for the Aristotelians, recall, it was the actualization of potential transparency) into an independent physical entity also allowed Kepler to come up with a mathematical-physical law for light's decline with distance. Light, he reasoned in his *Optics*, expands as the surface of a sphere and there is the same 'amount' of light at the source, the center of that sphere, as on the surface of the sphere. Since the surface of a sphere is proportional to the square of its radius ($A=4\pi r^2$ in our terms), the power of light must be declining by the same proportion: *the intensity of light is inversely proportional to the square of its distance from its source.*

Making *astronomy* physical meant dealing purely with physical objects – real bodies, real distances and real motions – and establishing mathematical physical laws similar to this inverse square law of light. In 1609, Kepler published the results of this project in a book proudly titled *Astronomia Nova*, and introduced the idea with the spectacular diagram in Figure 7.5, right. This was a first of its kind: as we saw in previous chapters, astronomical diagrams always displayed the perfect circles to which the changing positions of the planets were reduced. Like no astronomer before him, Kepler was introducing a real, physical trajectory: the path that a planet was actually supposed to travel in the heavens. But it wasn't an actual trajectory: he calculated and plotted on paper the path Mars *would have taken* if Tycho or Ptolemy were correct – if the Earth was stationary – and asked his readers: does this make any sense to you? What seems exceedingly orderly as a set of lines, angles, deferents and epicycles turns out to be unbearably complex – "a pretzel," Kepler calls it – when thought of as real motion. No mechanism can produce such a convoluted motion (remember that Tycho showed that there could be no real spheres in the heavens); even if the planets were smart enough to navigate on their own, there is no point of reference for them (remember that the real center is empty); and there's no space left for any other planet under the fixed stars (represented in Kepler's diagram by the outer circle of the signs).

From a practicing astronomer's point of view, a physically sound astronomy meant that all calculations should have the real Sun as their point of reference. This wasn't how it used to be done: astronomers from Ptolemy to Copernicus to Tycho would calculate where the Earth (and later the Sun) *would have been* had it been properly stationary in the heavens, and relate the planetary positions to this fictional point – the 'mean earth' (or 'mean sun' for the Copernicans). Kepler would have none of this: whatever it was in the Sun that caused the planets to move around it, it was *the real Sun* that caused it, not a calculated one. All angles and motions should therefore refer to the real position of the Sun, as established from observations. This was a difficult demand. There was no precedent for such practice, and when he tried to convert Tycho's model to a straightforward Copernican model (rejecting Tycho's compromise) with the real Sun, Kepler could not make the data fit. Yet one small, unexpected benefit of this way of calculation convinced Kepler that he was on the right track: Mars was known for its 'libration'; a 2° oscillation around the ecliptic. When his orbit was calculated around the real Sun, the libration disappeared – it turned out to be an artefact of the fact that Mars' orbit was oblique to the Earth's.

Kepler then did something that for a committed Copernican may seem very strange: he calculated the equant point of Mars' orbit. Copernicus, you may remember, rejected equants vehemently: they were meaningless points introduced only for convenience. But Kepler was already asking of astronomy a different question: not to assume and produce perfect circular order, but to establish a mathematical order that would make physical sense. With this question in mind, the equant became very significant – a point around which a real physical process was happening. Kepler's next move demonstrates this line of thought: he tried to put the Sun at this very point. He didn't know, he admitted, what exactly caused the planets to move (he had some speculations, to which we'll return below), but since they moved around the Sun, it was almost certain that the Sun was the main cause of their motion. And if the Sun was what moved the planets, it was reasonable to assume that they moved uniformly around it. Still – the data didn't fit.

But Kepler again made an unexpected discovery: when Mars was closest to the Sun (in its *perihelion* – remember that the orbits were eccentric), its velocity was highest, and when farthest (*aphelion*) – slowest. This suggested to him what he called "the vicarious hypothesis": that the planets' velocity is inversely proportional to their distance from the Sun. It made good sense when thinking about the Sun as moving the planets – the closer they were to the Sun, the more powerful was its influence and vice versa. But again, the hypothesis didn't fit the data and, again, an unexpected discovery suggested itself: although a straight inverse proportion didn't work, a somewhat more complex proportion did. *The line between the Sun and the planet (the 'radius vector') covered equal areas in equal times* (Figure 7.6).

This is what came to be called 'Kepler's Second Law' – the law of areas – though he had established it first. We are bound by traditional historians' titles, but for Kepler it wasn't really a law of nature, but a way to approximate orderly planetary motion, similar to the approximations used by traditional astronomers. Yet he was clearly abandoning their ancient rules: it was no longer simple uniformity of angular velocity that he was seeking, but uniformity in a physical relation to the Sun. And from the Area Law, an

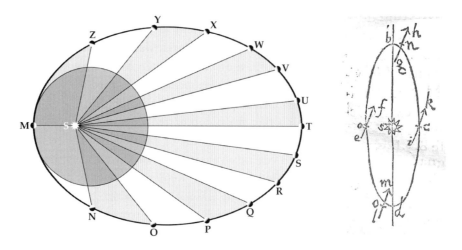

Figure 7.6 Kepler's two 'laws' formulated in his 1609 *Astronomia Nova*. The first in order of discovery – the Area Law (known as the Second Law): *The Radius Vector sweeps areas proportional to time.* That is: if the times by which the planet travels from N to P, P to Q, Q to R, etc. are equal, the areas of the curved triangles SNP, SPQ and SQR will be equal. When he first formulated it, Kepler was still thinking in terms of a circular orbit with the Sun just off-center. His failure to make the data fit well enough into the Area Law convinced him that the orbit must be 'squeezed' – an oval. This led to the so-called First Law: *The planet moves in an ellipse with the Sun in one focus.* The ellipse is the curve created by a point moving around two foci, with the sum of its distances from both being constant. The circle, then, can be thought of as a special case of an ellipse, in which the distance between the two foci is zero. In 1619, in the *Harmonices Mundi*, Kepler would add the 'Harmonic Law': *The squares of the planets' Periods are Proportional to the Cubes of their Radii.* On the right is Kepler's illustration of his physical hypothesis of the motion of the planets and the reason why they converge to and rescind from the Sun, creating the ellipse (which he draws quite exaggerated). The arrows designate the magnetic polarity of the planets, which causes them to be intermittently attracted and repelled from the Sun.

even more spectacular break with those rules followed. Near the aphelion and perihelion (the apsides), the law indeed worked very well, but away from them, it was as if the orbit needed to be 'squeezed' for the areas to fit. Kepler later narrated that he stared at his calculations for years until it dawned on him what he was seeing: the orbit was *not* circular! It was oval! Maybe an ellipse? For a skilled seventeenth-century mathematician, the ellipse was a clear 'second best' after the circle in terms of simplicity and orderliness. He would know well the geometry of conic sections: the circle is the result of slicing a cone parallel to its base; an ellipse – of slicing it obliquely (if the section was parallel to the side, the result is a parabola; still more oblique – hyperbola).

So Kepler tried to construct a model of an elliptical orbit with the Sun at its center, and the fit was still not satisfactory. He moved the Sun to the ellipse's focus – the next point of significance – and here it was, Kepler's (so-called) First Law: *The planet moves in ellipse with the Sun in one of its foci* (Figure 7.6). Some ten years after the *Astronomia Nova*, in his *Harmonices Mundi* (*Harmonies of the World*) of 1619, Kepler would add his Third Law – again, a mathematical law relating physical motions and distances: *The squares of the planets' Periods are Proportional to the Cubes of their Radii.*

It's hard to overstate the novelty of Kepler's law of ellipse. We saw in Chapter 3 that the idea that the Earth moves had been entertained – albeit rejected – in the past, but no astronomer ever thought that the heavenly bodies moved any other way but circularly (or spherically). This circularity reflected the eternal and unchanging character of the heavens and their strict distinction from Earth and the terrestrial realm, a distinction assumed by pagans and Christians alike; it was ingrained in cosmological, natural-philosophical and theological beliefs; and it was embedded in all astronomical practices. Kepler's account of how difficult it was for him to recognize that his model called for an oval orbit demonstrates that: although he was consciously and explicitly searching for a new astronomy, it was nearly impossible for him to turn away from the most fundamental assumption of the old one.

With his laws, and especially the Law of the Ellipse, Kepler reshaped astronomy. The heavens were now no longer exempt from physical laws and mathematics was no longer shunned from dealing with real physical bodies. Kepler willingly admitted that he didn't know how this physics worked. What was the *virtus motrix* – the force from the Sun moving the planets? Could it simply be light? Mathematical considerations led him to reject this idea: light expands three-dimensionally and declines (we saw) by the square of the radius, whereas all the planets are on one plane, so the motive force seems to expand only two-dimensionally and decreases by some other law. A book published in 1600 – *De Magnete* (*On the Magnet*) – provided another hypothesis. In it, the English physician William Gilbert (1544–1603) suggested that the Earth was a loadstone, a huge magnet, an idea he supported with many experiments and explanations of curious phenomena. If the Sun can also be thought of as a magnet, it could be that as it rotates it pulls the planets along. When the poles of the two grand magnets are opposite (north pointing to south and vice versa), the planet is drawn nearer the Sun; when they align (north to north), it is repelled (Figure 7.6, right). This hypothesis proved a dead end, but the way to a physics of the heavens was paved.

Galileo and the Telescope

The Telescope

In the May of 1609, as Kepler was completing his *Astronomia Nova*, a mathematics lecturer at the University of Padua (then belonging to the Republic of Venice) by the name of Galileo Galilei (1564–1642) heard that one Hans Lipperhey had applied for a patent for a so-called 'spyglass.' It was an instrument comprising a cardboard tube with a convex lens on one side and a concave one on the other, which, when put to the eye, magnified the image seen through it three times. The application, in the Netherlands the previous October, was rejected. By June, Galileo had managed to copy the invention. By October, he had an eight-factor instrument, and towards the winter a twenty-factor, which he proceeded to aim at the Moon. He didn't know that in England Thomas Harriot (1560–1621) was doing exactly the same thing, but he must have guessed that he was not the only one thinking along these lines, because he worked obsessively, and in June 1610 published the *Sidereus Nuncius* (the title's meaning is purposefully ambiguous, between *Starry Messenger* and *Message*).

Copernicus suggested that the Earth moved; Tycho claimed that the heavens changed; Kepler reshaped optics as the physics of light and astronomy as the physics of planets – but they all did so in esoteric publications, aimed at scholars and experts. Galileo's book was marvelously different: one didn't need to know any natural philosophy or mathematics to understand it; all one needed was to look and see (Figure 7.7): the Moon was just like the Earth. It was a whole sphere, had mountains and seas, and – as both Galileo and Kepler assumed – inhabitants. There were also very many more stars in the heavens – the Milky Way comprised such stars – and one more world-changing discovery that the book announced, to which we'll return momentarily.

Climbing the Walks of Life

Sidereus Nuncius delivered the instant success that Galileo was craving. That a 45-year-old professor would spend his days grinding lenses and his nights observing stars seeking such fame and fortune testifies again to the Renaissance culture of knowledge, in which a skilled, talented, ambitious, brazen individual could dramatically improve his social and economic status, especially if he could bridge the social and epistemological gap between *knowing-how* and *knowing-that*. Brunelleschi and Tartaglia were our examples in Chapter 5; Galileo's father Vincenzo (c. 1520–1591) is another excellent one. A lute player of humble origins, he taught himself

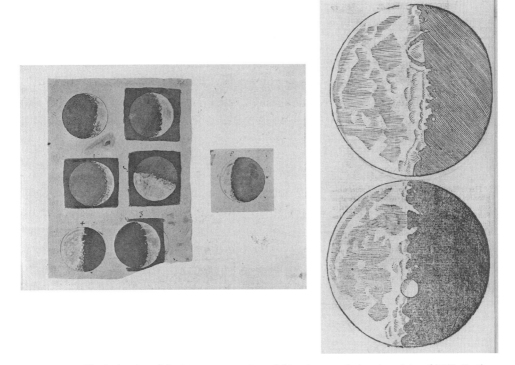

Figure 7.7 Galileo's drawing of the Moon as seen through his telescope during the winter of 1609. On the left: an aquatint engraving of his drafts. On the right: the images published in the *Sidereus Nuncius*. Note his sophisticated use of black-and-white shading (chiaroscuro) to create a sense of three-dimensionality.

musical theory and acquired a name and enough fortune to marry into minor (and penniless) nobility with a book on musical theory, in which he relentlessly and venomously assaulted the canonical theory as detached from real musical practices.

Galileo thus inherited a better starting point in life than his father, but also his ambitions, drive and artisanal skills (and a considerable musical talent). Born near Pisa before his father's career moved the family to Florence, he returned to the University of Pisa to study medicine, but stayed in the lower faculties and taught mathematics there from 1589. Two years later, he moved to Padua, his salary rising slightly from 160 Scudi to 160 Ducats a year. In 1599, he invented a military compass and dedicated it to the Venetian Senate to have his salary doubled and his contract extended for six years. When Paolo Sarpi (1552–1623), Galileo's friend and minor patron, arranged for the spyglass to be presented and dedicated to the Senate in 1609, Galileo's salary was doubled again and he was tenured for life.

This was not the last step in Galileo's climb up the walks of life, using his technical skills, academic knowledge and a smart maneuvering of the cultural structures sanctioning them: his next move would take him out of the university. Through all of his promotions, his academic position remained in Padua as it was in his early days in Pisa: a lecturer in astronomy and mathematics; bound to the lower faculties, prohibited from voicing ideas about the structure of the cosmos. This was the institutional side of the Thomistic compromise (see Chapter 5) that allowed the Church to continue sponsoring natural philosophy and the university: as long as the disciplines and faculties were kept apart – cosmology and theology untouched by what was discussed in astronomy and natural philosophy – all could continue uninterrupted. But Galileo was much too ambitious to accept these limitations – like Kepler, he had in mind a new understanding of the world, commensurate with his mathematical skills and talents, which we'll discuss in Chapter 9. To realize that vision, to become *a philosopher*, he needed someone to support him, financially and politically; someone who would provide the intellectual freedom that the university did with its *universitas* status and the institutions of *disputatio* and *ex hypothesi*, but without the boundaries by which the university protected them (Chapter 4). In late Renaissance terms, he had to find a powerful patron.

The exchange between the patron-prince and his (more rarely, her) client – humanist, artist or natural philosopher – was quite straightforward: the former provided protection and support and the latter – glorification. But these mercenary relations were clothed in many rituals and careful gestures; the client had to submit himself to the patron and offer some gift; but what can one offer the prince who, by his very station, has everything? A moonless period during his arduous observations provided Galileo with a solution. Not wanting to waste the clear nights, he started watching Jupiter, then in retrograde, hence bright and full (it is a good exercise in traditional astronomy to work out why this has to be the case). He reports how he delightedly saw three fixed stars – only observable through his telescope – nicely aligned to Jupiter's east (Figure 7.8 – the diagram on the top left under the second line). Observing the little constellation the following day, he found that Jupiter had moved to their east, and four stars were now visible (the diagram to the left of the first one). But this was strange – in retrograde Jupiter should have been moving west. After convincing himself that neither he nor his ephemerides were wrong, he continued watching, until it dawned on him: it was not Jupiter moving against the backdrop of faraway fixed stars, but small bodies moving around Jupiter – Jupiter had moons!

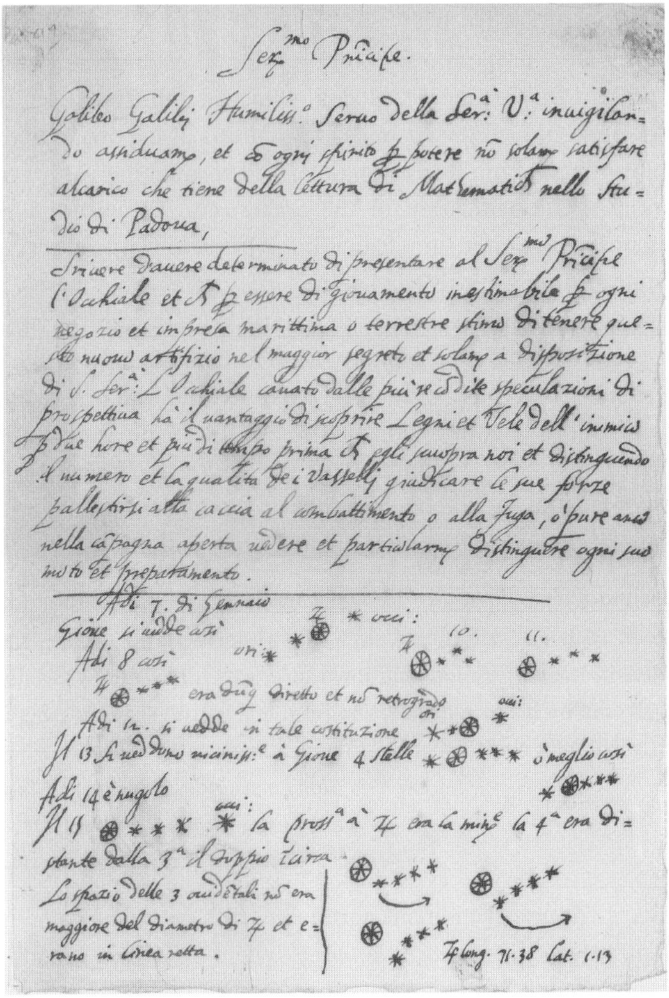

Figure 7.8 At the bottom of the page, under the line, are Galileo's notes from around the turn of 1609, documenting his discovery of the moons of Jupiter. Note how they begin (third row on the left) with an attractively simple pattern – a line – that draws Galileo's attention, become confusing over the following night, and then regain order as he understands that it is these bodies that move in relation to Jupiter, not the other way around. As is very rarely the case, the writings above the line are almost as momentous: they comprise Galileo's draft for a letter he sent on August 24, 1609, presenting his 'occhiale' to the Doge of Venice. There is no mention of astronomical observations in the letter:
"most Serene Prince, Galileo Galilei most humbly prostrates himself before Your Highness, watching carefully, and with all spirit of willingness, not only to satisfy what concerns the reading of mathematics in the study of Padua, but to write of having decided to present to Your Highness a telescope that will be a great help in maritime and land enterprises. I assure you I shall keep this new invention a great secret and show it only to Your Highness. The telescope was made for the most accurate study of distances. This telescope has the advantage of discovering the ships of the enemy two hours before they can be seen with the natural vision and to distinguish the number and quality of the ships and to judge their strength and be ready to chase them, to fight them, or to flee from them; or, in the open country to see all details and to distinguish every movement and preparation."

The Copernican Tribute

Moons were a proper present for a prince. These moons were a particularly apt present for Cosimo de' Medici, who had Jupiter written into his family myth, who enjoyed the cosmic connotations of his own name, and who had four brothers whom he very much liked to imagine as orbiting him. Sarpi negotiated the dedication of the *Sidereus* to Cosimo and the naming of the moons of Jupiter the Medici Stars. In turn, Galileo received his wish: a position for life as the Medici court *philosopher* – not merely astronomer or mathematician – and a very hefty salary. With the confidence and authority that came with his new position, Galileo embarked on a journey of professional success. Kepler, still imperial astronomer and before actually looking through the telescope, published his enthusiastic support in the form of a short book – *Conversations with the Sidereus Nuncius* – embracing the telescope as an invaluable addition to the project of establishing physical astronomy supported by instrumental observations:

> Therefore let Galileo take his stand by Kepler's side. Let the former observe the moon with his face turned skyward, while the latter studies the sun [with his *camera obscura*] by looking down at a screen (lest the lens injure his eye). Let each employ his own device, and from this partnership may there some day arise an absolutely perfect theory of the distances.
>
> *Kepler's Conversation with Galileo's Sidereal Messenger*,
> Edward Rosen (trans. and ann.) (New York: Johnson
> Reprint Corporation, 1965 [1610]), p. 22

Galileo's triumphal moment continued with a trip to Rome in the summer of 1610 to visit the *Collegio Romano*, where he gained the full support of its illustrious leading astronomer, Christopher Clavius (see Chapter 5). He was then received by Prince Federico Cesi (1585–1630) and elected a member of his *Accademia dei Lincei* (more about academies in Chapter 9), which from then on dedicated itself to promoting Galileo's work.

No one could observe the Earth move, but the *Sidereus* provided, for the first time, powerful empirical evidence – which anyone could see with their own eyes – that it was a planet like any other. The Moon looked like the Earth, and there was at least one more planet sporting moons; the Earth clearly wasn't the only possible center of motion in the cosmos. And the telescope would soon provide more evidence. Later in 1610, Galileo observed that Saturn had moons as well (in 1616, he would decide it was a ring) and in December – that Venus exhibited the whole range of phases, like the Moon, and as Copernicanism predicted (see Figure 7.9 for an explanation of why). In 1612, it turned out that even the purported new center of the world – the Sun – was neither perfect nor static: turning the

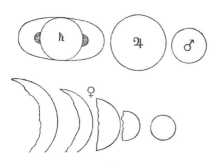
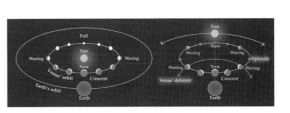

Figure 7.9 The phases of Venus. On the left: Galileo's drawing of the planets as observed through the telescope – among them Saturn with its ring – from his 1623 *Il Saggiatore* (*The Assayer*), a polemical text in which Galileo argues for the complete superiority of telescopic observation to the naked eye, despite the additional mediation that it presents between observer and object. Venus, at the bottom, is shown in a full set of phases, like the Moon, and the diagram on the right explains the importance of this observation as evidence for Copernicanism: if the Sun and Venus orbit the Earth, as in the top diagram, and their orbits just happen to synchronize, Venus should always appear as a crescent, because the Sun should always light it from behind (at most slightly obliquely). If, however, the Earth and Venus orbit the Sun and Venus always appears close to the Sun because our orbit is further away, as in Figure 7.2, Venus should present the whole spectrum of phases: from the new crescent, when it's between us and the Sun; through half spherical, when we see it at 90° to the Sun; to full, when it's on the other side of the Sun.

telescope "down at a screen (lest the lens injure his eye)," Galileo observed the Sun's moving spots. The Jesuit Christoph Scheiner (1573–1650), who had already observed them, boldly claimed the spots were small planets orbiting close to the Sun. Galileo, however, convincingly argued that they were *on* the Sun: they changed their shape and moved slower when near the edges. This suggested that they were moving with the spherical surface of the Sun, traveling in a near-straight line when close to the center but approaching or receding from us when closer to the sides. Even the Sun was a purely physical body, best understood when looked at with an artisan's eye: three-dimensional, spotted and rotating around its poles.

The Galileo Affair: The Church Divorces Science

Yet as Galileo's fame – and that of the type of science he represented – was growing, well beyond the bounds of the scholarly world, trouble was looming. In 1611, gossip and allegations began that he was teaching that the Earth moves, against the word of the scriptures; in 1614, he was denounced from the pulpit in Florence; and in 1615 the Inquisition initiated an inquiry into the question.

Copernicus' hypothesis had been public knowledge for a century; it was formulated by a devout clergyman, dedicated to the pope, and had been a matter of respectable debate within Church-sponsored universities: what changed to make it a cause for theological anxiety?

The answer is complex. We've discussed the reasons as to why monotheistic believers were comfortable with the notion that the Earth is static in the middle of the cosmos, even as it came from a pagan like Aristotle. The Bible, though containing no astronomy, does present a handful of verses that could easily be read this way (*Genesis* 1:14–18; Psalm 104:5; Job 26:7; Isaiah 40:22; Joshua 10, where Joshua makes the Sun stand still, was popular sermon material). Copernicans were happy to offer interpretations of these verses which allowed for the motion of the Earth – primarily suggesting that they were written to suit the human perspective, and in a different era this may have been acceptable, because Catholic scholars always took pride in their capacity for inspired interpretation. But the defensive orthodoxy instigated by the Reformation made this difficult: the Council of Trent, which had concluded fifty years earlier, specifically prohibited "distorting the Scriptures in accordance with [one's] own conceptions ... interpreting them contrary to that sense which the Holy Mother Church ... has held or holds." Moreover, enthusiasts like Bruno certainly alarmed the Church, suggesting that Copernicanism might develop into a new heresy. It also must have registered with Church authorities that Galileo, by nature and new status, was not restricted by scholarly humility. As a courtier, he was expected to be boisterous and seek attention with controversy and spectacular claims. These he increasingly made in Italian rather than Latin, making them available to the common folk, whose hesitations about issues of cosmology, Biblical interpretation and religious authority the Church could ill-afford. But perhaps what was making Copernicanism unacceptable when presented by Galileo was the tremendous power of the visual. With the telescope, the motion of the Earth and the earthliness of the planets were no longer issues solely for astronomers and enthusiasts like Bruno: they were there for everyone to see.

The First Procedure: Reason vs. Revolution

The investigation of the Galileo affair was charged to Cardinal Roberto Bellarmine (1542–1621) – a Jesuit, an important theologian and the person who had condemned Bruno to the stake some fifteen years earlier. Perhaps still traumatized by the outcome of that procedure – the Church was loath to execute its clergy – he handled it delicately. When it was completed, some eighteen months later, Galileo requested and received an official declaration that he was neither suspected nor indicted of any wrongdoings, but the (indirect) correspondence between him and Bellarmine demonstrates how far the New Science had veered away from traditional scholarship.

"It seems to me that Your Paternity and Mr. Galileo," Bellarmine wrote to Paolo Foscarini, Galileo's ally,

> Are proceeding prudently by limiting [yourself] to speaking suppositionally and not absolutely, as I have always believed Copernicus spoke. For there is no danger in saying that, by assuming the earth moves and the sun stands still, one saves all the appearances better ... and that is sufficient for the mathematician.
>
> Finocchiaro, *The Galileo Affair*, p. 67

We can recognize what Bellarmine had in mind: the institution of arguing *ex suppositione* and Osiander's preface to *De Revolutionibus*. But whether he was genuinely commending Galileo and Foscarini or gently threatening them, Bellarmine was wrong: the Copernicans of Kepler and Galileo's generation were no longer willing to limit themselves to the traditional role of the mere mathematician, and Galileo's answer was that of a physicist:

> ... since we are dealing with the motion or stability of the earth or of the sun, we are in a dilemma of contradictory propositions (one of which has to be true), and we cannot in any way resort to saying that perhaps it is neither this way nor that way. Now, if the earth's stability and the sun's motion are de facto physically true and the contrary is absurd, how can one reasonably say that a false view agrees better than the true one with the phenomena?
>
> Finocchiaro, *The Galileo Affair*, p. 75

For Galileo, the Earth either moved or did not, and if the evidence was for the former possibility, it made no sense to hold the latter.

Bellarmine and Galileo agreed that the scriptures were beyond doubt and that "it is not the same to show that one can save the appearances with the earth's motion ... and to demonstrate that these hypotheses are really true in nature" (this is the way Galileo put it! – Finocchiaro, The *Galileo Affair*, p. 85). Their conclusions, however, differed in a way that would turn out dangerously, despite the inquisitor's goodwill.

Bellarmine didn't force Galileo to give up Copernicanism or to adopt a literal interpretation of the relevant verses. He even carefully considered the Copernican arguments for the possibility that we move without experiencing it, although he found them unconvincing. What Bellarmine demanded of Galileo was to take into consideration what "a dangerous thing" it would be to play around with scriptural interpretations, and "whether The Church can tolerate giving Scripture a meaning contrary to [that of] the Holy

Fathers and to all the Greek and Latin commentators," who are "all agreeing in the literal interpretation that the sun is in heaven and turns around the earth with great speed, and that the earth is very far from heaven and sits motionless at the center of the world" (Finocchiaro, *The Galileo Affair*, p. 67). A good scholar, in Bellarmine's view, would take it all into account – the theological and political implications as well as the intellectual price: to refute 1,500 years of arguments would require an extremely powerful counter-argument. Still, Bellarmine added:

> If there were a true demonstration that ... the earth circles the sun, then one would have to proceed with great care in explaining the Scriptures that appear contrary, and say rather that we do not understand than that what is demonstrated is false.
>
> Finocchiaro, *The Galileo Affair*, p. 67

If Copernicanism can be demonstrated powerfully enough, Bellarmine stressed, the Church would have to reinterpret the scriptures after all.

This is the most telling of Bellarmine's notes. He was the inquisitor, and had no need to cater to Galileo or offer any kind of concession to Copernicanism. That he did, shows that for him – and the Church – the affair was neither about defending dogma nor enforcing authority, as the common myth has it. Rather, Bellarmine was defending *reason*. From his point of view, Galileo was another Bruno: an enthusiast in a time of many heresies, unwilling or incapable to consider reasoned arguments, completely unwavering in his conviction that his is the only way to *Truth*. This was what Bellarmine emphasized: the scriptures are divine, but interpretation is human, so if reason truly and forcefully points that way, the Church would reinterpret the problematic verses to align with the new scientific claims. What it *wouldn't* do, is "say ... that what is demonstrated is false," because, as Augustine put it, "in [reason] we are made unto the image of God."

Bellarmine was no wide-eyed champion of humanist values. He was a powerful emissary of a domineering institution, and he wasn't defending only human reason, but also the Church's privilege to represent it. He wasn't only stressing that the Church would abide by a "true demonstration," but also that it retained the right to decide what the criteria for such a demonstration were, and when they're met. But neither was Bellarmine completely wrong in his assessment of Galileo, who had only this terse retort about truth:

> Nor can one or should one seek any greater truth in a position than that it corresponds with all particular appearances.
>
> Finocchiaro, *The Galileo Affair*, p. 85

"Appearances," for Galileo, meant quite simply what could be observed through his telescope: the Truth, for him, was what *he* saw. And what about those verses? "The motion of the earth and stability of the sun could never be against Faith or Holy Scripture," Galileo says, "if this proposition were correctly proved to be physically true by philosophers, astronomers and mathematicians." The significance of Galileo's claim is clear and outrageous: it is him and his allies, the "philosophers, astronomers and mathematicians," and they alone, who will decide the truth of Copernicanism, and then,

> If some passages of Scripture were to sound contrary, we would have to say that this is due to the weakness of our mind, which is unable to grasp the true meaning of The Scripture in this particular case.
>
> Finocchiaro, *The Galileo Affair*, p. 81

Bellarmine's worries were very well-founded: there was no reasoning with Galileo. He was a revolutionary.

Galileo's Trial

Although Galileo received the clearance he requested, the relations between the Church and Copernicanism – and by extension, the New Science – didn't emerge from the 1615–1616 investigation unscathed. The mounting pressure from below forced the Inquisition to set, in February 1616, a committee of theologians who deliberated for less than a week before unanimously concluding that the motion of the Earth is "foolish and absurd in philosophy, and formally heretical since it explicitly contradicts in many places the sense of Holy Scripture." The unrest was disturbing enough that the pope himself instructed Bellarmine to summon Galileo and instruct him "to abandon completely the ... opinion that the sun stands still at the center of the world and the earth moves, and henceforth not to hold, teach, or defend it in any way whatever, either orally or in writing" (Finocchiaro, *The Galileo Affair*, p. 146). All books concerning Copernicanism were banned, and *De Revolutionibus*, too practically valuable for this harsh treatment, was sent to be 'corrected,' while its uncorrected version remained on the index of prohibited books until 1758.

One would have thought the message was as clear as could be, but fifteen years later, Galileo, now in his late 60s and by far the most famous scientist in Europe, decided to invest all his reputation and authority into an unmitigated support of the Copernican hypothesis. In 1630, he completed, in Italian, *The Dialogue concerning the Two Chief World Systems, Ptolemaic and Copernican*, in which his own Copernican views were presented by the character of the wise *Salviati*; the intelligent challenges and evidence provided by the clever and attentive *Sagredo* (both names were tributes to real friends); and the Aristotelian counter-arguments, including those favored by the pope, were delegated to the mouth of *Simplicio*, whose name suggests the level of sophistication that Galileo allotted him.

What could have made Galileo blatantly defy an unequivocal episcopal ruling? He may have become entangled in – or thought he could maneuver – the delicate legal distinctions between the prohibition to believe, to argue for and to teach a position. He would later claim that he understood the 1616 decree as disallowing only the former but not the latter, that he forgot the original formulation and that he intended to give all sides their due, none of which is convincing – nor was to his prosecutors. Or he may have thought that his fame, the Medicis' patronage and his friendliness with the Florentine cardinal Mafeo Barberini, now Pope Urban VIII, made him untouchable. If that were the case, he was politically naïve. The Medicis didn't, and couldn't have been expected to, take on the Holy See to save one client, venerable as he was, and the pope's friendship couldn't have survived Galileo's mockery. As a supreme political leader, he simply couldn't afford it. Most pertinent to our interests, Galileo may have thought that he found the "positive demonstration" Bellarmine was demanding: his 'theory of the tides'; the idea that it is the rotation of the Earth that makes the oceans rise and fall by (what would be called later) centrifugal force. This idea would remain popular with Copernicans for a generation, until the advent of gravitation theories, but never made much impression on anti-Copernicans.

Whatever the reasons for Galileo's confidence, it was clearly ill-founded; indeed, many of his long-time patrons and advisors – Bellarmine, Sarpi, Cesi, Cosimo – were no longer alive. The publication of the *Dialogue* immediately met with great difficulty – it could not be published in Rome, under the auspices of the *Lincei*, and it took declarations of questionable veracity to enable it to pass the Inquisition censor in Florence. When it *was* finally published, in February 1632, it was enthusiastically received by philosophical peers and disciples, but complaints immediately started piling in. Within months its sale was banned – first temporarily, as a special committee of experts was set again and the case was transferred to the Inquisition. In September it met, presided over by the pope, and decided to summon Galileo to Rome to stand a real trial. He tried to evade the summons, but was finally forced down in January 1633. After the first interrogation, he reached a deal which didn't satisfy the pope and was interrogated again – this time with torture instruments presented to him (the second stage of torture use, the first being mentioning them, the third – using). Unsurprisingly, he agreed to confess, and on this basis was found "vehemently suspected of heresy" – strong enough to sentence him to lifetime imprisonment, soon to be mitigated to house arrest – but not, as full heresy would have mandated, to send him to the stake.

Conclusion

The story of Galileo's trial plays an extremely important role in the way modern science presents itself to itself as well as to wider culture. For centuries, the image of the brave scientist wielding reason against dogmatic political power has served to initiate students and to argue the cause of many modern values. As such, it has raised great curiosity among more sophisticated historians, keen to reveal the actual story behind the myth and unearth the making of this crucial moment in the history of science in particular and modernity in general. Let's quickly recount some of their insights.

The trial was a series of miscalculations on both sides. Galileo forced a substantial inquiry into the content of the book – which was doomed since it had to be handed over to the theologians – when it could have been resolved as a procedural lapse by obtaining proper authorization from the censor. Indeed, it perhaps should not be looked at primarily through a legal lens: once he left the university to join the court, Galileo subjected himself to its rituals and theatrics, and 'the fall of the favorite' was a script he was bound to be subjected to, especially at the pinnacle of his fame. This does not mean that the issues were not serious: Galileo's challenges to orthodoxy were much wider and more complex than the motion of the Earth. His matter theory made the Eucharist difficult to conceive of, and worse: the idea of the maculate Moon interfered with the worship of Mary. In this popular cult, the perfection of the Moon was allegorically identified with Mary's immaculate virginity, and Galileo's allies weren't shy of flaunting the discrepancy anymore than he was hesitant in arguing over Biblical interpretation with Church theologians. See (Figure 7.10) how Lodovico Cardi (known as Cigoli, 1559–1613), the friend who first warned Galileo of the gathering discontent, installed the spotted Moon of the *Sidereus* drawings (see especially '1' in Figure 7.7, left), where every viewer at the time would have expected the brilliant, silvery white befitting Mary's flawless divinity.

Finally, historians have shown that Galileo found himself embroiled in a political, educational and theological struggle in which he was far from the main actor. Truly at stake was the enormous popularity of the Jesuit

Figure 7.10 *The Assumption of the Virgin*; a 1612 fresco at the Cappella Paolina at the Basilica di Santa Maria Maggiore in Rome by Galileo's friend Lodovico Cardi (commonly known as Cigoli). Note the Moon under Mary's feet: not only is it "maculate," but it's a defiant copy of Galileo's Moon's drawing from the *Sidereus Nuncius*. Cigoli, a fellow Tuscan, was an important ally to Galileo in the following year, during the early stages of the affair that would culminate in his trial two decades later, reporting to him about the dangerous opposition to his Copernicanism developing in their home region and the people and intrigues involved.

Order and the price it was willing to pay for this popularity. In the eyes of the older, more orthodox orders, especially the Dominicans, the Jesuits were compromising core Christian beliefs and practices. Their adoption of the New Science was in line with their adoption of Mandarin attire in China or shaman rituals in the Amazon, all much too close to outright heresy. Being wrapped up in a powerful educational program (see Chapter 5) made these habits particularly dangerous. Galileo's relations with the Jesuits saw ups and many downs, but at that moment what determined his personal (mis-) fortune was his affiliation with the losing side in the grand struggle within the Catholic Church of the Counter-Reformation.

Galileo's personal lot ended up less than horrendous. He spent the last decade of his long life in the Medicis' Villa Arcetri near Florence, playing the lute as he gradually lost his eyesight. He still worked, and in 1638 published – in Leiden this time – his enormously impactful *Discourses Relating to Two New Sciences* (whose main teachings will be discussed in Chapter 9). In the story of the construction of the cathedral which is science, however, his trial was as dramatic a chapter as any: it signaled the final collapse of the Thomistic synthesis. For centuries, the Catholic Church maintained its position as the leading European institution of knowledge through that delicate intellectual and institutional compromise between monotheism and pagan science. This compromise allowed it to sponsor the study of the scriptures on the one hand and the Book of Nature on the other as complementary parts of one grand project. Already put on the defensive by the Reformation, the tensions built into this compromise could not withstand the pressure of the great novelties of the New Science and the radical, uncompromising approach of its practitioners. There would still be Christian scientists and science would still be used for (and against) Christian belief, but by prosecuting the most famous philosopher of its time, the Church determined and declared that there would no longer be Christian science.

Discussion Questions

1. Where do you think the most dramatic departure from traditional astronomy occurs? With Copernicus declaring that the Earth moves? With Tycho demonstrating that there is change in the heavens? With Kepler calculating that the planetary orbits are not round? With Galileo showing mountains of the Moon?
2. What can one make of the underlying Aristotelianism of Copernicus or the theoretical conservatism of Tycho? Are they agents of change or revivers of ancient traditions?
3. Compare the sketch of Galileo's and Kepler's lives to that of people of earlier eras – from Chapters 4 or 2, for example. Is there an interesting lesson to be learnt from the differences? Does the difference in biographies have bearings on how we should look at the knowledge they produced?
4. Astronomers always used instruments. Is there something unique to Kepler's camera obscura and Galileo's telescope, or are they just incremental improvements?
5. Looking at the details of Galileo's trial – does it still seem like a momentous event in the making of modern science or just a series of miscalculations and political intrigues? Or are the two interpretations contradictory?

Suggested Readings

Primary Texts

Copernicus, Nicolaus, "Commentariolus," in Edward Rosen (trans. and ed.), *Three Copernican Treatises* (New York: Octagon Books, 1971), pp. 55–65.

Kepler, Johannes, *Somnium (The Dream)*, Edward Rosen (trans. and ed.) (Madison, WI: University of Wisconsin Press, 1967), pp. 11–21 ff.

Galileo, Galilei and Roberto Bellarmine, "Correspondence" and "Considerations Concerning the Copernican Hypothesis," in Maurice A. Finocchiaro (ed.), *The Galileo Affair* (Berkeley, CA: University of California Press, 1999), pp. 67–86.

Secondary Sources

Biographies:

Caspar, Max, *Kepler*, C. Doris Hellman (trans. and ed.) (New York: Dover, 1993).

Heilbron, John L., *Galileo* (Oxford University Press, 2010).

Thoren, Victor E., *The Lord of Uraniborg: A Biography of Tycho Brahe* (Cambridge University Press, 1990).

On the Copernican revolution:

Kuhn, Thomas S., *The Copernican Revolution: Planetary Astronomy in the Development of Western Thought* (Cambridge, MA: Harvard University Press, 1957).

On Tycho and his project:

Christianson, J. R., *On Tycho's Island: Tycho Brahe and His Assistants, 1570–1601* (Cambridge University Press, 2000).

On Rudolf's court:

Evans, Richard J., *Rudolf II and His World* (Oxford: Thames & Hudson, 1997).

On Kepler's defense of Tycho's project:

Jardine, Nicholas, *Birth of History and Philosophy of Science* (Cambridge University Press, 1984).

On Kepler's optics:

Chen-Morris, Raz, *Measuring Shadows: Kepler's Optics of Invisibility* (University Park, PA: Pennsylvania State University Press, 2016).

On Kepler's cosmology:

Field, Judith V., *Kepler's Geometrical Cosmology* (London: Athlone Press, 1988).

Hallyn, Fernand, *The Poetic Structure of the World: Copernicus and Kepler*, D. M. Leslie (trans.) (New York: ZONE BOOKS, 1993).

On Kepler's astronomy:

Stephenson, Bruce, *Kepler's Physical Astronomy* (Princeton University Press, 1994).

On the Accademia dei Lincei:

Freedberg, David, *The Eye of the Lynx: Galileo, His Friends, and the Beginnings of Modern Natural History* (University of Chicago Press, 2002).

On Galileo's telescope:

Van Helden, Albert, "The Telescope in the Seventeenth Century" (1974) 65 *Isis* 38–58.

Reeves, Eileen, *Galileo's Glassworks: The Telescope and the Mirror* (Cambridge, MA: Harvard University Press, 2008).

On Galileo's discoveries:

Reeves, Eileen, *Painting the Heavens: Art and Science in the Age of Galileo* (Princeton University Press, 1997).

On the cultural impact of the visual instruments:

Alpers, Svetlana, *The Art of Describing: Dutch Art in the Seventeenth Century* (University of Chicago Press, 1983).

Clark, Stuart, *Vanities of the Eye: Vision in Early Modern European Culture* (Oxford University Press, 2007).

Edgerton, Samuel Y., Jr., *The Mirror, the Window, and the Telescope: How Renaissance Linear Perspective Changed Our Vision of the Universe* (Ithaca, NY: Cornell University Press, 2009).

Panofsky, Erwin, *Galileo as a Critic of the Arts* (The Hague: Martinus Nijhoff, 1954).

On Galileo's trial:

Biagioli, Mario, *Galileo Courtier: The Practice of Science in the Age of Absolutism* (University of Chicago Press, 1993).

Feldhay, Rivka, *Galileo and the Church* (Cambridge University Press, 1995).

Redondi, Pietro, *Galileo Heretic*, R. Rosenthal (trans.) (Princeton University Press, 1987).

8

Medicine and the Body

Harvey and the Circulation of the Blood

Harvey in Padua and London

In 1599, when Galileo was starting to make a name for himself at the University of Padua, a young Cambridge graduate, William Harvey (1578–1657), arrived there to study medicine. Padua was one of the two most prestigious medical faculties in Europe (the other being Leiden), and known to admit non-Catholic students, providing them with special colleges, so an Englishman aspiring to a medical career in the metropolis was all but obliged to make the journey.

The eldest son of a yeoman and carter from Kent, Harvey represents, much like his senior Tuscan contemporary, the new opportunities for fame and fortune that Europe was now making possible through education. Returning to England in 1602, Harvey set up a successful practice, aided by his marriage to Elizabeth Browne in 1604, daughter of a wealthy and well-connected physician. In 1607, he was elected a fellow of the Royal College of Physicians, by then an almost 100-year-old guild assembling and accrediting mainstream, university-educated doctors; and, in 1615, he was appointed the 'Lumleian Lecturer,' responsible for the annual lecture on anatomy to the College fellows. He carried out the task for the next forty years, until a year before his death, thus wielding enormous influence. This influence extended beyond the medical profession: most of the early members of the Royal Society were physicians, so their concepts of careful, literally hands-on empirical research – the Lumleian Lectures included not only dissections of corpses, but also vivisections (live dissections) – were shaped by Harvey's teaching (in which he indeed profusely praised Bacon, 'Lord Verulam'). In 1618, Harvey became James' court physician, and in 1625 – Charles' (the latter's gruesome death was beyond his vaunted medical prowess). This position took him both around England, on the king's many hunting expeditions, and back for three years through war-devastated Europe: Harvey was keenly aware of death, having seen much of it in his lifetime. The civil war forced him to retreat with the king to Oxford, again putting him in the mentoring role of the would-be Royal Society founders, and with the fall of the monarchy he retired and spent his last years in London. His impact on anatomy and physiology paralleled the impact his senior Paduan, Galileo, had on astronomy and mechanics.

Harvey's Heart and Blood

Harvey's main interest was generation – the process through which parents produce new living creatures or offspring – which he investigated with an inordinate number of observations of eggs and fetuses, though, unimpressed with the new stress on instruments, he very rarely used the microscope. In Harvey's account, the egg is an essential stage in any organism's development, whether plant, or viviparous or oviparous animal. In all, the development of the offspring begins as a shapeless mass in a small vessel – like the egg or the seed – already independent of the parent, but lacking any structure. There is no 'little hen' inside the egg; the evolution of the chick (and any fetus) is a gradual process of structuring: first appears a red point which will become the heart, then a red web which will become the blood vessels and so on.

This was novel and interesting, but not something that Harvey's contemporaries found very difficult to accept: Harvey's account of generation incorporated much of the Aristotelian concept of form operating on matter. Where Harvey found much more resistance and ended up having a much deeper impact was in a little book titled *On the Motion of the Heart and the Blood* (*Exercitatio Anatomica de Motu Cordis et Sanguinis in Animalibus*, 1628). The heart, Harvey claimed there, was but a pump, drawing the blood from the extremities of the body into itself and pushing it outwards again.

According to the common – though contested – theory that Harvey would have studied in Padua, the blood was produced in the liver from ingested food in the form of 'chyle.' It was then propelled through the veins by its *pneuma* – a term referring both to air and 'vital spirits' (not unlike what we mean by calling alcohol 'spirit'). On its way to deliver nutrition to the organs, the blood passed through the right ventricle of the heart, to be heated and loaded with more pneuma. A small portion of it was sent to support the lungs, through which the heart released excess heat and harmful substances and absorbed new pneuma from the outside. When the heart contracted, these 'sooty vapors' were expelled to the lungs and could be exhaled; when it expanded, air was drawn in. The remaining blood passed to the left ventricle through invisible pores in the septum – the fleshy wall separating the two sides of the heart – from which it continued through the arteries to the body, again driven by the vital spirits that also created the arteries' throbbing motion. Although the blood in the arteries was created in the heart enriching the blood arriving through the veins with pneuma, the two kinds of blood were fundamentally different: the venous blood provided nutrition, and the arterial – vitality; and each was flowing from a different organ towards the periphery.

Harvey painted a completely different picture. The blood circulates, he claimed, and not by vital powers it carried, but by the pulsating motion of the heart. When it contracts (systole), the heart drives the blood from the right ventricle through the arteries; when it expands (diastole), it draws it back through the veins to the left ventricle. At the extremities, the 'used' blood moves from the arteries to the veins through minute, invisible capillaries. These capillaries were a bold hypothesis – they were first observed only in 1661, by Marcello Malpighi (1628–1694), who was exploring frogs' lungs using a microscope. There were no pores in the septum, argued Harvey: the blood in the left ventricle is sent through the pulmonary artery to be refurbished in the lungs, returning to the right ventricle through the pulmonary vein.

Harvey's Way into the Body

Harvey's arguments are a fine example of the methods and approaches of the New Science, particularly exciting in their application to the human body. He begins with a quantitative consideration:

> First, the blood is incessantly transmitted by the action of the heart from the vena cava to the arteries in such quantity that it cannot be supplied from the ingesta.
>
> William Harvey, *On the Motion of the Heart and the Blood*,
> Robert Willis and Alexander Bowie (trans.) (London:
> GlobalGrey, 2018 [1618]), ch. 9, p. 46

Harvey actually calculates an estimation of how much blood goes through the heart: he measures the volume of the left ventricle of a cadaver and multiplies it by the number of heartbeats, and concludes that the volume of blood the heart moves in one hour is larger than the volume of the whole body.

To that Harvey adds a series of simple but powerful experimental arguments that he encourages the reader to try at home: "tie the arteries, [and] immediately the parts not only become torpid, and frigid, and look pale, but at length cease even to be nourished." Partially release the knot, and the blood in the thicker and deeper arteries returns to flow, and the veins *under* the ligature swell: apparently, the blood in the periphery is struggling

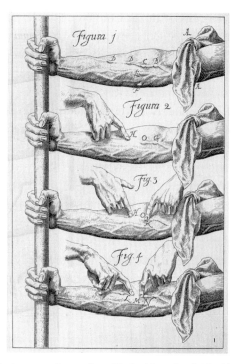

Figure 8.1 Harvey's ligature experiments from his *De Motu Cordis*. The blood in the deeply set arteries continues to flow down (away from the heart), but the blood in the shallower veins stops (Figura 1). The bumps on the veins are unidirectional valves, allowing only an upward flow: if you force the blood down, towards H (Figura 2), the vein under the valve O remains empty. If the blood is forced up, towards O (Fig. 3), and the finger at H is removed, the blood returns to fill the vein. The diagram, which is a clearer version of Harvey's original one but very loyal to the source, is after a drawing by S. Gooden for the Nonesuch edition of *De Motu Cordis* (London, 1928).

to make its way up. Trying it on the arm as in Figure 8.1, one can see that the blood in the veins is only allowed up: the "knots or elevations" (Harvey, *On the Motion of the Heart and the Blood*, p. 66) turn out to be unidirectional valves (recognized but wrongly identified by Harvey's Padua mentor Fabricius, to whom we'll return later). Such valves are not needed in the arteries, through which the blood is forcefully pushed away from the heart and towards the extremities, but are necessary in the veins. First, to ensure that used blood doesn't return to the body, and secondly to ensure that blood doesn't flow from the bigger veins to the smaller ones, rupturing them.

Harvey supports these homespun experiments with many expert observations: dissections of human cadavers and vivisections of animals. He can report on the size, weight and motions of organs; on regular blood flow in vessels and on squirts of blood when they're opened; on the regular pace of the heart and on its change as blood is removed. The body, according to Harvey, is a wonderful natural machine, which can be investigated like any other part of nature. Concerning *the role* of the blood, Harvey can offer little beyond the ancient concepts of vitality and nutrition, and it would take over two centuries for his analysis of the motion of the heart and the blood to make any impact on medical treatment. But neither fact should obscure the radical novelty of his way of analysis and the type of arguments and evidence he marshals in its support. To understand this we need to look at what it was that Harvey studied in Padua.

Harvey's Curriculum

Modern physicians acquire their medical skills at universities, and so did Harvey. But it is not self-evident that this is where such knowledge should be obtained. Originally, the university was a community of learners, committed to abstract learning (Chapter 4). Taking on training professionals in practical disciplines like medicine (and law), the university reshaped itself as an institution and reshaped the relations between 'knowing-how' and 'knowing-that' (Chapter 1).

Medicine had always presented a challenge to the dichotomy between these two types of knowledge. It comprised both the art of healing and the theory of the body, so it was too practical for the pretense of pristine knowledge befitting the free man, and too important and interesting to leave in the hands of the bound: Figure 8.2 illustrates how entrenched yet complicated it was to maintain this distinction. Once adopted by the university, medicine came to express, at least partly, the institutional divisions and hierarchies which we found in natural philosophy and magic. It is true that to be taught, knowledge had to be standardized and written, and what could not tended to disappear; texts incorporated into the curriculum were canonized, and those which weren't were lost more often than not. Yet these texts – first as manuscripts and then as books – were rich with material and practical knowledge that the physician required but that we don't usually associate with the university. Moreover, Padua taught Harvey and his peers how to put their hands on the body: how to open it and what to expect within. Medicine therefore presents an opportunity to add nuance to the relations between *technē* and *episteme* (Chapter 2), whose investment into the cathedral of science has interested us throughout the book.

Figure 8.2 John Banister delivers an anatomy lecture (*The Visceral Lecture*) at the Barber-Surgeons' Hall, London, 1581. The Company of Barber-Surgeons was allowed by Henry VIII to dissect the corpses of four executed criminals a year, and these lectures became compulsory for its members. Banister himself became a member in 1572, and a licensed physician a year later, receiving a medical degree (MB) from Oxford. The painting, which Banister commissioned for the frontispiece of his *Anatomical Tables*, captures well the ambivalence towards the body that made his position – both a practical surgeon and a learned physician – rare and difficult. Standing at the center, he gestures towards the cadaver, perhaps laying his hand on it, perhaps not. The *assistants*, in contrast, definitely *are* touching the dead body: the one in front of the operating table and the one to Banister's left (our right) are holding probes, and the next one to his left is holding a scalpel. The bearded senior colleague on the far right of the painting, on the other hand, definitely isn't touching – he is only pointing. Finally, Banister distances himself somewhat from the gore of the body (note that he also has a bloody probe under his right hand) and celebrates his erudition by directing his pointer at the skeleton with Realdo Colombo's *De Re Anatomica* on the stand, opened at Book 11 – "de Visceralibus."

The Learned Tradition

Hippocrates and the Hippocratic Corpus

The core of the medical canon, like that of the other branches of the learned tradition, lay in texts from the classical Greek period. The fundamental concept of the body that Harvey learned in Padua – and did much to dismantle – was formulated some two millennia earlier, in a corpus of about sixty texts, written around 440–340 BCE and ascribed to Hippocrates of Cos, who allegedly lived around 460–377 BCE. Hippocrates himself may have been only a legend, and the texts – collated around 250 BCE – vary from very theoretical to very practical. Together, however, they present a coherent naturalistic tradition, with a clear conception of the human body, both healthy and sick, of the medicine to sustain it, and of the physician deserving to serve it.

The human body in the Hippocratic Corpus was fluid. It was material, and in line with what the Greeks – especially Plato – thought about matter; it was in constant flux, and disturbingly so: constantly changing, unstable and unruly. The Hippocratic physician had little interest in structures: organs and their distinct function; the different kinds of blood vessels. He was apparently oblivious even to muscles, despite the famous enchantment that their outer appearance had on his contemporary sculptor. For him, the body was made of liquid substances: humors. Their number and descriptions varied from one Hippocratic text to another, but not the fundamental understanding of their import. Life was the flow of these humors: they related the parts of the body to one another and the body to the cosmos – the food it provided, its seasons, its climates and the influences of its celestial bodies. In the healthy body, the flow was balanced, but rarely was it in such harmony. Not only because of its material mutability, but also because, unlike that of other animals, the human body wasn't well suited to life on Earth. It needed to be covered from the cold, its food to be cooked, and its humors regimented; it needed constant medical care, as Plato himself put it:

> Suppose you asked me if it was enough for the body to be the body, or whether it needed something else. I would reply: "It certainly does need something else. That's the reason why the art of medicine has come to be invented, because the body is defective, and therefore not self-sufficient. So the art of medicine was developed to provide it with the things which were good for it."
>
> Plato, *The Republic*, G. R. F. Ferrari (ed.), Tom Griffith
> (trans.) (Cambridge University Press, 2000), p. 17

Hippocratic medicine, then, was the reasoned, organized know-how – the *technē* (Chapter 2) – by which the "defective" body was brought into balance: balance between its humors and between them and the cosmos. This balance was always precarious, dependent on a myriad of varied, personal factors; so the Hippocratic physician, responsible for the physical and mental health of each of his patients, relied much less on medical theory that on information gathered from each one of them. Particular, personal inquiry was necessary: what did the patient eat, how they slept, etc. But the answers weren't sufficient. Patients tended to misremember or simply lie because they were embarrassed or didn't follow the prescribed regimen, and, moreover, they didn't know what happened inside their body. All outer signs of inner processes were therefore crucial: the color of patients' urine and the smell of their feces; the pace and rhythm of their pulse.

The humors were ever flowing and balanced precariously, so Hippocratic physicians would almost invariably take a mild approach to treatment: careful diet; hydrating with mild solutions of barley, honey, vinegar and the like. 'Dietetics,' however, comprised much more than food: health was to be achieved by a regimen of exercise, sleep and sex – balance and self-control befitted and benefitted the body as well as the mind of the free man (Chapter 2). Physicians did know how to invade the body if necessary: the Hippocratic texts explain how to set a broken bone, open and clean a wound or remove a stone with a fine catheter. But aggressive intervention didn't fit their understanding of the way the body worked nor their notion of their role as healers – nicely captured by the epithet 'first, do no harm' (which appears in the text *On Epidemics*, but is not, as is usually believed, part of the famous Hippocratic oath). And yielding a knife on a regular basis also didn't cohere with their idea of themselves as free, scholarly men – it was instead appropriate for slaves or "practitioners," as the oath calls them: "I will not cut persons ... but will leave this to be done by men who are practitioners of this work."

The one Hippocratic remedy that to the modern eye does appear radical is bloodletting. We'll return to this therapeutic procedure below, as it became increasingly popular in the centuries after the Hippocratic Corpus was assembled and remained so for two millennia. What is worth stressing here is that this very popularity – in both learned and folk western medicine and in Chinese and Indian medicine as well (though with different historical trajectory) – suggests how much sense it made. The blood was the only humor that could be approached in a controlled way from outside, so removing some of it from different parts of the body was the only way in which the physician could directly manipulate the quantities of the humors and their distribution throughout the body. This could rid the body of excesses and defects and restore the balance between the rest of the humors. Moreover, bleeding wasn't haphazard: blood was taken from particular points to affect particular parts of the body: "One had to bleed the right elbow for liver pain, and the left for spleen problems, because the vein in the right elbow ran to the liver, whereas that in the left elbow led directly to the spleen" (Kuriayama, *The Expressiveness of the Body*, p. 202).

There was some theory behind the localization: blood vessels (the Hippocratic texts don't differentiate between veins and arteries) carried not only blood but also *pneuma* (the spirit or breath of life mentioned above) through the body and to its organs. But the locations of the bleeding points suggest that they were related less to theory and more to practices of pain relief. A curious and telling fact is that the Hippocratic maps of bleeding points are very similar to Chinese maps of points of acupuncture from the same era (as discussed in Chapter 1, Chinese medicine, like other systems of non-Western knowledge, is regrettably out of our scope). It could be that the relations between the Greek and the Chinese cultures at the time was more intimate than we usually assume. But it's much more likely that the similarity reveals that acupuncture developed from bleeding, while bleeding developed from practical, trial-and-error attempts to relieve pain, a role acupuncture fulfills successfully to this day.

Galen and the Systematization of Medicine

The Hippocratic Corpus seems to have been consolidated in the middle of the third century BCE, probably in the Library of Alexandria, and our knowledge of medical thought and practice in the last Hellenistic centuries before the Christian era is almost exclusively mediated by the reports of later writers. One reason may have been the burning and destruction of the library (Chapter 4). Less dramatically, it was probably the towering figure of Galen (Claudius Aelius Galenus, c. 130–c. 200/c. 216) who made his predecessors appear obsolete to his successors, in much the same way Aristotle, some five centuries earlier, shaped subsequent views of his contemporaries and precursors. Otherwise, Galen can be compared to his contemporary Ptolemy:[1] much like Ptolemy's in astronomy, Galen's work made his name synonymous with medicine until the early modern era. This was partly the result of the sheer volume of Galen's writing – more manuscripts of his texts survive than all other Greek medical writers put together. But the status of Galen's work also stemmed from his empirical and philosophical innovations and from his ability to synthesize and systematize traditional ideas into a coherent and explanatory framework. Everything in medicine and physiology that came after him was read through his eyes.

We sketched Galen's biography in Chapter 4 as an example of a life lived through the period in which the Hellenistic world was giving way to the

---

[1] Ptolemy was already an old man when Galen arrived in Alexandria, but it's still exciting to think of them chatting on the stairs of the *forum*.

Roman one: born in Hellenistic Asia Minor, Galen moved to Alexandria and then to Rome, worked in Latin and wrote in Greek. Galen was a hybrid scholar in other respects as well: he came from a wealthy family but worked hard for his living; he was proficient in theory and practice alike and comfortable in emperors' courts as well as in the stables of gladiators. Their open wounds, which he was celebrated for healing, he treated as "windows into the body."

But if Galen's prolific output obscured many of his contemporaries, his own polemical mentions of them, despite being often dismissive, reveal something about their thought which is crucial to the history of science as a whole. That is: physicians of Late Antiquity were engaged in a rich and sophisticated debate about the very nature of their knowledge and the proper and efficient way in which it should be pursued. It was not just a part of the philosophical debate of the Greek golden age on the nature of Truth and how to attain it (Chapter 2), but also a practical and well-informed discussion couched in categories parallel to the ones we've been using throughout the book.

Medical thinkers, according to Galen and Celsus – a medical encyclopedist of a century earlier – belonged to one of three groups. The Empiricists shunned theory. They distrusted the ability to know the inner workings of the body, and preferred to work by analogy and induction, prescribing for each malady a therapy that appeared to have worked on similar symptoms in the past. The Dogmatists, or Rationalists, did believe in theory, although they adopted different theories, according to the school they chose and the type of evidence they found convincing. But even if their theories differed, they agreed that general claims about the nature and operations of the body are not only possible to formulate and defend, but are also very useful, as they enabled reasoning from the symptoms of a disease to its causes and devise proper remedies to undo those causes.

The third group were the Methodists, who took the name to boast the medical 'method' they claimed they could teach anyone within six months. They had a basic theory, founded on an atomistic idea of the body rather than Hippocrates' flow of humors, although their concept of health as balance was similar: atoms stream around the body through pores – in the healthy body, the stream is free and smooth; disease is its obstruction. The training of the physician could therefore be brief because he didn't need to absorb a complex theory of the body nor know much about the complex particularities of the patient's body, history, behavior and diet.

We've often discussed the relations between epistemological values and social-political standings, so it shouldn't be surprising that Galen abhorred the last option. A person of his standing couldn't allow such a populist approach, rooted in rural healing, which rendered moot the erudition and sophistication he stood for. For him, the Methodists' theory was simply wrong and their 'method' a sham. This was an attitude towards folk practices that learned medicine would maintain throughout history, with a few crucial exceptions which we'll discuss below (Harvey's interests in the knowledge of breeders and midwives is one such exception). Between the Empiricists and the Dogmatists, Galen was somewhat conflicted. No theory was preferable to a wrong one, leading, as in the Methodists' case, to dangerous treatment, but Galen did believe in his ability to attain real certainty in his own theoretical claims. This could be achieved by using Aristotle's logic, and in particular syllogism – the fundamental mode of argument taught by Aristotle. Here is an example of such use: Galen's demonstration that the rational, 'ruling' soul – *hēgemonikon* – is in the brain:

> where the source of the nerves is, there is the hēgemonikon;
>
> but the source of the nerves is in the brain;
>
> the hēgemonikon, then, is in the brain.
>
> Quoted from https://plato.stanford.edu/entries/galen/

The reader might wonder how Galen could think that an argument like this, clearly based on empirical observations and contestable assumptions, may lead to certainty – perhaps he was using 'certainty' in a less restrictive way than we do. But his readers in the next 1,500 years hardly perceived a problem, and in the universities of the Middle Ages he was taught as the prime example of a brilliant application of the Aristotelian tools in the acquisition of natural knowledge.

The most influential aspect of this application was Galen's fundamental theory of the body. Galen adopted and canonized the single text in the Hippocratic Corpus – *On the Nature of Man* – that specifies exactly four humors: blood, phlegm, yellow bile and black bile. This quartet allowed him to release the Hippocratic concept of body from its bond to Platonic ideas about the constantly fluctuating matter, and frame it instead in the symmetries of Aristotelian natural philosophy. In this Aristotelian interpretation of the Hippocratic ideas, the four humors correspond to the four elements and, like the elements, Galen conceives them less as physical, flowing liquid substances, and more as realizations of the fundamental properties – cold and hot; dry and wet. Each humor he relates to an organ: blood to the heart; phlegm to the brain; yellow bile to the liver; and black bile – the spleen. Each is dominant in a different personality type: the blood in the music- and fun-loving sanguine; the phlegm in the slow and lazy phlegmatic; yellow bile in the ill-tempered and querulous choleric; and black bile in the moody melancholic (Figure 8.3). Galen can then further map the humors and their corresponding organs and personalities onto the

Figure 8.3 The four personalities, or temperaments, from an undated and unsigned medieval manuscript. Top left: the sanguine – fleshy, pinkish, optimistic and passionate; top right: the choleric – hardy, ruddy, irritable and violent; bottom left: the phlegmatic – fat, pale, slow and lazy; bottom right: the melancholic – lean, anxious, stubborn and sensitive.

seasons of the year and of human life, onto climates and planets (whose astrological characters we discussed in Chapter 6): the northerners living in a cold climate are phlegmatic; teenagers are sanguine, etc.

Galen thus describes the healthy human body as varied and changing yet stable and balanced – just like nature as a whole. Dis-ease, as the English word still captures, is conceived as a loss of this balance. For example: when the cold and wet winter enhances the cold and wet phlegm (which was related to the element of water), it could travel from the brain to the sinuses and the lungs, causing those maladies characterized by coughing, like bronchitis and pneumonia. Spring increases the amount and flow of blood (the humor aligned with air, hence warmth and moistness), so one could expect high fevers and nose bleeds. Choler (yellow bile) and black bile are associated with harm, and are held culpable in more serious diseases. Some fatal conditions (probably what we'd call hepatitis) come with yellowed eyes and swollen liver – signs of excess of yellow bile; the excrements and puss of people dying of the Black Death (the Bubonic Plague) are indeed black, and the inner organs of people dying of severe melancholy (likely malaria) are also badly darkened – both clear signs of the effects of the black bile.

Much of the power of Galen's writing came from his ability to produce a coherent, explanatory theory, tying all these aspects and relating them to the teachings of the main philosophical schools. Method, we saw, he drew from Aristotle and fundamental ideas about the material body – from Plato via Hippocrates. His concept of life was adopted from the Stoic school (Chapter 4): the body is alive, moving and sensing because it comprises three types of *psychē*, all created within it by transformations of the *pneuma* – the spirit or soul of the world. Galen gives these philosophical conceptions concrete physiological underpinning: the liver produces the vegetative or desirous soul, responsible for basic living processes – nutrition and generation – and distributes it through the veins. The heart creates the spirited soul, responsible for motion and the passions, which is carried through the body in the arteries. The brain, we saw, is the site of the rational soul, which is transmitted through the nerves, and here Galen manages to synthesize Plato's idea of *logistikon* with the competing Stoic concept of *hēgemonikon*. The arguments Galen provides to support his claims are varied. Some are common tropes – for example, everyone knows that great emotions are felt at the heart and may flush the face with blood. Some evidence was gathered from careful anatomical observations of the bodies of pigs, monkeys, sheep and goats (the human body was not to be desecrated). Some was based on bold experiments – for example, severing the nerves of a pig one by one, Galen showed that he could stop the poor pig from struggling while it kept squealing and vice versa.

From a practical, medical point of view, Galen's most lasting legacy may have been bloodletting. He trusted this procedure much more than the Hippocratic writers – enough to recommend continuing it, sometimes, even after the patient fainted, and even in cases of blood loss. This didn't follow from an underestimation of the crucial role of blood: Galen, we saw, assigned blood a crucial life-carrying role. But since he had a convincing theory of how blood was constantly produced in and consumed by the body, removing some of it (and over a few treatments – quite a bit) didn't appear to him as dramatic as it seems to us. Moreover, Galen developed a sophisticated method of diagnosis by pulse-taking and a careful clinical manual of locations and quantities of blood to be let, and his trust in bloodletting was well supported not only theoretically, but empirically. He fiercely defended the Hippocratic localization of bloodletting, which physicians of his time were for some reason losing trust in, claiming to have confirmed it by "extensive research" (quoted by Kuriyama, *The Expressiveness of the Body*, p. 207). He was also clearly impressed by the apparent healing powers of menstruation – women seemed immune to many diseases afflicting men, like epilepsy and gout. As we noted in discussing Hippocrates, the millennia-long belief in bloodletting, and definitely Galen's, wasn't based on negligence or superstition.

Muslim Learned Medicine

Learned medicine resembled the other branches of the learned tradition also in the subsequent fate of its Hellenistic canonical corpus. As we discussed in Chapter 4, it was the prize purchase of the great Abbasid translation project. Tellingly, the great Hunayn (whom the Latins knew as Johannitius) not only translated almost all of Galen's work, including some texts which were lost in the original Greek and survived only in his Arabic translation. He also compiled an introduction to Galenic medicine (*Medical Questions*, to which we'll return below) and authored a book on the health of the eye: *Book of the Ten Treatises on the Eye*.

The role assigned to Galen by Muslim learned physicians, as it would be by the later Christians, was only comparable to that of Aristotle. They found in Galen not only practical medical instructions and a fundamental account of the body, but also an overall system of sound thought, research and argumentation. Yet, as with Aristotle, this is not to say that they followed his teaching blindly.

A good example of the sophisticated critical approach of these Muslim Galenic philosophers is provided by one of the earliest and most influential among them: the Persian Muhammad ibn Zakariya al-Razi (865–925), known in Latin Europe as Rhazes. His biography adds an example of their social and cultural place as people of practical and theoretical knowledge. Al-Razi studied in Baghdad, the imperial capital, and was apprenticed in a new kind of institution – the charitable Islamic hospital or *bimaristan*. He was called back to his hometown Ray (Rey in today's Iran) by the Persian ruler Mansur ibn Ishaq to run a similar institution, and later his reputation took him back to Baghdad to establish a new *bimaristan*. Al-Razi chose its location, so it's told, by hanging slabs of meat in various parts of the city to see which would take the longest to rot, indicating the freshest air.

Al-Razi was a Galenist not only in accepting most of Galen's theories and incorporating many of Galen's case studies into his own comprehensive manual of medicine (*al-Hawifi'l-tibb* or *Continens Liber*, Latin translation). Following Galen's example, he fashioned himself a medical *philosopher*, committed to determining the proper balance of theory and observation, as the story of the establishment of the hospital conveys: whether the anecdote is true is less important than the claim it makes about al-Razi's commitment to a careful and sophisticated empirical approach. At the same time, he wasn't shy of correcting Galen on empirical details. His most famous correction is the distinction between measles and smallpox, which in the Galenic tradition were categorized together as rashes:

> The physical signs of measles are nearly the same as those of smallpox, but nausea and inflammation are more severe, though the pains in the back are less. The rash of measles usually appears at once, but the rash of smallpox spot after spot.
>
> Cited in Porter, *The Greatest Benefit to Mankind*, p. 97

Nor was al-Razi shy of directly criticizing Galen on theoretical principles, and he dedicated a treatise – one of some 200 he composed – to such a critique: the very influential *Doubts about Galen* (*Shukuk 'ala alinusor*).

This persona of the physician-philosopher was emulated a century later by an even more influential Persian – Abu Ali al-Husayn ibn 'Abdallah Ibn Sina (980–1037). We already mentioned Ibn Sina in Chapter 4, when we touched on his role in the monotheistic adoption of pagan science: his impact in Christian Europe, under the Latinized name Avicenna, was rivaled only by Averroes and Alhazen. The stories about Ibn Sina's childhood and youth – again, whether accurate or exaggerated – tell of the great admiration for learnedness developed in Islamic culture under Abbasid rule. They also demonstrate that this admiration didn't come at the expense of respect for practical knowledge. Ibn Sina was the son of a tax collector from around Bukhara (in today's Uzbekistan), and allegedly taught himself the whole Qur'an (by heart) by the age of 10, and the theory and practice of medicine by 16. While practicing medicine from that tender age, he became an expert in every branch of all available theoretical knowledge: Indian arithmetic, Islamic jurisprudence, and all translated Hellenistic science and philosophy, including Aristotle, Euclid, Ptolemy and the Platonists.

Ibn Sina led an itinerant life in the inner Asian regions of the Muslim realm, carried by political upheaval and a quest for knowledge, knowledge by which he made his living as a physician to the Emir, a master of jurisprudence, a court philosopher and an astronomer. He left 270 or so works, about forty on medicine, many composed on horseback, in prison or in hiding, and many remaining in active use for centuries – both in the original Arabic (a handful in Persian) and in their Latin translations. Among all this, his *Canon of Medicine* (*Al-Qanun fi al-Tibb*), completed in 1025, still stands out. The *Canon* comprises about a million words, covering medical knowledge ranging from Hippocratic antiquity, through Galen and his disciples, to Ibn Sina's own immediate predecessors and contemporaries. It's meticulously arranged in five books: the first including fundamental humoral theory, anatomy, hygiene and etiology; the second, *materia medica*, or plants and other substances serving as remedies; the third and the fourth, diagnosis, prognosis and treatment of specific diseases; the fifth, the preparation and use of compound drugs. The *Canon* was still the most important book in Harvey's medical education in Padua at the turn of the seventeenth century, and remained the core of the medical curriculum a century later.

The human body portrayed in the *Canon* is that described by Galen, with one important exception. For Ibn Sina, a devout Muslim who adopted a Neo-Platonic interpretation to his monotheism (see Chapters 1, 5 and 6), it was crucial to maintain a strict separation between the material body and immaterial soul. But in the *Canon* the body itself is not completely material. The idea that the body is permeable to the influences of the immediate and remote environment, we saw, can already be found in the Hippocratic Corpus. But in Ibn Sina the four humors come to relate to the world in a way we discussed in Chapter 6: they represent, symbolize and echo the seasons, the planets and the configurations of objects. The analogy between the *microcosmos* and the *macrocosmos*, as it would come to be called in a couple of centuries, was crucial for the function and temperaments (see Figure 8.2 again) of the Galenic body as Ibn Sina analyzed it, as well as to health and disease. This meant that the disciplines studying these symbolic relations – alchemy, astrology, the interpretation of dreams and the art of talismans (Chapter 6) – were essential for Ibn Sina's medicine.

The theoretical medicine of the *Canon* is particularly interesting because Ibn Sina carefully shaped it as a paradigmatic Aristotelian natural philosophy. The philosopher aspired to *scientia*, which Aquinas, after Ibn Rushd (Chapter 5), formulated as thus: "certitude of knowledge which is gained by demonstration" (quoted by Charles Lohr, *The European Image of God and Man* (Leiden: Brill, 2010), p. 260). Medicine was too complex and its subject matter too varied to achieve this level of certainty, but Ibn Sina did his best to put it in syllogisms, similar to Galen's own which we saw above, and he carefully structured his accounts of the bodily phenomena according to the four causes (Chapter 2). The material causes were the body's humors, pneuma and vital spirits; the formal causes were the temperaments, faculties and organs. All external agents, affecting changes on the body, would count as efficient causes: food and drink, exercise and sleep, medications and poisons; and the final cause was the body's healthy balance. The unique popularity of the *Canon* was therefore based not only on its use as a compendium of medicine, but also as a model of proper Aristotelian inquiry.

Christian Learned Medicine

The learned role played in the Muslim realm by 'professionals' and courtiers like Hunayn and Ibn Sina was fulfilled in Christian Europe, as discussed in Chapter 4, by people of the clergy – first in the monastery and then in the university – and medical knowledge was no exception. It was also the clergy who saw to the collation of the text of the Hellenistic-Muslim tradition and their translation to Latin, and at the center of that project stood, as discussed in Chapter 4, Gerard of Cremona and the Toledo College of Translators. Yet the fundamental practical nature of medical knowledge, even that delivered in books, did make for important differences in the ways this knowledge was adopted, assimilated and taught.

The main portal for medical texts was the eleventh-century work at the Benedictine monastery of Monte Casino, concentrated around the intellectual skills and commercial connections of Constantinus Africanus (c. 1020–1087), which we discussed in Chapter 4. It was sponsored by Alphanus (d. 1085), the Bishop of Salerno (some 150 kilometers south-west of Monte Casino), who was one of its former monks. Alphanus traveled to Constantinople in 1063, where he acquainted himself with Galenism and summarized it to the Latin scholars in his *Premnon Physicon*. Alphanus' own approach was philosophical, but the legacy of the Salerno school of medicine was a true hybrid of learned and practical medicine.

Defining this legacy are two books. One is Constantinus' *Liber Ysagogarum* (or *Isagoge – Book of Introductions*) – his Latin version of Hunayn's *Medical Questions* – which concentrated on Galen's concept of the 'six non-naturals.' The non-naturals were those factors of health that were not part of the body's own nature: air, food and drink, elimination and retention (which for Hunayn meant both digestive and sexual), exercise and rest, sleep and wakefulness, and states of mind. This was a very practical set of ideas for healthy conduct – still easy to abide by. But it also powerfully relayed the theoretical Galenic-Hippocratic conception of the body, mediated by Ibn

Sina, in which these ideas were embedded: a body permeable to the cosmos and to its immediate environment, with the humors expressing the seasons and the elements; the vital spirits relating body and soul, etc.

The second book – *The Salerno Health Regimen* (*Regimen sanitatis salernitanum*) – was composed anonymously some 200 years later and brought this hybrid of theory and practice even closer to home. It was written in verse, so it made the Galenic stress on hygiene, exercise, diet and temperance, as well as the theoretical framework it embedded, available not only in the formal educational context – to which we'll presently turn – but also in the big household, the manor and the parish.

In the thirteenth century, the newly established universities began to incorporate medicine into their educational duties, and it was this synthesis of medical *know-how* and *knowing-that* coming out of Monte Casino and the Salerno School that they picked up as their basic curriculum. It came encapsulated in six texts, known together as the *Articella* – the little art (of medicine): the aforementioned *Isagoge*; Galen's *Tegni* (or *Techne*); Hippocrates' *Aphorisms* and *Prognostics*; *De Urinis* (*On Urines*) of Theophilus Protospatharius; and *De Pulsibus* (*On Pulses*) by Philaretus. The latter two authors were seventh- to eighth-century Byzantine physicians who had been translated by Alphanus; little is known about them apart from that, which further demonstrates how powerfully this knowledge transcended physical, cultural and linguistic boundaries. Indeed, texts by Muslim and Jewish authors – on diets, on fevers, etc. – were continuously translated from Arabic and added to the *Articella*. The titles hint at the fundamentally practical nature of these texts, although from the middle of the twelfth century they came bound together and commentated upon by the Salerno scholars – first Bartholomaeus (d. 1192), then Maurus (1130–1214). These commentaries not only stressed the Galenic theory of the body that made sense of the remedies and regimens, but also how this theory was embedded in the Aristotelian natural-philosophical framework. They became the basis for a thriving commentary tradition adjoining the *Articella* in the following centuries within which university scholars would discuss philosophical-theoretical ideas about the human body and its place in the cosmos.

Yet despite very harsh criticism from physicians of Harvey's era, this literary and studious leaning didn't define the output of the Salerno school, as demonstrated by another corpus coming out of the school in the twelfth century. It was a collection of three texts about medicine for women, by a woman – *On the Conditions of Women*; *On Treatments for Women*; and *On Women's Cosmetics* – called *Trotula*, namely *Little Trota*. 'Trota' was the name of the woman who reportedly wrote the second of the three texts, and about whom nothing else is known with certainty (although heroic stories abound). Indeed, the essentially practical nature of medicine, even in its learned form, allowed women to have a significant presence in it throughout the Middle Ages. The abbess Hildegard from Bingen (1098–1179) provides another good example. She is famous for her religious visions and her music, but her medical writings were no less copious and influential. The most famous of them, the *Book of Simple Medicine* (*Liber simplicis medicinae*), includes a vast array of healing techniques employing herbs, tinctures and precious stones; instructions for diagnosis and prognosis; and theoretical discussions of the human, in particular female, body.

Nor did university medical education ever veer too far from practical considerations. A medical degree was an arduous undertaking, requiring seven years (including the basic education discussed in Chapter 4) to achieve a Bachelor of Medicine (BM) and ten to become a medical doctor (MD). Beyond absorbing the *Canon* and the *Articella*, students striving towards the MD at the medical faculties in places like Bologna and Paris had to serve some time as apprentices to practicing doctors. From the beginning of the fourteenth century they were also expected to attend a dissection. The patients clearly appreciated this thorough education, with its strong basis in the liberal arts and natural philosophy: MDs were those who'd serve as physicians to lords and kings. Their duties included overseeing the health, diet and regimen of the whole royal household, as well as giving reasoned, learned explanations for their advice. But MDs were all men: women (and non-Christians) were prohibited from the university, so its appropriation of learned medicine meant that they were now excluded from these more prestigious and lucrative branches of practice.

The Healing Tradition

The Leechbook

Of course, only a small portion of the population – in Europe or anywhere else – had access to learned medicine. Most people relied for their health on the advice of physicians whose knowledge resembled the mason's much more than the scholar's. As we've discussed throughout the book, this knowledge is a crucial ingredient of science's cathedral, but is by its very nature much more difficult to recover. Yet it is not impossible: images, recipe books, instruments and indirect accounts allow some glimpses into the medical concepts and practices outside the institutions of knowledge. And just as the learned medical tradition was quite practical, the practical, healing tradition had access to and interest in knowledge delivered in writing. Practical physicians were probably partly literate, especially if they were members of the clergy who took up medicine as a religious duty of benevolence. *The Leechbook of Bald* is both an example of the practical, semi-learned medical tradition and a good source of information concerning actual medical practices, especially in the medieval Christian realm, but as we shall see – also well beyond. Comprising three books, it is written in Old English ('leech' means 'physician' in Anglo-Saxon), and was compiled around the turn of the tenth century for someone called Bald by someone called Cild: the phrase "Bald owns this book, which he directed Cild to write" appears towards the end of Book II, one of very few in Latin (Figure 8.4).

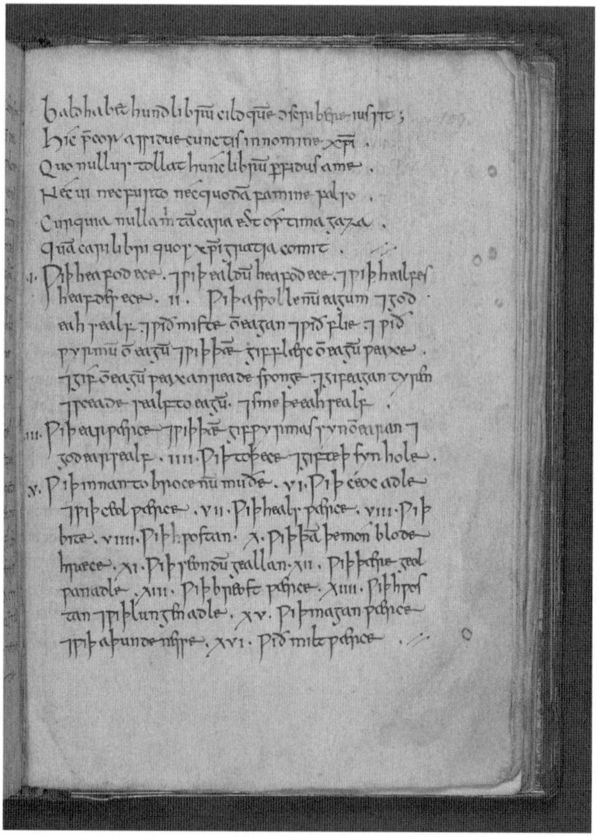
Bald habet hunc librum cild quem conscribere iussit ;
Hic precor assidue cunctis in nomine Christi .
Quo nullus tollat hunc librum perfidus a me .
Nec vi nec furto nec quodam famine falso .
Cur quia nulla mihi tam cara est optima gaza .
Quam cari libri quos Christi gratia comit .

Figure 8.4 *The Leechbook of Bald*: "Bald habet hunc librum, Cild quem conscribere iussit" (uppermost line). Note the beautiful calligraphy in a language we hardly associate with learning, suggesting that the distinction between 'high' and 'low' knowledge may be less pronounced than is assumed, especially for medicine.

Practical Remedies

As a practical aid to treatment, the *Leechbook* has much more interest in remedies – 'leechdoms' – for ailments than in theories of the healthy body. For example, humors are mentioned, but not their related organs and their balance – the fundamental concepts of the Hippocratic-Galenic theory. Moreover, to be useful, the ailments needed to be described specifically and the remedies needed to be particular and local – they needed to be available. So, the "leechdoms for all infirmities of the head," with which the book begins, contain, for example:

> A wort has been named murra, rub it in a mortar as much as may make a pennyweight, add to the ooze a stoup full of wine, then smear the head with that and let the patient drink this at night fasting …
>
> For head ache, take willow and oil, reduce to ashes, work to a viscid substance, add to this hemlock and carline and the red nettle, pound them, put them then on the viscid stuff, bathe therewith.

And among "Leechdoms for mistiness of the eyes" we find:

> … take juice or blossoms of celandine, mingle with honey of dumbledores, introduce it into a brazen vessel, half warm it neatly on warm gledes, till it be sodden. This is a good leechdom for dimness of eyes.

And so on.

Yet the locality of the remedies within the healing tradition doesn't mean that the knowledge represented by the *Leechbook* extends only to headaches. In fact, the book offers "Leechcrafts and wound salves and drinks for all wounds and all cleansings in every wise, and for an old broken wound, and if there be bone breach on the head, and for a tear by a dog; and a wound salve for disease of the lungs, and a salve for an inward wound," for example:

> For broken head … take wallflower and attorlothe and pellitory and wood marche and brownwort and betony, form all the worts into a wort drink, and mix therewith the small cleaver and centaury and waybroad … and if the brain be exposed, take the yolk of an egg and mix a little with honey and fill the wound and swathe up with tow, and so let it alone; and again after about three days syringe the wound, and if the hale sound part will have a red ring about the wound, know thou then that thou mayest not heal it.

Like the Hippocratic-Galenic tradition, the Leechbook is wary of wielding the knife – unsurprising, given how difficult it is to stop bleeding – but when necessary it doesn't hesitate:

> ... if the swarthened body be to that high degree deadened that no feeling be thereon, then must thou soon cut away all the dead and the unfeeling flesh, as far as the quick, so that there be nought remaining of the dead flesh, which ere felt neither iron nor fire ... If thou wilt carve off or cut off a limb from a body, then view thou of what sort the place be, and the strength of the place, since some or one of the places readily rotteth if one carelessly tendeth it.

And it is even confident enough to provide instructions for cosmetic surgery:

> For hair lip, pound mastic very small, add the white of an egg, and mingle as thou dost vermillion, cut with a knife the false edges of the lip, sew fast with silk, then smear without and within with the salve, ere the silk rot. If it draws together, arrange it with the hand; anoint again soon.

Learned Resources

The local and practical nature of the *Leechbook*'s remedies is hardly surprising: the physician was supposed to heal rather than explain, so his knowledge needed to be effective and his means readily available. More surprising is that the healing tradition was in fact quite attentive to the learned medical tradition, and showed interest in both abstract ideas and remote, inaccessible remedies.

Some of the ideas did have immediate practical implications. Bloodletting is recommended in the *Leechbook* – for example, "against pocks, a man shall freely employ bloodletting and drink melted butter." It is not recommended as often as it is in Galenic medicine, but the book does attempt (not in nearly as many details) to follow Galen's instructions regarding its proper application. It considers, for example, "[o]n what season bloodletting is to be foregone" and "how a man shall forego bloodletting on each of the six fives in the month, and when it is best." It's not easy to say by what route these Hellenistic teachings trickled into ninth-century England – remember that the *Leechbook* dates before the establishment of the universities and the great Christian translation project. But many of the leechdoms are clearly translations from the Greek, presumably through Latin, and names of teachers are mentioned – Oxa and Dun, presumably learned physicians.

Perhaps most surprising and telling, though, is that the *Leechbook* shows interest in remedies which are clearly impractical because they are very far away and thus almost impossible to obtain:

> ... scamony for constipation of the inwards, and ammoniac drops for pain in the milt, and stitch, and spices for diarrhoea, and gum dragon for foul disordered secretions on a man, and aloes for infirmities, and galbanum for oppression in the chest and balsam dressing for all infirmities, and petroleum to drink simple

for inward tenderness, and to smear outwardly, and a tryacle, that is a good drink, for inwards tendernesses, and the white stone, lapis Alabastrites, for all strange griefs.

None of these ingredients was available in England: "All this Dominus Helias, patriarch at Jerusalem, ordered one to say to King Alfred." Arriving from Jerusalem, the knowledge dispatched by Helias is a glimpse into an earlier global age: he is recommending plants and minerals from the Syrian desert, Mesopotamia and Persia; through the vast Eastern reaches of the Muslim realm.

Practitioners

Bald the leech (assuming that his book was for his own perusal), then, possessed medical knowledge concentrating on the local and practical, but ranging also to the remote and esoteric. Other practitioners of the healing traditions had different levels of interest and use for abstract and theoretical knowledge.

Apothecaries

Closest to the learned physician in cultural and intellectual standing was the apothecary, who gathered and dispensed *materia medica* and prepared medications. Most of the apothecary's knowledge was local and particular: his ingredients came from the nearby paddock, forest and quarry. Apothecaries affiliated with institutions – monasteries or hospitals, and later faculties of medicine – often had their own *herbaria*. There they would grow fragrant plants to drive away *miasmas* – vapors carrying epidemics and recognized by their foul smell – and medicinal herbs. These plants and minerals were the 'simples,' which the apothecary could provide to be used as-is, or mix to create compound remedies.

But as Bald's exotic ingredients remind us, from very early on – perhaps as far back as this book is concerned – knowledge and goods traveled, if with difficulties, across Asia, North Africa and Europe – especially *materia medica*. This was true even if the medical theories making sense of the ingredients and procedures were often lost when crossing linguistic and cultural boundaries. This became particularly true with the dramatic globalization of commerce from the fifteenth century onwards: from Malabar, European apothecaries imported the calumba root; nutmeg, mace and cloves from Moluccas; hell-oil from Madagascar; from Mexico, jojoba and physic nut; guaiacum from the Bahamas and many, many more.

The distant and esoteric origins of many of the apothecary's ingredients meant that it was almost necessary for him to be literate, at least to some degree and at least in the vernacular. This was required in order to communicate with merchants of exotic materials and to consult with learned physicians, and it also gave him access to two definitive textual resources. The first was the book *On Medical Material* of the first-century Greek physician Pedianos Dioscorides (Figure 8.5). In it, the apothecary found the description of some 1,000 plants and minerals, almost exclusively from Dioscorides' native Asia Minor, carefully arranged according to their medical properties. *On Medical Material* remained available in the original and in Arabic and Latin translations throughout the Middle Ages (the Latin title – *De materia medica* – is the origin of the term we've been using), finally finding its way to the vernacular and remaining in use until the nineteenth century. Dioscorides' book was the most popular, but not the only one of its kind. Apothecaries in the Muslim realm, for example, could also make use of the *Compendium on Simple Medicaments* of the Andalusian physician Diyāʾ Al-Dīn Ibn al-Baytār (1197–1248), which contained three times the number of medical substances as Dioscorides' book. Surprisingly, neither Ibn al-Baytār's *Compendium* nor his commentary on Dioscorides were known in the Christian world, although his *Treatise on the Lemon* was.

The other learned resource of importance to the apothecary was Galen's theory of simples, which he could find either in translations of Galen's own *On the Powers of Simple Remedies* or in its elaborations by Ibn Sina and his Arabic- and Latin-writing disciples. This was the theory of elementary substances and their medical efficacies, which gave sense to the use of the simples and direction to compounding them. The theory explained the properties of the simples in terms of Aristotelian elements and their fundamental properties, and also explained why the powers of some simples could not be analyzed in this way. These powers were 'occult' in the way we discussed in Chapter 6, namely: they arose from properties that belonged to the unique natural essence of the substance and couldn't be further analyzed – they just were.

Witches

Occult properties, you may recall, were important for magical thought, and magic indeed provided another theoretical framework for the practitioners of medicine. 'Wise women' like Matteuccia or Walpurga (Chapter 6) may have been only marginally literate, but what they did know about the magical ideas ingrained in their healing made sense of their practices and related them to learned medicine.

The fundamental concepts of the learned medical tradition did lend themselves to the magician's approach. The Galenic body, we saw, represented its environment, both immediate and remote, and was open to its influences. These representations and influences were the symbolic relations assumed

Figure 8.5 A drawing of a blackberry from Dioscorides' *On Medical Material* in the original Greek. Images like this were added later, and can be found in various editions. This one is from the *Codex Medicus Graecus*, which is a particularly interesting manuscript for its longevity and its continuous use. It was created in 515 in Constantinople for the Byzantine princess Anicia Juliana and contains 435 (!) mostly naturalistic illustrations, including no fewer than 383 of them full page, but it was used for centuries as a textbook in Constantinople's imperial hospital, and intensively so, as the annotation in cursive Greek attests (including βάτος – batos, or bramble – at the top). It was still in the hospital's records as a working text in 1406 (!) when it was restored, re-bound and augmented with a table of contents and scholia. After the fall of Byzantium to the Ottomans, it came to the use of Moses Hamon, the Jewish physician of the Sultan, and the knowledge was still relevant enough that the name of each plant was added in Arabic (which can be seen among the branches: عُلَّيْق وَاتُوْش – wa'atoosh ullaik) and Hebrew (written in very small Rashi script, bottom right: בטוש – batosh). The *Codex Medicus* was the basis for many copies of Dioscorides, including one for the pope in the middle of the fifteenth century, and only a century later did it take on its apparent original intended use as an object of marvel, when purchased for the Holy Roman Emperor in Vienna, where it has been since, coming to be known as *The Vienna Dioscorides*.

and manipulated by magicians, and the macrocosm-microcosm analogy we discussed above captures these common grounds of magic and medicine. Bald, practical in approach but familiar with learned sources, dedicates in his *Leechbook* significant space to remedies for magically induced harms, like "Leechdoms against every pagan charm and for a man with elvish tricks; that is to say, an enchantment for a sort of fever, and powder and drinks and salve." He's also keen on magical remedies for natural maladies, for example:

> For joint pain; sing nine times this incantation thereon, and spit thy spittle on the joint: "Malignus obligavit; angelus curavit; dominus salvavit."

Surgeons and Barbers

Most independent of learned considerations were those who put their hands directly on the body: drew blood, set bones, cleaned and sutured wounds, pulled teeth, leeched, cupped and administered enemas. Not all of them were illiterate: the guilds knew to differentiate between educated surgeons, semi-educated barber-surgeons, and simple barbers who learned by apprenticeship to wield razor and scalpel, but not much more. Yet for all, knowledge of the body was primarily practical (recall the porous boundaries between the groups expressed in Figure 8.2).

Opening the body safely, we said, was difficult – it was hardly possible to relieve pain or stop serious bleeding – and the traditional hierarchy between *know-how* and *knowing-that* placed even erudite surgeons below their more theoretically leaning learned physicians. Yet despite the Hippocratic aversion to radical treatment, surgeons, from Antiquity through Harvey's time, didn't lack ambition, confidence or practical capacities. Tracing practitioners' *know-how* is difficult, as we noted about the mason and the witch, but the great medical compendia (including the *Leechbook*, we saw) included treatises on surgery, from which we can also construe some of their skills and methods. The most influential of these treatises, both in its original Arabic and in the Latin version, was Book 30 (the final one) of *al-Tasrif* (*The Recourse*) by the Cordovan physician Abu'l-Qasim Khalaf ibn Abbas al-Zahrawi (Albucasis, 936–1013). Al-Zahrawi, providing some 200 illustrations of medical and dental instruments (Figure 8.6), described not only cauterizing, suturing, draining abscesses, setting fractures, etc., but also obstetrical and dental procedures and operations for removing kidney stones. Particularly representative are his unwavering instructions on how to remove a cataract:

> Patient has to be seated directly in front of the sunshine, while his normal eye is completely covered ... Then put the needle to the border of iris adjacent to the corneoschleral limit. Now push and twist the needle until getting inside the

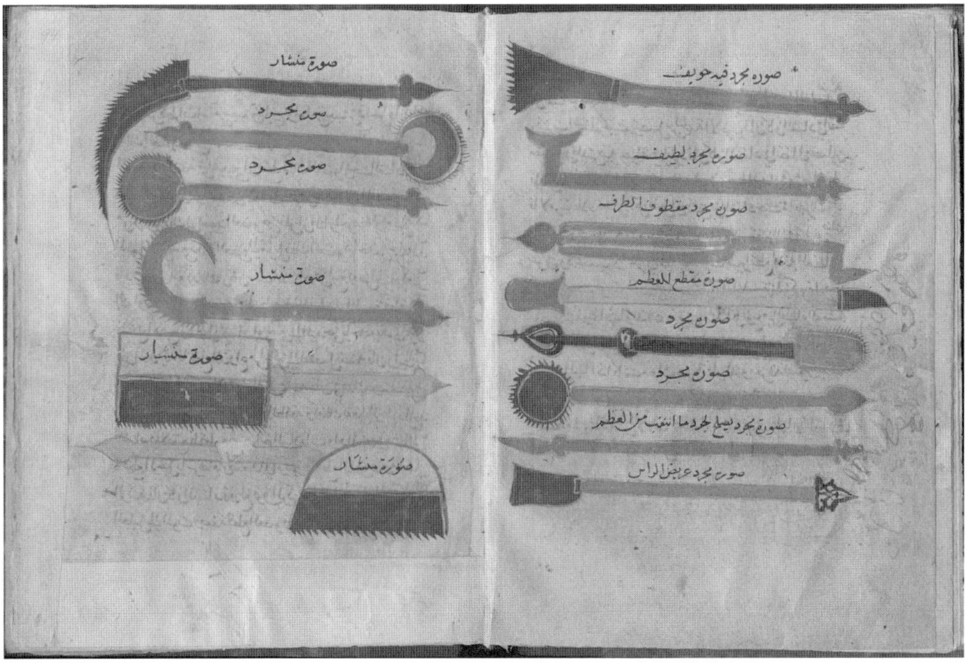

Figure 8.6 Surgical instruments from a manuscript of al-Zahrawi's (Albucasis) *Al-Tasrif*. As the variety of saws depicted serves as a reminder that despite the precision of the optical procedure quoted, much of the surgeon's work was very brutal, especially without pain relief or efficient means to stop bleeding. The combination of skills as both medical scholar and hands-on surgeon was thus rare, and al-Zahrawi is indeed aware and proud of that.

eyeball then you feel that it is gone into an empty space. You will see the needle in the centre of the pupil because of corneal transparency. Now hold the needle where cataractous lens has been formed and push the needle down a few times. If all parts of the lens are discharged, patient can see while the needle is still in his eye. If not, push down the needle once more, then bring back the needle to its original place in the anterior chamber and then turn the needle gently and pull it out. Now dissolve Turkish salt in water and wash the eye then put a clean cotton pad coated with white egg, rose water and oil on the eye and cover both eyes with the pad.

Quoted from A. Zargara and A. Mehdizadeh, "Cataract Surgery in Albucasis Manuscript" (2012) 24(1) *Iranian Journal of Ophthalmology* 75

How many patients were willing to risk such a delicate operation? How many surgeons dared to perform it? Did they gain their knowledge from such manuals or only by apprenticeship? Questions like this are very hard to answer, but for the cathedral of science this quote adds an important course (layer of masonry). It is a concept of the body quite different from the one we found in the Hippocratic-Galenic tradition. It is a body with clear, independently operating physical parts and with strict boundaries which can only be penetrated by violence. This alternative concept would be an essential resource for Harvey and the anatomical tradition he embraced in Padua, as well as for new ideas about health and disease, and we'll return to it shortly.

Midwives

Of all who nursed and healed the human body, midwives deserve special consideration. First, because in childbirth, they tended to a healthy bodily process, if often difficult and dangerous. This meant that they had physi-ological and functional knowledge of the body shared by neither learned physicians nor other healers – who attended to the body mostly in distress. Secondly, because midwifery was the one discipline – not just of medicine but of knowledge in general – dominated by women. As noted, women were relatively welcomed in medicine, but men wrote almost all the books (Trota and Hildegard are the exceptions that highlight the rule), including books on women's medicine. There were men-midwives, but when men tried to gain control over the lucrative practice of midwifery, women suc-cessfully defended their relative independence.

It was a partial independence. The medical guilds were closed to women just as the university was, and midwives don't seem to have established one of their own. Moreover, being a mysterious, feminine process, childbirth was an object of nervous attention by city, court and especially the Church. This was because the child, not yet baptized, was particularly susceptible to witchcraft (*Sleeping Beauty* documents this fear well). So as the fasci-nation with and anxiety about witchcraft mounted in the late Middle Ages and through the Renaissance, the authorities appropriated the licensing and supervison of midwives.

But these restrictions also had a liberating effect on midwives: they gave them a clear sense of professional identity and integrity, as well as the legal means to protect them. The self-confidence that this identity bred is nicely captured by Jane Sharp's dedication of her 1671 *The Midwives Book* "to the midwives of England":

> Sisters.
>
> I Have often sate down sad in the Consideration of the many Miseries Women endure in the Hands of unskilful Midwives; many professing the Art (without any skill in Anatomy, which is the Principal part effectually necessary for a Midwife) meerly for Lucres sake. I have been at Great Cost in Translations for

all Books, either French, Dutch, or Italian of this kind. All which I offer with my own Experience. Humbly begging the assistance of Almighty God to aid you in this Great Work, and am

Your Affectionate Friend Jane Sharp.

Sharp, *The Midwives Book*, Forward

All that is known about Jane Sharp is what she tells her readers: that she's had thirty years of experience as a midwife. But we can conclude, obviously, that she could read and write (in a free and sure-handed style), and less obviously, that she indeed possessed and made competent use of the library she brags about: she refers to and quotes with great accuracy all the important (male) authorities, both ancient and contemporary (many of these probably came via the translations and compendia of Nicholas Culpeper from the 1650s and 1660s). For Sharp, "[t]he Art of Midwifry is doubtless one of the most useful and necessary of all Arts, for the being and well-being of Mankind," and

As for [midwives'] knowledge it must be two-fold, Speculative; and Practical, she that wants the knowledge of Speculation, is like to one that is blind or wants her sight: she that wants the Practice, is like one that is lame and wants her legs.

Sharp, *The Midwives Book*, p. 1

So Sharp establishes her own authority by demonstrating her deep acquaintance with the writings of the "speculative" – learned – physicians, and fearlessly debating them when they do not agree with her "practical" expertise. "Anatomists have narrowly [hardly] enquired into this secret Cabinet of nature" – the womb – she laments, so those "best Learned men," from Hippocrates through Galen to Colombo, cannot agree on even the most fundamental facts, like "how many Cells are in the womb." So, she declares – and indeed follows suit – "Let their ignorance or disputes be what they will to no purpose, I shall satisfie all by true experience, which cannot be contradicted": that is, her "own experience."

The New Medicine and the New Body

Medical practitioners' trust in their skills and knowledge, which Sharp powerfully expresses, resonated among learned physicians. This was a development with deep cultural roots: since the rise of Humanism (Chapter 5) and through the sixteenth and seventeenth centuries, the demand that all knowledge be based on "true experience" was becoming increasingly common. It also had crucial significance for the shape which the cathedral of science would assume (we will return to this idea in Chapter 9), and in medicine, with its essential practical bent, this demand found strong adherents even earlier than in other branches of the budding New Science. Humanists from the Italian Petrarch (1304–1374) to the Dutch Erasmus (1466–1536) – both intellectual leaders of their respective centuries – expressed their resentment towards the medieval intellectual traditions with deep suspicion towards learned medicine. Sharp was therefore already acquainted with learned physicians turning to her midwife "sisters," as well as to "tramps, butchers and barbers," for their practical experience, importing with it, perhaps unwittingly, their concepts of the body and of disease.

Paracelsus and the Alchemical Body

Priding himself for learning from "tramps, butchers and barbers" was the most celebrated of these physicians: Philippus Aureolus Theophrastus Bombastus von Hohenheim (1493/4–1541), who brazenly called himself Paracelsus, that is – "beyond Celsus" (the classical medical encyclopedist mentioned above). A son of a Swiss physician with some suspect claims to noble descent, Paracelsus attended the medical faculties of Basel and Ferrara, apparently graduating neither. Very unimpressed, according to his own account, with what was taught there, in 1517 he opted instead to enlist as a military surgeon in the Venetian army. In that capacity, he traveled for the next seven years through much of Europe and as far as Constantinople, gathering practitioners' and artisans' knowledge, as well as a reputation for his own skills. His attempts at an urban medical practice were frustrated by the German peasants' revolt (he barely escaped the gallows), but in the mobile Renaissance society this reputation for practical medical skill proved efficacious. In 1526, apparently vouched for by Erasmus himself, it gained him a position as town physician and professor of medicine in his *alma mater* Basel, where he made a point to lecture in German and in an alchemist's leather apron. Add to this image the spectacle he performed by publicly burning editions of Ibn Sina and Galen (imitating, as in his insistence on the vernacular, another contemporary revolutionary – Martin Luther) and it's easy to understand why his tenure lasted all of two years. He returned to the life of an itinerant medic, wandering through German-speaking middle Europe, intermittently gaining popularity and losing his license. Throughout these journeys he wrote incessantly – mostly in German – on subjects ranging from the treatment of syphilis, through surgery and alchemy to Hermetic philosophy, allegedly dictating his teachings while drunk to ensure their authenticity (at the price of their coherence). He ended his tumultuous life in Salzburg, aged 48.

The colorful details of Paracelsus' personality and biography may be slightly embellished, but they testify to the complexity of the task he took upon himself and the unconventional intellectual tools he employed. "When I saw that nothing resulted from [doctors'] practice but killing and laming, I determined to abandon such a miserable art and seek truth elsewhere," he declared. That meant a new theory of the body, and to frame it – a new account of the world in which the body dwells. The "elsewhere" in which he sought truth turned out to be the hermetic tradition (Chapter 6) and magical practices.

The healthy body of the Hippocratic-Galenic tradition was a temple: a tranquil and permeable whole, seeking balance within itself and with the cosmos. Paracelsus', in comparison, was a bustling site of activity: an alchemical workshop. Powerful transformations were taking place inside it: digestion, respiration and generation comprised fermentation and distillation, calcination and sublimation. In this respect (though not in others, as we'll immediately see), the body's boundaries were clear and physical: materials, coming from the outside, were transmuted into life processes on the inside. Its inner organs were clearly separated – like the instruments of the alchemist – and each was controlled by its own *archeus*; a kind of life spirit or governing principle. This 'internal alchemist,' as Paracelsus sometimes referred to it, was responsible for the proper operations of the healthy organ as well as for the malfunction of the diseased one. Disease, for the traditional learned physician, was imbalance; for Paracelsus, it was a foreign invasion: an external, evil agent, corrupting the *archeus*.

For us, used to chemical biology, to microbes and viruses, Paracelsus' ideas may sound familiar: the body is an enclosed system of processes; disease is the effect of an independent entity, existing outside the body, subverting its operations when infiltrating inside it. But as we stressed on similar occasions, historical ideas appear 'modern' because we ended up adopting them, not because the people elaborating them were anticipating us. Paracelsus' case is particularly interesting in this regard, because he drew his resources from *magic* – both practical and learned – so his influential medical writing is one of the most important conduits of this kind of thought into modern science.

According to Paracelsus, disease is the effect of an autonomous, specific entity, invading from outside and disrupting a specific part of the body. But here the 'modern,' physical resonance of his analysis gives way to the Hermetic resources he draws on. In Paracelsus' world, growing things – which for him, the committed alchemist, meant also minerals and metals – receive their life from spirits emanating from the stars. In their role as mediators between matter and spirit, these astral emanations have direct influence on the *archei*, which fulfill a similar role within the body. And with this mediation these spirits also carry the marks of human debasement – the original sin and the fall – so occasionally they project this corruption onto the operation of an *archeus*, turning, for example, fermentation into putrefaction. The effects of such corruption are disease.

Alchemy provided Paracelsus with a model for the internal operations of the body and resources for a concept of disease, and it also provided a concept of healing and medication. In traditional medicine, balance could be restored by negating the cause of the imbalance: cholera, for example (as its name suggests), was caused by an excess of yellow bile, which relates to fire, so it needed to be treated by cooling and moisturizing. According to Paracelsus, on the contrary, like healed like: a kidney stone was cured by stones, such as crab's claw or lapis lazuli, and arsenic cured arsenic poisoning. This was because symptoms signaled which *archeus* was suffering corruption – the *archeus* of the organ in which the disease resided. That *archeus* required support from the parts of the macrocosm to which it related by affinity and representation. It also explained why every substance could be both poison and medicine, depending on the dose and the disease.

The Paracelsian physician, therefore, was a careful observer of those aspects of nature that interested the magician. He had a keen eye for hidden signs, by which things signaled their relations to other things and their capacity to influence them, because "there is nothing that nature hasn't marked, and through these marks one can discover what is hidden in the marked."[2] Orchids looked like genitalia, walnuts like the brain. These similarities between parts of the macrocosm and parts of the microcosm – the body – couldn't be accidental; they signaled affinities, and these affinities were relevant for treatment. The Paracelsian physician was also an alchemist, because "[n]othing has been created as *ultima materia* – in its final state" ("Alchemy, Art of Transformation," in Oster, *Science in Europe*, p. 100). Alchemy was the art of perfecting; of elevating materials from the state of *prima materia* in which God had left them to "the nobler pure, and perfect" (ibid.) state in which they could be used medically. Just as the farmer grows fruits out of seeds, the alchemist transmuted base metals into silver and gold, and simple minerals and herbs into medicines.

Paracelsus found in alchemy even more: a new understanding of the cosmos as a whole. The Aristotelian cosmos, the basis for Galenic medicine, had four elements configured in harmonious symmetries. In their stead, Paracelsus installed the *Tria Prima* – the three active principles of the magician: Salt – the unburnable; Sulphur – the combustible; and

---

[2] "Nichts ist, was die Natur nicht gezeichnet habe, und durch die Zeichen kann man erkennen, was im Gezeichneten verborgen ist." Paracelsus, *Der Mystiker Paracelsus*, Gerhard Wehr (ed.), Kindle edn. (Wisbaden: Marixverlag, 2013).

Mercury – the metallic and volatile. Like the inside of the body, nature, according to Paracelsus, was a site of incessant, intense change. It didn't strive to harmony but to constant separation: it began when God separated light from darkness, continued with the separation of the three principles, from which separated the elements and so forth, to the separation of every animal from its mother. The alchemists, we saw in Chapter 6, harnessed different processes of separation – distillation, digestion, fermentation – in their quest to gain control over the transmutation of metals. But this is where Paracelsus' fascination with their theories and practices came to an end: for him alchemy was in the service of medicine.

Iatrochemistry

It's difficult to fully gauge the legacy of Paracelsus. The term 'Paracelsian' became quite common in the century following his death, but it's not likely that many of the physicians adopting it could carefully follow his difficult and less than consistent philosophy. For them, it primarily connoted the release from the shackles of traditional learned medicine and its institutions and the demand for practical and efficient treatment. Within the traditional institutions of learning, the title 'Paracelsian' was used mostly pejoratively, to connote the charlatanism of magic. Those who were more genuinely curious came to know Paracelsus' work mostly second-hand, and mostly through summaries like that offered by Peter Severinus (Peder Sørensen, 1540/2–1602) in his *Idea medicinae philosophicae* (*The Idea of Philosophical Medicine*, 1571). Severinus was an emblematic learned physician – a professor at the University of Copenhagen and a physician to the Danish court – and in renditions like his, Paracelsus was domesticated: the Hermeticism (Chapter 6) was put in a Platonic framework; the medical philosophy was reconciled with Hippocratic medicine; and the role of alchemy was reduced to the production of useful drugs.

Ironically, given Paracelsus' volatile personality, drunken visions and furious rants,[3] it is this practical aspect of his teachings that seems to have survived best, under the title *Iatrochemistry*. Mostly rejected in the

---

[3] One that sums up his attitude can be found in his "Credo": "I do not take my medicines from the apothecaries; their shops are but foul sculleries, from which comes nothing but foul broths. As for you, you defend your kingdom with belly-crawling and flattery. How long do you think this will last? ... let me tell you this: every little hair on my neck knows more than you and all your scribes, and my shoebuckles are more learned than your Galen and Avicenna, and my beard has more experience than all your high colleges." Paracelsus, *Selected Writing*, Jolande Jacobi (ed.), Norbert Guterman (trans.) (Princeton University Press, 1979 [1951]), p. 6.

universities, which remained loyal to the Aristotelian framework and the Galenic medicine within it, the search for new chemical remedies found a home in the courts of monarchs and nobility, where the promise of effective treatment outweighed worries about academic substance. Learned physicians could accept the idea of distilling essences from medicinal herbs and minerals – this was an ancient practice – but were apprehensive of its magical connotations, which were intellectually suspect and legally dangerous. The required compromise was provided by *pharmacopeia* books, like *Alchemia* (1597) and *Alchymia triumphans* (1607) by Andreas Libavius (Libau, c. 1540–1616) – an impressive intellectual who taught history and poetry at the University of Jena and oversaw medical disputations there. His approach exemplifies not only the state of European medicine around the turn of the seventeenth century, but the combination of excitement and trepidation by which scholars welcomed the New Science in general.

Libavius was happy to teach his readers alchemical practices and dictate recipes for transmutations of metals as much as for distillations medicines, but he was careful to warn against all superstitions and call for a proper *chymia*, free of mysticism and secrets. A generation later, the German professor Daniel Sennert (1572–1637) went even further in this attempt to reconcile Paracelsianism with traditional, mainstream academic medicine. He introduced iatrochemistry to the medical curriculum at the University of Wittenberg, and in his writings, which remained influential through the eighteenth century, tried to give it an atomistic grounding (with the aim to give credence to both). Yet the magical ancestry of these ideas wasn't easy to ignore: the *Pharmacopoeia Londinensis* that the Royal College of Physicians published in 1618 (already with Harvey at the helm) still included remedies made from dried viper, foxes' lungs, live frogs, wolf oil and crabs' eyes.

Van Helmont and the Invading Disease

The most important advocate of iatrochemistry, Jan Baptist van Helmont (1579–1644), demonstrates its complex relation to magical thought: he found himself in trouble not for defending magic, but for attempting to naturalize it, hoping to make use of magical practices without the danger of their associated beliefs. The worst came in 1621, when he volunteered his views about the possible efficacy of the weapon salve: an ointment that could heal wounds when applied to the instrument that caused the wound. This is possible, argued van Helmont, but not because of some occult affinity between the weapon and the wound. Rather, it was a magnetic attraction between the blood in the wound and that remaining on the weapon that did the healing. The censures at Louvain University in the Spanish Netherlands found this idea (and similar attempts to naturalize miracles) very dangerous, because they blurred the all-important distinction between divine, natural and demonic magic (see Chapter 6 for a discussion of the role of this distinction), and he was put under house arrest.

Van Helmont resented being regarded as a Paracelsian, but it's easy to see why both his biography and his medical philosophy made both historians and contemporaries see him in that light. The son of a Brussels council member, van Helmont studied medicine in Louvain, struggling, like Paracelsus, with boredom and frustration. Unlike the Swiss, he did return from periods of travel to complete his MD in 1599, and, confined for many years by these house arrests, had little opportunity to gather knowledge from far and wide, and thus concentrated instead on experimenting from within his secluded household. Van Helmont survived, but was only released from confinement two years before his death, so almost all his writings were published posthumously, mostly by his son Franciscus Mercurius.

In the next two chapters we'll explore how the New Science fashioned itself in the seventeenth century: as an experimental, instrumental, mathematical, worldly and state-sponsored venture. Together with Paracelsus, van Helmont represents an alternative route that was available in the rebellion against Aristotelianism: a medicine-based natural philosophy (not to be confused with science-based branches of medicine, like biochemistry) whose later versions came to be known as 'vitalism.' This natural philosophy was as fiercely empirical and as committed to knowledge relevant to the human condition in this world as the mathematized natural philosophy that came to define the New Science. But unlike the *mathematical* objects and forces with which Kepler, Galileo and their disciples populated the universe, Paracelsus and van Helmont found a cosmos of *living* forces. And accordingly, whereas the physicalist tradition adopted mechanics as the fundamental science, the vitalists found it in chemistry.

For van Helmont, the vitality of the cosmos was such a strong assumption that it made Paracelsus' analogy between macrocosm and microcosm superfluous, and he rejected it as empty words, just as he rejected Aristotelian elements and qualities and Galenic humors. He did admit the three active Paracelsian principles – Salt, Mercury, Sulphur – but not as elementary; they arose in chemical operations. Water and air were the only true elements, and all nutrition was water, as van Helmont demonstrated experimentally and quantitatively: a willow gained in weight, over five years, almost exactly the overall weight of the water it was given in this time, hence it fed only on that water. Everything, whether mineral, vegetable or animal, evolved from seminal seeds, directed by their own *archeus*, and so did disease. Each disease was caused by a particular entity – *ens* – existing independently from the body, invading it from the outside and planting its seed within, confounding the *archeus* and diverting it to act at its service rather than the body's.

Just as with Paracelsus, most of the physicians who called themselves 'Helmontians' didn't seem to appreciate or even understand the complexities of his natural philosophy. But by the second half of the seventeenth century, especially in England, where Franciscus Mercurius settled, a significant number of well-known physicians were proud to identify themselves as van Helmont's disciples. As with the Paracelsians, this mostly meant a rebellion against traditional learned medicine and its powerful institution – the Royal College of Physicians. But it also meant a devout – 'Christian' – commitment to the welfare of patients, and a claim to expertise in the treatment of contagious diseases.

Iatrochemistry, with its concept of disease as an independent entity, was much better equipped than traditional learned medicine to explain contagion. Whereas the Galenists needed to ascribe the spread of disease in a particular place to the corruption of water and air, the Helmontians could make sense both of the peculiarity of the disease and how it traveled from one patient to another. Such an account was provided by Paracelsus' contemporary Girolamo Fracastoro (c. 1476/8–1553) in his 1530 treatise on syphilis (a term he coined): *Syphilis, sive morbi gallici* (*Syphilis or the French Disease*). Fracastoro seems to be the first iatrochemist to employ the concept of a seed of a disease, which he further developed in his 1546 *De contagione et contagiosis morbis* (*On Contagion and the Contagious Disease*) and that van Helmont would adopt and adapt. Van Helmont's version of the concept was quite esoteric, involving the imagination as an agent of transmission of both disease and cure, but it gained him and his disciples a reputation as the most proficient in treating epidemics.

In 1665, the devotion and the expertise came to a test together. The Bubonic Plague broke in London in the winter, lasting for a year, killing some 200,000 of its inhabitants and driving anyone who had the means to leave out of the city. Most well-established Royal College physicians were among those who left, but the leading Helmontians stayed, among them George Thomson (c. 1619–1676), who had this to say about their behavior:

> ... although I could enjoy my ease, pleasure, and profit in the Country, as well as any Galenist; yet I would rather chuse to loose my life, then violate in this time of extreme necessity, the band of Charity towards my neighbour, and dedecorate that illustrious profession I am called to, in hopes to save myself ... according to that infamous, and insidious advice which Galen hath given his Disciples.
>
> Cited in Wallis, "Plagues, Morality and the Place of Medicine," p. 16

This "Charity" cost many Helmontian physicians their lives, most famous among them George Starkey (see Chapter 6), the great alchemist of American descent. But it was also curiosity that lured them into staying in afflicted London. They had an elaborate theory of the plague that Thomson summarized thus:

> I pass to the Essence and Quiddity of the Pest; which is contagious, for the most part very acute, arising from a certain peculiar venomous Gas, or subtile Poyson, generated within, or entering into us from without: At the access, or bare apprehension of which, the Archeus is put into Terror, and forthwith submitting to the aforesaid Poyson, invests it with part of its own substance, delineating therein the perfect Idea of Image of this special kind of Sickness distinct from any other.
>
> Cited in Miller, *The Literary Cultures of Plague*, p. 63

Unwilling to pass up the opportunity to examine it with their own eyes, they risked their lives performing autopsies of victims. Thomson himself survived dissecting a just-deceased servant of one of his patients, but others weren't so fortunate. In fairness, although some College physicians treated the Helmontians' deaths with more glee than compassion, others found their own in a similarly heroic style. The famous Samuel Pepys (1622–1703), from whose diary we learn much about the daily goings-on in this crucial time and place for the rise of early modern science, laments, on Friday, August 25, 1665, the death of his personal physician, Alexander Burnett. Burnett, a Royal College fellow, was one of an eclectic group of some ten men, comprising Fellows, surgeons and apothecaries, who

> ... in a consultation together, if not all, yet the greatest part of them, attempted to open a dead corpse that was full of the tokens, and being in hand with the dissected body, some fell down immediately, and others did not outlive the next day at noon.
>
> Cited by Munk, http://munksroll.rcplondon.ac.uk/Biography/Details/665

Among the victims was William Johnson, a chemist of the College and a moderate Helmontian – the lines between the camps were sometimes blurred, and were definitely not respected by the plague.

The Rise of Anatomy

The Paracelsians and Helmontians were challenging learned medicine ethically and politically; they even purported to establish their own "Noble Society for the Advancement of Hermetick Physick" to compete with the Royal College of Physicians. And they were challenging its fundamental principles and practices with ideas developed from that "Hermetick" tradition. In these ideas lay a new concept of the human body, with much clearer internal and external boundaries: disease came from without and affected each organ separately and differently. The body's boundaries were to be respected, and van Helmont had a particularly dim view of the traditional practice of trespassing them through bloodletting: "a bloody Moloch presides in the chairs of medicine," he decried.

At the same time, a similar challenge to the permeable, balanced body of traditional – Galenic – medicine was coming from the heart of the establishment: from the medical faculties at the universities, where interest and proficiency in anatomy was growing at a fast pace.

Early Questions and Limits

Physicians have always been interested in anatomy. Barbers needed to know how to locate blood vessels to open for bleeding or to avoid damaging while cleaning wounds; surgeons needed to know their way within and around organs while removing a cataract or opening a fistula; and learned physicians were naturally curious about their subject matter – the human body. Yet there were clear limits to the usefulness and availability of anatomic knowledge. Practical physicians, much like the modern GP, could reach and treat only organs close to the body's surface. For the learned physicians, the humoral concept of the body provided little motivation and guidance concerning the various internal organs: the body operated as a balanced whole, so studying each organ on its own was fascinating but potentially misleading. The inside of the body could only be observed when it was no longer alive, and even that was rare and difficult: the dignity of the human body was carefully observed by both pagan and monotheistic cultures, although for different reasons.

Curious anatomists didn't always perceive the inability to peer into the human body as a great hindrance: what distinguished humans from animals was their *soul*; their bodies seemed analogous. Galen, for example, learned much about human anatomy by dissecting sheep, goats, pigs and apes. Yet this analogy between humans and animals caused confusions, which could not be rectified without comparison with human bodies. Galen, for example, assigned much importance to the *rete mirabile*; a network of nerves and blood vessels found at the base of many species' brains. Its position and complexity suggested to him that it was the place where the vital spirits emanating from the heart turned into the animal spirits distributed by the brain. When dissections of the human body became common, however, it turned out this structure wasn't to be found.

These uncertainties disturbed anatomists, and debates about structures and functions within the body had raged since Antiquity: whether sensation occurred in the brain or the heart; how many lobes the liver had; how many compartments to the womb; and so forth. Some of these questions were practically important: when bleeding, for example, it was useful to know whether the arteries contained physical air, *pneuma* or blood. But physicians healed individual patients, and mostly, individual bodies were different enough from each other that physicians needed to literally feel their way around each. All they could gain from anatomical theory was the general knowledge of relative positions, like the knowledge represented in the diagrams in Figure 8.7. Images like these are not failed attempts at a

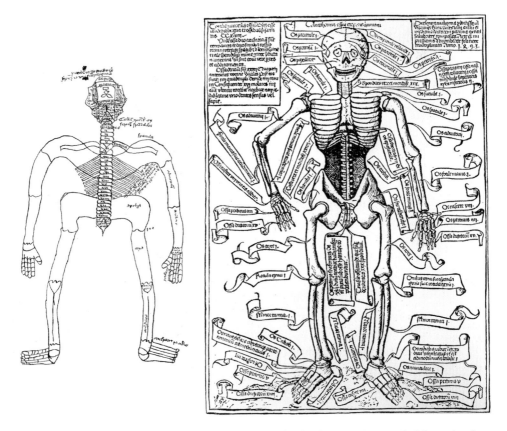

Figure 8.7 Two diagrams of skeletons from Johann Ludwig Choulant's 1852 *History and Bibliography of Anatomic Illustration*. The image on the left is from the fourteenth-century Latin *Munich Manuscript Codex*; the one on the right is the first printed image of the skeleton, from Schedel's 1493 *Nuremberg Chronicle* (see Figure 1.12). Neither is attempting a realistic depiction, and the later one, despite the details allowed by print, demonstrates this point even more strongly. Rather than a model, the drawing explicitly follows the theory of Richard Helain – the dean of the Faculty of Medicine in Paris at the time – about the bones, their total number (which he set at 248) and their configuration, assisted by the name tags, and prints from this woodcut were apparently used as teaching devices.

realistic depiction of the body. Rather, they are maps – mnemonic and educational devices – and in this role their schematic appearance is a benefit. They represent the anatomical knowledge required of the physician without the burden of unnecessary visual details.

Humanists and Artists

In the grand cathedral of science, realistic, detailed human anatomy and physiology came to define the import of learned medicine. But for the reasons we just discussed, the new interest in anatomy didn't originate with institutional medicine. Tellingly, it didn't originate with the disciples of Paracelsus and van Helmont either, both of whom didn't think much of "dead anatomy," as Paracelsus called it. For them, we saw, the living body wasn't an assemblage of static structures, but a site of active processes and constant change.

The new curiosity about human entrails arose from two sources that did not share an admiration of novelty but a rhetorical nostalgic tone: an admiration for the great achievements of the Hellenistic glorious past. The first source was what has been called 'medical humanism': the search for and translation of classical medical texts around the turn of the sixteenth century. It culminated in the discovery and translation of the first part of Galen's *On Anatomical Procedures*, which aroused so much admiration it made humanists replace contemporary procedures of dissecting human cadavers with Galen's allegedly more rational ones. The practice of starting dissection from the internal organs, for example, because they were most vulnerable to putrefaction, gave way to Galen's order of dissection, which followed the structural order of the body: starting from the bones, moving to the muscles, nerves, blood vessels and only then the abdominal organs.

If humanists were drawn to anatomy because of Hellenistic texts, artists became infatuated with it because of Hellenistic art. The new attention to the spectacular exemplars of Greek and Roman art in Rome, Pompeii and elsewhere in Italy convinced fifteenth-century artists that emulating such marvels of corporeal harmony required a thorough knowledge of the human body, inside and out; the body's internal structures had to be meticulously studied in order to understand its external appearance. The most famous outcome of these studies is Leonardo da Vinci's (1452–1519) anatomical illustrations, which he worked on from the late 1480s with the unfulfilled intention of producing a comprehensive anatomical atlas (Figure 8.8); but he wasn't alone. Leonardo's younger contemporaries Albrecht Dürer

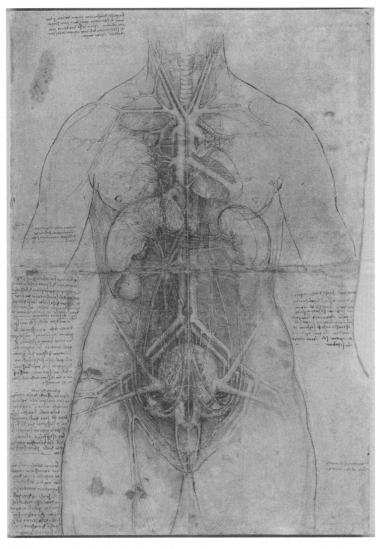

Figure 8.8 Leonardo da Vinci's drawing of the cardiovascular system and principal organs of a woman from c. 1509–10, now in the Royal Collection in Windsor. Although this marvelous image is much more realistic than those in Figure 8.7, it isn't only a drawing but an anatomical study: it doesn't present the cadaver in front of Leonardo – no open abdomen, etc. – but a recreated living body. It reflects careful attention to structural details like spatial relations between organs, and requires more than one 'sitting,' so even as a drawing it would have to have been a composite of different cadavers or different states of the same one. But in fact, much of the knowledge conveyed in this image comes from traditional accounts of the body that later observations discarded, like the understanding of the structure of the liver that appears also in Figure 8.9 and the spherical shape of the uterus. Moreover, the only documented autopsy conducted by Leonardo (in the winter of 1507–1508) was of an old man and, indeed, the external organs in this drawing – the arms, shoulders and pectorals – are masculine.

(1471–1528) and Michelangelo (1475–1564) were as proficient in human anatomy, and Leonardo's teacher Andrea del Verrocchio (1435–1488) had all his students study flayed cadavers. Paradoxically, the new admiration for the human (mostly male) body came with a relaxation of the anxiety about its posthumous dignity: in 1482, Pope Sixtus IV formalized this by informing the University of Tübingen that, provided a body comes from an executed criminal and its remains are given a Christian burial, there was no objection to dissecting it.

Indeed, even if the original urge came from outside, with aesthetic rather than medical motivations, it was in universities like Tübingen where the new curiosity about human anatomy turned into an innovative project. Leonardo's drawings, for example, as detailed and realistic as they are, remain completely loyal to Galen's descriptions of the internal organs (they also remained unpublished until the nineteenth century, so couldn't inform his contemporaries and immediate successors). The medical faculties were primed to take the challenge: from the middle of the thirteenth century, they needed to prepare their graduates to carry out public autopsies for forensic purposes, and around 1315, Mondino de' Luzzi (c. 1270–1326), who held the chair of medicine in Bologna, turned this into part of the curriculum there. Mondino used his empirical studies to compose the *Anatomia mundini*, which served as the fundamental textbook on dissection until the sixteenth century, first as a manuscript and from 1478 in print – in some forty editions.

Back to the University: Vesalius and the Padua School

As the interest in human anatomy spread over the next century, other universities adopted the spectacle of public dissection (Figure 8.9) as a combination of pedagogical and promotional means. Its most important center after Bologna became Padua, where Harvey was educated.

The tradition developed there, already famous by the time of Harvey's arrival, is most associated with the name of the Belgian anatomist Andreas Vesalius (Andreas van Wesele, 1514–1564), yet, like his contemporary Copernicus, Vesalius wasn't self-consciously a revolutionary. His early medical education was in Paris, under the tutelage of the medical humanist Jacobus Sylvius (Jacques Dubois, 1478–1555), whose trust in Galen was such that he insisted that any dissected corpse exhibiting a deviation from the master's teaching must be a negligible exception. When religious hostilities drove Vesalius back to Louvain, he was taught dissection by Guinther von Andernach (1487–1574), a translator of Galen's *On Anatomical Procedures* into Latin. In later years, Sylvius and Vesalius developed disparaging views of each other, but when the latter arrived in Padua in 1537 to complete his

Figure 8.9 Vesalius' anatomical spectacle. On the left: the frontispiece of his 1543 *De humani corporis fabrica*. "The frontispiece of the *Fabrica*," wrote Roy Porter (*The Greatest Benefit to Mankind*, p. 181), "presents the dreams, the programme, the agenda, of the new medicine. The cadaver is the central figure. Its abdomen has been opened so that everyone can peer in; it is as if death itself had been put on display. A faceless skeleton [above the cadaver] points towards the open abdomen. Then there is Vesalius, who looks out as if extending an invitation to anatomy. Medicine would thenceforth be about looking inside bodies for the truth of disease. The violation of the body would be the revelation of its truth." On the right: his *Tabulae Anatomicae* from 1538. Much less famous than the later images from the *Fabrica*, these sketches are in a sense more telling. They are drawn in Vesalius' own hand, and demonstrate the original commitment to Galenism that he was gradually shedding with the development of his project of autopsy: the liver (on the left) is depicted as the origin of the blood vessels and has five lobes.

MD, he was committed to re-establishing Galen's authority. Upon completion, Vesalius was immediately offered Padua's professorship of surgery and anatomy, and took to the task (Figures 8.9 and 8.10).

Yet Vesalius' loyalty to Galen's *empiricism* turned out to be deeper than to the Alexandrian's anatomy. He continued to take Galen's teaching as the default, but whenever his autopsies (literally meaning: 'see for oneself') suggested that the great master was wrong, apparently led astray by his non-human subjects of dissection, Vesalius took it as his duty to correct him. In Vesalius' hands, empirical, detailed accuracy became the core of anatomy, and a new tradition was created, in which every generation proudly corrected the errors of its predecessors. Realdo Colombo (Realdus Columbus, c. 1515–1559), for example, who studied under Vesalius and taught some of his teacher's classes in the 1540s, turned Vesalius' anti-Galenic argument against him. In his own *De re anatomica* (published posthumously in 1559), Colombo demonstrated that many of Vesalius' own anatomical claims were based on dissecting animals rather than humans. Credit was fervently sought, and expertise became necessary: Colombo himself concentrated on the heart and the sexual organs, and demanded recognition for the identification of the pulmonary cycle and the clitoris.

The new anatomy was an exemplary instantiation of the New Science, but it never forsook its roots in painters' studies. On the one hand, teaching in art academies would become a common employment for leading anatomists – Colombo himself was a close collaborator with Michelangelo. On the other, Vesalius gave a new meaning to the 'see for yourself' dictum by instituting visual presentation as a fundamental part of the discipline. His relatively modest and conservative *Six Anatomical Tables* (Figure 8.10) was followed in 1543 by *De humani corporis fabrica* (*On the Structure of the Human Body*, Figure 8.9), containing more than 250 illustrations, in

Figure 8.10 The anatomical theater in Palazzo Bo of the University of Padua, designed by Fabricius, constructed in 1599 and still in place, although the wooden structure is now considered too fragile to allow standing on. The standing-benches are steep and narrow, so the viewers are shoulder to shoulder. The cadaver would be prepared on a stretcher in the lower level that can be seen through the opening at the bottom. The stretcher would be raised to the floor where it would be dissected.

the talented hand of Jan Stephan van Calcar. It corrected some of Galen's errors – like getting rid of the five-lobed liver and the *rete mirabile* – but its seven books were very much arranged according to Galen's logic: from the bones, through the muscles and tendons, then the blood vessels, nerves, organs and generation, the heart; and, finally, the brain. Published in the same year as Copernicus' *De revolutionibus* – 1543 – *De fabrica* is as emblematic of the new age of science.

Conclusion: Tradition, Innovation and the New Body

While Vesalius' images were taking the anatomy spectacle outside the university walls for display, his disciples were carefully marshalling it within them. Restoring seriousness to public dissection was important: the frontispiece of Colombo's *De re anatomica* is much less extroverted than that of *De fabrica* – Colombo is dissecting there amid only a small group of somber men in academic gowns – and Rembrandt's 1632 *The Anatomy Lesson of Dr. Nicolaes Tulp* is even more somber. More important yet was the establishment of a physical place and a standardized procedure for the anatomical dissection. This was provided in 1599 by Hieronymus Fabricius (Girolamo Fabrizio, 1537–1619), who designed the anatomical theater – still to be viewed in Palazzo Bo of the University of Padua (Figure 8.10) and a model for all such theaters around Europe. William Harvey arrived in Padua at the very year of its completion, and although the education he received there still had much of Galen and Ibn Sina, the anatomical theater embedded the new practices he was taught and the new concept of the body emerging from them: a body of clear structures and processes; separated from the world materially and connected to it causally; to be studied by a violent infringement of its boundaries. Harvey's heart and circulating blood were prime representatives of this body.

Discussion Questions

1. What is the significance of the relations between medicine and magic? What does it say about our interest in and our understanding of our body?
2. How different is the holistic conception of the body at the basis of pre-modern medicine from the one we are familiar with? Was rejecting this conception a necessary price for the advantages of modern medicine? Was it a reasonable price?
3. The advent of early modern and then modern medicine was predicated on inflicting a great amount of pain on other animals. Is this an important fact? Does it have bearing on the knowledge gained?
4. Pre-modern physicians and their patients described their pains and maladies in very different terms than we would, referring to very different symptoms and causes. Is it reasonable or useful to say that they actually sensed their bodies differently?
5. Does the role of art in the development of medical anatomy give modern medicine an aesthetic dimension?

Suggested Readings

Primary Texts

Galen, *On the Natural Faculties*, Arthur John Brock (trans.), The Loeb Classical Library (London: William Heineman, 1652), pp. 3–17.

Galen, *Galen on Anatomical Procedures*, Charles Joseph Singer (trans.) (Oxford University Press, 1999), pp. 1–9.

Bald and Child, "Leechbook" in Oswald Cockayne (ed.), *Leechdoms, Wortcunning, Starcraft* (London: Longman, Green Reader & Dyer, 1865), Vol. II, pp. 3–17, 27–31, 57–59, 323–335.

Paracelsus, "Alchemy, Art of Transformation" in Malcolm Oster (ed.), *Science in Europe, 1500–1800* (New York: Palgrave Macmillan, 2002), pp. 99–107.

Sharp, Jane, *The Midwives Book, or the Whole Art of Midwifry Discovered* (London: Simon Miller, 1671), frontispiece and pp. 153–162.

Secondary Sources

General history of medicine:

Porter, Roy, *The Greatest Benefit to Mankind: A Medical History of Humanity from Antiquity to the Present* (London: Fontana Press, 1999).

Ancient medicine:

Kuriyama, Shigehisa, *The Expressiveness of the Body and the Divergence of Greek and Chinese Medicine* (New York: ZONE BOOKS, 1999).

Nutton, Vivian, *Ancient Medicine* (London: Routledge, 2004).

von Staden, Heinrich, *Herophilus: The Art of Medicine in Early Alexandria* (Cambridge University Press, 1989).

Medieval medicine:
Siraisi, Nancy, *Medieval & Early Renaissance Medicine: An Introduction to Knowledge and Practice* (University of Chicago Press, 1990).

Other medicines:
Lo, Vivian (ed.), *Medieval Chinese Medicine* (Oxford: Routledge Curzon, 2005).

Winterbottom, Anna and Facil Tesfaye (eds.), *Histories of Medicine and Healing in the Indian Ocean World*, Vol. 1: *The Medieval and Early Modern Period* (New York: Palgrave Macmillan, 2016).

On the weapon salve and the relations between medicine and magic in general:
Harline, Craig, *Miracles as the Jesus Oak: Histories of the Supernatural in Reformation Europe* (New Haven, CT: Yale University Press, 2011).

Moran, Bruce T., *Distilling Knowledge: Alchemy, Chemistry and the Scientific Revolution* (Cambridge, MA: Harvard University Press, 2005).

On the changes to medicine in the age of global commerce:
Cook, Harold John, *Matters of Exchange: Commerce, Medicine, and Science in the Dutch Golden Age* (New Haven, CT: Yale University Press, 2007).

On the London plague and its significance:
Miller, Kathleen, *The Literary Culture of Plague in Early Modern England* (New York: Palgrave Macmillan, 2016).

Wallis, Patrick, "Plagues, Morality and the Place of Medicine in Early Modern England" (2006) 121(490) *English Historical Review* 1–24.

9

The New Science

Galileo's Mechanical World

Leaving the protective confines of the university for the glamour of the court ended up costing Galileo his freedom, but it wasn't merely a reckless career move. Galileo did have a powerful idea of himself as a *philosopher*, making unreserved claims about the make-up of nature – claims that his university rank as a mathematician didn't allow. Like Kepler, however, he didn't intend to forsake mathematics for philosophy. His project, rather, was to submit the philosophy of nature to mathematics: to claim mathematical foundations for nature and thus designate the mathematician as its most skilled and authoritative interpreter. The mathematical disciplines Kepler reshaped in his quest to fulfill this ambition, as we saw in Chapter 7, were optics and astronomy. For Galileo, the aim was his own discipline of expertise – mechanics.

The Aristotelian Theory of Motion and Its Discontents

Mechanics was (and still is) the mathematical science of solid bodies in motion. To turn it into the foundation of natural philosophy meant turning these elements – bodies and motion – into the foundations of nature. Moreover, it meant giving them – especially motion – a new meaning: a meaning that would serve the tools and skills of the mathematician rather than the analytical tools of the traditional, Aristotelian natural philosopher. This was indeed an ambitious undertaking. Motion, as we discussed above (especially in Chapter 2), was at the core of the Aristotelian philosophy of nature, and it had a rich and precise meaning.

Nature was change – the acorn changed into an oak and the child changed into an adult – and the epitome of change was change of place; local motion (in fact, Aristotelians often used 'motion' as a synonym for all motion, but for simplicity we'll now reserve 'notion' for local motion). Every change required a cause – deciphering these causes was the main task of the natural philosopher – and so did local motion. The cause could be internal – not in a spatial sense, but belonging to the moving body's nature – in which case the motion was natural. The heavy elements belonged at the center of the cosmos, so bodies consisting primarily of them tended to move down naturally, and vice versa for the light elements. Or the cause could be external – a stone could be thrown up or air could be forced down with a bellow – and then the motion was forced, or violent. A body moving naturally carried its cause of motion with it, as it were – the heaviness of the stone, which was the cause of its moving down, was an inseparable property of the stone. So the naturally moving body would only come to rest when it arrived at its natural place or as close as possible to it; the stone would fall until reaching the center of the earth or more likely – hitting the ground. Violent motion, by contrast, would stop immediately when the force causing it ceased to affect the body. Again: change – motion – could only happen for a cause; once the cause ceased, so did the change; once the moving force ceased, so did the motion.

Natural and violent motions were thus metaphysically opposites, so they could not mix – a body moved either naturally or violently, never both. Acceleration and deceleration were self-explanatory: a heavy body falling would naturally get faster as it neared its proper place and vice versa. The velocity of the fall depended on the body's weight: the heaviness was the cause of the motion and an effect is proportional to the cause. Velocity was the measure of the effect. Conversely, velocity was inversely proportional to resistance of the medium in which it moved, hence motion only made sense in a medium (details in Chapter 2).

Aristotelian theory of motion was as well entrenched as any in the history of science, but Galileo didn't need to take on this set of concepts all alone. Neat as it was, it had some serious deficiencies which were noted already in antiquity and deliberated upon in the Middle Ages, and a number of these deliberations were particularly relevant to Galileo's project.

Buridan and Impetus Theory

A particularly acute deficiency of the Aristotelian theory was that it made little sense of the most familiar phenomenon of motion: the flight of projectiles. In simple words: the fact that when we throw a stone or shoot an arrow, it keeps on moving *after* it's left the hand or the bow. According to the Aristotelian analysis, we saw, this shouldn't happen: a thrown stone or shot arrow are in violent motion, so it needs a continuous external cause in order to continue moving. But once a projectile leaves the hand or the bow it is no longer pushed by anything: why then does it keep moving?

Aristotle tried to resolve the mystery by suggesting that it was the air, pushed by the bowstring, that kept pushing the arrow, as it was spread by the arrow's head and curled behind its rear end to prevent a vacuum from being created behind it. To many of Aristotle's own disciples, this was clearly an unsatisfactory explanation. First, the air, in Aristotle's own analysis, was supposed to be the resisting medium – how could it provide both propulsion and resistance? Think of a boat, which keeps moving after its sails are dropped – clearly the water is not what pushes it, because it is the same water that would soon bring it to a halt. Secondly, the same question asked about the arrow and the boat could be asked about the air and water themselves: how do *they* keep moving after detaching from the original cause of their motion? And if instead of a stone or an arrow we think of an object sharp on both ends – like a spear – how can the air move it by pushing against a sharp end? And what about circular bodies and motions, like a potter's wheel, which also continues to spin on its axel for some time after the potter has left it? In the fourteenth century, a professor at the University of Paris, Jean Buridan (c. 1300–c. 1358), collated these arguments to conclude that Aristotle's theory of violent motion had to be wrong. In its stead, he suggested, we need to assume that the moving body – hand, bow or wind – delivers some force to the moved body – stone, arrow or boat. This force, he argued, remains in the moved body and continues to propel it after the two have separated: after the stone has left the hand, the arrow has left the bow or the wind has stopped. He named this force 'impetus.'

'Impetus' was a very fertile concept. It explained not only why projectiles keep on moving, but also why they slow down and stop: this is because the impetus diminishes with time and resistance. It also explained why falling bodies – moving in natural motion – accelerate: the inner force propelling them – their gravity – continuously adds impetus to the falling body, which accumulates and speeds it up. Impetus could even be used to make sense of the unceasing, uniform motion of the heavenly bodies: all God needed to do was to provide them with the appropriate amount of circular impetus at creation, and because there is no resistance to their motion, it continues in perpetuity.

The science-educated modern reader may be tempted to read modern concepts like momentum and inertia into Buridan's theory, but note: in spite of his criticism of Aristotle, Buridan's fundamental understanding of motion is still Aristotelian. For him, motion always needs a cause, and its speed is proportional to the force, or the cause, and inversely proportional to resistance. Some of these ancient commitments Galileo would keep, others he'd discard; sometimes consciously, sometimes not.

The Mysteries of Free Fall
Not only violent motion presented difficulties to Aristotelian natural philosophy, but also natural motion. The paradigm of such motion was the falling body: moving because of its own heaviness, its velocity was assumed to

be proportional to its weight (and inversely proportional to the resistance of the air or water it fell through). But the relations between weight and motion were far less clear than one would have expected. A good example of these difficulties was the phenomenon of the lever, or the balance: the weight of a body put on a balance is proportional to its distance from the center of the balance – the fulcrum. This is an empirical fact that we can tell was well known in Antiquity, because there are many ancient remains and images of the *steelyard*, an instrument based on this principle (Figure 9.1, right). In the Aristotelian world, heaviness – *gravitas* – was the striving of a heavy body towards the center of the Earth, which was its proper place; there was no reason why horizontal displacement, parallel to the center, should change its weight. Aristotelians tried to solve the riddle in terms of motion (as captured in the diagram in Figure 9.1, left) by treating the balance as a compass. They divided the body's circular motion around the fulcrum into a natural component (down) and a forced component (towards the fulcrum), and reasoned that the higher "positional weight" was an expression of the higher ratio of natural-to-forced motion the farther the body was from the fulcrum. This was an ingenious solution, but also a true strain of Aristotelian theory: natural and forced motions were not supposed to be mixed, and weight was supposed to explain motion, not the other way around. More than anything, these attempts demonstrate the difficulty of the Aristotelian analysis in handling the relations between the static and dynamic expressions of weight – on a balance and in free fall.

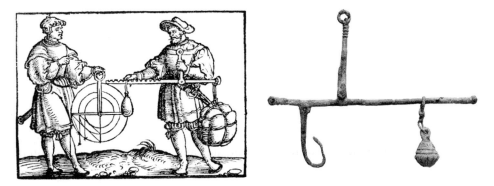

Figure 9.1 The steelyard and the principle of the lever. The image on the left is a woodcut from Gaultherus Rivius' *Architecture ... Mathematischen ... Kunst* (Nuremberg, 1547). This sixteenth-century diagram stresses that the practical tool can be mathematically analyzed. The image on the right, however, is a Roman steelyard, preceding it by more than a millennium. It demonstrates that the principle of the lever was understood well enough to be embodied in a material, artificial device well before it was formulated and formalized.

So it was in free fall that Galileo found a particularly soft spot to assault the Aristotelianism theory of motion, using arguments mostly formulated by sixteenth-century scholars of mechanics, especially Giambattista Benedetti (1530–1590). Common lore and Galileo's own rhetoric credit his turn to empirical observations, but these arguments are strictly abstract – a series of thought experiments. According to the Aristotelian theory, Galileo points out, the velocity of a falling body should be proportional to its weight. This means that a big cannon ball, weighing 1,000 drachms (Galileo's favorite unit, about 4 grams), should fall 100 times faster than a small musket ball weighing 10 drachms. The big ball may be falling somewhat faster, but had anyone ever seen one metal object, no matter how large, fall 100 times faster than another metal object, no matter how small? Even worse is the following insight: attaching a light body (a) to a heavy one (b) makes the combined body (c) heavier than (b), hence, according to the theory, it should fall faster. But (a) was supposed to be slower than (b) – how can the addition of a slower-moving body make the faster-moving one accelerate?

The complementary Aristotelian assumption – that the fall is inversely proportional to resistance – didn't make any more sense. A piece of wood falls through air and rises in water.

But if the velocity of the fall is inversely proportional to resistance, no resistance can change the direction of the falling body – only its velocity. If water is ten times more resistant than air, the velocity of the fall through water should be a tenth of the velocity through air; if it's 100 times more resistant – the velocity should be a hundredth. But no division will change the direction of the fall.

Galileo's conclusion from this consideration was simple: to calculate the velocity of a falling body, resistance should be *subtracted* from the driving force, not divided by it ($V \propto F - R$ rather than $V \propto F/R$). If the resistance is greater than the force, the direction of motion *will* change, as happens to the wood that falls through air and rises through water.

This seemingly minor mathematical alteration had far-reaching metaphysical consequences. $V \propto F - R$ means that resistance could be zero – with zero resistance, the body would move at the velocity determined solely by the original force: $V \propto F$. But a body can suffer zero resistance – no resistance at all – only if it moves through empty space. As you may recall (Chapter 2), this was exactly Aristotle's argument against the possibility of void or vacuum: in a void resistance is zero, and assuming that velocity is inversely proportional to resistance, he argued, that means that through a void bodies would move in infinite velocity ($F/0 = \infty$). Infinite velocity is absurd (it means being in more than one place simultaneously), Aristotle continued, hence a void is absurd, and motion necessitates medium. Galileo's new formula annulled this consideration: in direct opposition to Aristotle, motion through a void *was* possible!

Galileo's Resources
Archimedes and the Simple Machines

This non-Aristotelian concept of resistance hints at the resources Galileo drew on in his assault on Aristotelian natural philosophy and in developing his alternative. One crucial resource came from Antiquity, but only became fully available to Europeans in the middle of the sixteenth century: the works of Archimedes (287–c. 212 BCE).

Archimedes presented a world quite different from Aristotle's. Aristotle populated the terrestrial realm with natural substances moving in straight lines towards or away from their proper places. The elements of Archimedes, instead, were *simple machines* (Figure 9.2) – idealized mechanical devices like the inclined plane, the wedge and the lever, which turned forces into motions and vice versa. Instead of Aristotle's causal stories, this world was best described by mathematical proportions – the law of the lever, namely the ratio between weight and distance from the fulcrum, is one example. These were mathematical laws, but they were neither the wild mathematical

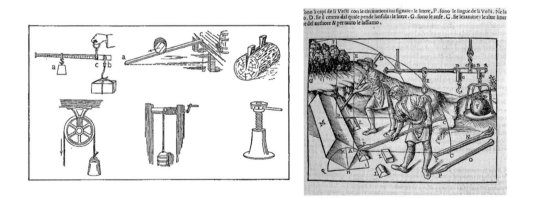

Figure 9.2 The simple machines. On the left, clockwise from upper left: lever; inclined plane; wedge; screw; wheel and axle, pulley. The diagram is from John Mills, *The Realities of Modern Science* (New York: Macmillan, 1919), and it expresses the idea, going back all the way to Archimedes, that geometrical configurations of matter have real, physical effects: the more pulleys are engaged in lifting a body, the lighter it is; the farther a body is from the fulcrum of a lever, the heavier it is. The idea that these machines are both practical implements and abstract principles is nicely captured by the image on the right – an engraving from a drawing by Cesare Cesariano, from a 1521 Italian translation of Vitruvius' *De Architectura* (see Chapter 1). The principle of the lever is embodied both in the two rods used to roll the hewn stone M on the left and in the steelyard on the right.

speculations of the Pythagoreans, abhorred by Aristotle, or the very rigorous mathematics of Euclid, admired by Aristotle's disciples. Euclidean mathematics recognized only constructions with ruler and compass and one motion at a time – that's how its theorems about angles, circles, triangles and so on were proven. Archimedes replaced this rigor with rich flexibility, defining a plethora of curves created by a combination of motions. These were tellingly known as 'mechanical curves' – both because they were produced by motions of solid bodies – the subject matter of mechanics – and because they were, again, idealized machines: they could do work in the world. The screw, which Archimedes famously invented, is a good example of those curves: mathematically, it's produced by a point which moves simultaneously around a given point in one plane and away from it in a line perpendicular to this plane. Mechanically, it pumped water from the Syracusan ships trying to fend off the Roman invaders.

Archimedes' simple machines injected mathematics directly into the world. This concept of mathematics' import in the cosmos was very different from Kepler's dream of perfect divine order, but it fulfilled a similar function with similar success. It made sense of the new physical aspirations of the traditional mixed sciences (see Chapter 7), and elevated the mathematician to the role of natural philosopher. Archimedes also offered Galileo a model for analyzing the complex relations between weight and fall in a different way to Aristotle's. The basis for this analysis was his principle of buoyancy, immortalized in the famous *Eureka* story. Archimedes, so the story goes, was charged by the king to determine if a crown presented to him was indeed made of gold. The challenge was to find if the crown weighed as much as the same quantity of gold should weigh, but because of the crown's complex shape there was no simple way to tell how much metal it actually contained – what its volume was. Seated in the bathtub, the story continues, Archimedes hit on the solution: when he dipped into the water, the water rose, because his body pushed it aside. Here was the Eureka: measure the volume of the water displaced, and you know the volume of the submerged body – Archimedes' or the crown's.

Beyond solving the king's riddle, Archimedes had thus answered the question why heavy bodies float in water: if the water displaced weighed more than a submerged body, the water carried that body. A boat could be made of metal, but if it also contained enough air (or other light materials), then the weight of its combined volume could be lower than the equivalent volume of water, and the boat would float. This was a very different approach to weight than Aristotle's. It was not the total weight of the body that mattered, but its *weight per volume* or 'specific weight.' This was because it wasn't the body's individual drive towards or away from the center of the Earth that determined which body would descend and which would ascend. Rather, *all bodies* were heavy, and it was the relation *between* them that determined which would sink and which would rise.

The principle of buoyancy was, therefore, also an excellent example of the idea of a world made up of simple machines: what was the floating body if not the lighter hand of a balance, moving up not because it was *light*, but because the other hand was *heavier*?

Tartaglia and the Symmetrical Trajectory

Archimedes provided Galileo with one set of resources with which to develop his mathematical-mechanical alternative to Aristotelian natural philosophy; Renaissance scholars like Tartaglia (Chapter 5) provided another set. As you may remember, his *Nova Scientia* promised a marked improvement of artillery practices through the theoretical analysis of the trajectories of cannonballs. This analysis was supposed to be fundamentally Aristotelian and Tartaglia begins the book declaring "that it is impossible for a heavy body to move with natural motion and violent motion mixed together." Yet for the practical use of the "bombardier," this strict distinction needed some amendment; after all, cannonballs do both – fly violently, driven by the force of the gunpowder, then fall naturally by the force of their gravity. So Tartaglia devised the idea he expressed in diagrams like Figure 9.3, top: the projectile moves first in straight, violent motion (H to K); then in circular motion (from K to M); then falls straight down in natural motion (M to N). His explanation as to why balls don't usually fall down straight from the sky, but rather touch the ground or their target obliquely, was that the point at which the trajectory meets the ground is often before the end of that circular part, for example at M.

Whether Tartaglia genuinely believed he found a compromise between theory and experience or was only bluffing, the Aristotelian doctrine didn't really survive his analysis. What was that circular part of the trajectory – KM – if not the forbidden mixture of natural and forced motion? But even more interesting was his further conclusion:

> ... truly no violent trajectory or motion of uniformly heavy body outside the perpendicular of the horizon can have any part that is perfectly straight, because of the weight residing in that body, which continually acts on it and draws it towards the center of the world.
>
> Tartaglia, *Nova Scientia*, in Drake and Drabkin, *Mechanics in Sixteenth-Century Italy*, p. 84

Look again at the diagram at the top of Figure 9.3, Tartaglia's reasoning becomes clear: where exactly should point K be? It's supposed to be the

Figure 9.3 The two contradictory ways in which Tartaglia analyzes cannon ball trajectories in his 1537 *Nova Scientia*. Top: one of the many diagrams of 'triplicate' trajectory reflecting the principle that: "Every violent trajectory or motion of uniformly heavy bodies outside the perpendicular of the horizon will always be partly

point at which the heaviness of the body would start curling its trajectory – but the body is always heavy. So the trajectory should start curling immediately when the ball leaves the cannon, which means, contra Aristotle, that heavy bodies move in mixed motions and curved trajectories! Tartaglia was undoubtedly convinced that this was indeed the case – look at the smoothly curved, symmetrical trajectories in the illustration of his book's frontispiece (Figure 9.3, bottom). In a free drawing, rather than a mathematical diagram, Tartaglia showed no hesitation in offering a clear alternative to the Aristotelian theory of motion he had started with.

Galileo's Investigations
The Parabolic Trajectory
The trajectories of projectiles were smoothly curved and symmetrical – but what was their shape? This was the question for the new mathematical natural philosopher. Mathematics was imprinted *in* nature – perhaps distributed by God through light, as Kepler imagined, or embodied in those simple machines, as Archimedes taught and Galileo learned. It was for the philosopher-mathematician to decipher the mathematical curves that nature drew. The question was to be asked empirically, so the answers could be surprising – mathematics was no longer committed to saving the phenomena with ideal, perfect abstractions. Led by this conviction through the first decade of the seventeenth century (as we saw in Chapter 7), Kepler found that in the heavens, the planets were drawing ellipses – not circles. In the last decade of the sixteenth century, Galileo set out to find what curves heavy bodies were tracking on Earth.

Like Tartaglia, and for the same regrettable reasons, the heavy bodies drawing Galileo's curiosity were cannonballs, and in the 1590s, still a young mathematics professor, Galileo began experimenting with them. Together with his older friend and patron Guidobaldo del Monte (1545–1607), he was

---

Figure 9.3 (Cont.)
straight and partly curved, and the curved part will form part of the circumference of a circle." Bottom: the frontispiece of the 1550 edition. Here the trajectories of the cannon and mortar balls are clearly drawn to look smoothly curved and symmetrical, reflecting his insight that: "truly no violent trajectory or motion of uniformly heavy body outside the perpendicular of the horizon can have any part that is perfectly straight, because of the weight residing in that body, which continually acts on it and draws it towards the center of the world." Whichever way he was truly leaning, this discrepancy reveals the tremendous difficulty of using mathematics – the art of perfect forms – to capture nature in its constant change. Tartaglia hints that this is the main challenge he is undertaking with the allegorical landscape in which he places the artillery pieces in the frontispiece. To enter the first realm of knowledge, all comers must pass through a gate guarded by Euclid. Inside they encounter Tartaglia, flanked by the muses of the various mathematical disciplines – those of the quadrivium first, and the more esoteric, like astrology, behind. Only those who traveled through this mathematical terrain can enter the inner sanctum of *philosophia*, guarded by Aristotle and Plato.

dipping balls in ink and rolling them on an inclined roof, investigating the lines they were drawing (Figure 9.4).

This was quite a new way to ask the question: it was an empirical question, but the objects were not asked to behave as they do in nature. Tartaglia was also studying his cannonballs empirically: debating with the duke's 'bombardiers' as to whether the best range for the cannon would be achieved at an elevation of 30° (as custom had it) or 45° (as he calculated); they wagered and tested and Tartaglia won (it is he who tells the story). The test consisted of actually shooting the cannon, attempting to

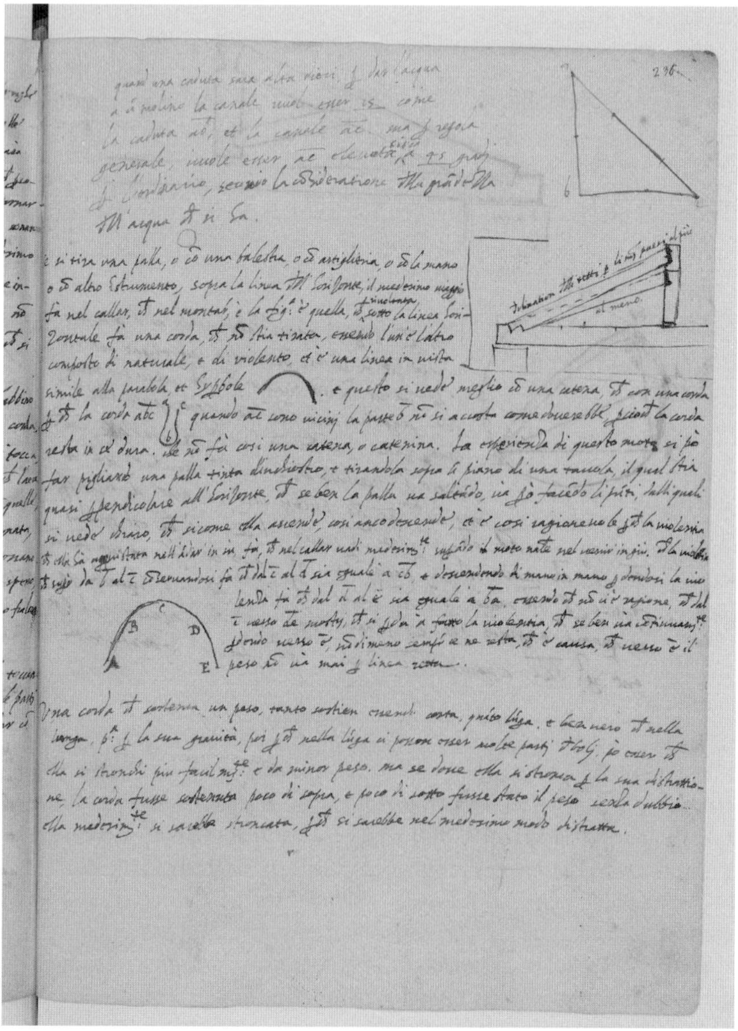

Figure 9.4 Guidobaldo's protocol of his experiments with Galileo. The roof is on the upper right; the curve of the rolled ball at the center, and its analysis as a parabola – bottom left.

keep all irrelevant factors (direction, amount of gunpowder, etc.) in check. Galileo and Guidobaldo, instead, created an artificial environment – balls rolling on a roof. It served their purposes, because the inked balls left a trace on the roof, a trace they could investigate. But this convenience came at the expense of the authenticity of the experiment: the assumption that slow-rolling balls truly represent fast flying ones was far from obvious. This reversal of hierarchy between natural and artificial would become a centerpiece of the new experimental study of nature, and we'll return to it below.

The experiments convinced Galileo and Guidobaldo that Tartaglia was right in claiming that the trajectory was symmetrical, but not if he really thought that it was a segment of a circle (the continuously curved trajectory on the frontispiece – Figure 9.4 bottom – indeed doesn't look circular). What *was* the curve then? They settled on the parabola. Why? Maybe it was a guess similar to Kepler's: Kepler decided that the planetary orbit – a closed curve – could not be a simple circle, and the next most-simple *closed* conic section was the ellipse. Galileo and Guidobaldo decided that the trajectory – an open curve – was not a circle, and the next simplest *open* conic section was the parabola.

But perhaps the parabola originated in a more educated guess: Galileo, practically inclined and curious, may have been familiar with masons' knowledge about the stability of the parabolic arch (the knowledge implemented by Brunelleschi building the Duomo – see Chapter 5). Yet why should the stability of a static curve intimate anything about the shape of a dynamic trajectory?

For Galileo, the answer would have come from Archimedes: it was because the two different phenomena actually represented the same mechanical principle. A trajectory is created by the combination of the weight of the cannonball and the force produced by the gunpowder explosion spontaneously balancing each other. Similarly, an arch is stable when the vertical weight of the stones and the horizontal forces of the structure balance each other. In other words: an arch is stable when its weight is distributed in the curve it assumes spontaneously – like a trajectory – otherwise the weight would stress the arch towards that curve and destabilize it. Hence, if a stable arch is a parabola, so should be the trajectory.

Galileo never reveals if this was his line of thought concerning trajectories, but he does provide hints that he was convinced that the parabola is the curve which will spontaneously result from a combination of weight and horizontal pull. He even thought that a freely hanging chain is also a 'simple machine' of the same kind – an inverted stable arch or trajectory – and should therefore also be a parabola. Here he was wrong: a chain line – a

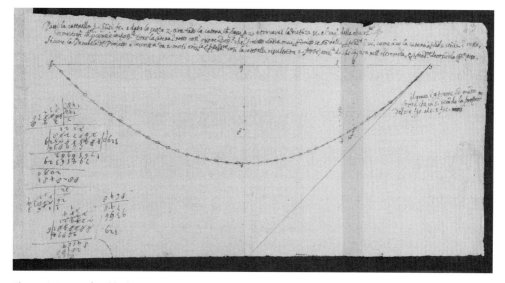

Figure 9.5 One of Galileo's attempts to reduce a chain line into a parabola. He drew the curve by hanging a chain against a piece of paper and marking through the links, and in the calculations on the left he tries to find a regular ratio between the empirically given curve, drawn by nature herself, and a similar parabola. Galileo MS72, 43 r.

catenary – is a more complex curve, but throughout his life he was hanging chains, marking them on the wall and attempting to reduce them to parabolas (Figure 9.5).

Pendulums, Inclined Planes and the Law of Free Fall

Galileo was constructing a mathematical philosophy of nature using ancient and Renaissance resources, and it was as different from Aristotelian natural philosophy as Kepler's astronomy was from Ptolemy's. The strict distinction between natural and violent motion was broken – they were mixed in the curved trajectories; as was the strict distinction between motion and rest – static and dynamic phenomena were explained in parallel ways. In place of the Aristotelian causal narratives came analysis in terms of mathematical structures. Symmetry in particular became a powerful tool in Galileo's hands. It was the feature common to the trajectory, the arch and the hanging chain, as well as the pendulum.

The pendulum was interesting exactly because it was so un-Aristotelian: an earthly motion that was circular, without a terminus, and seemingly could continue forever. Drawing on another of his resources – Buridan and his impetus theory – Galileo could make sense of this continuous motion. The 'impetus' gained by the pendulum on its way down should be sufficient to elevate it to the very same height on the other side, so it would now have enough impetus to return to its original height and so forth. Only the resistance of the air and the friction of the string should ever make the pendulum stop. Crucially, for Galileo, the symmetry was *not* created by the circular path of the pendulum, but by this exchange of forces and motions (Figure 9.6, left). This is why, even if a nail stopped the string along the way, the cord of the pendulum would wrap around it and deliver the bob (the weight at the end of the cord) to the same height.

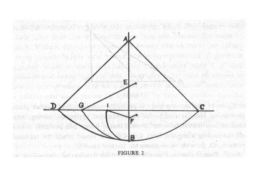

FIGURE 2

Figure 9.6 On the left: Galileo's stopped pendulum from his *Two New Sciences*. The pendulum oscillates around A and starts descending from C. It will always get to the same height of C along line CD, even if stopped, whether the stop is above this line at E, or even under it – at F or B. Galileo reasons that what determines how high the pendulum would ascend (ignoring friction) depends solely of the amount of *momento* – the amount of driving force – it receives during the fall. This amount of force would suffice to bring it to exactly the same height, independent of the slope in which it falls or rises. On the right: Galileo's pendulum clock design. The same line of reasoning convinced Galileo that the circular pendulum is isochronous – its period will remain the same (for a given length of the cord), regardless of its amplitude. This is because increased amplitude will increase its restoring force, and the consequent increase in velocity will exactly compensate for the lengthening of the trajectory. In 1641, already blind, he dictated the idea to his son Vincenzo (named after his father), who drew this design. The first working pendulum clock was designed by Christiaan Huygens (1629–1695) in 1658–1659. A few years later, he proved that the curve of isochrony is *not* the circle of Galileo's pendulum, but the *cycloid* – the curve drawn by a point on a rolling wheel. In pendulums of very small amplitude, however, the difference between the circle and the cycloid is negligible, hence the very long circular pendulums with very short amplitude of 'grandfather' clocks.

Within the Archimedean framework, this line of thought proved extremely fertile: the pendulum is yet another simple machine, and the descending bob is just like a ball rolling down an inclined plane – one of Archimedes' simple machines. Just as the bob will always ascend to the same height, even if the string is restrained by a nail, so the descending ball, if presented with a rising inclined plane at the end of its role, will roll up on it to its initial height – regardless of how steep either incline is. But what if the slope on which the ball ascends is very gentle? So gentle that the ball would *never* get to its original height? Then it would continue to roll, forever (if we ignore friction), because the force moving it – the impetus from the fall – would never be exhausted. It would, in fact, require an opposite force to stop it.

Galileo thus invented a new kind of motion: 'neutral motion,' with no designated beginning or end, neither accelerating nor decelerating, neither forced nor natural:

> A body subject to no external resistance on a plane sloping no matter how little below the horizon will move down in natural motion, without the application of any external force. This can be seen in the case of water. And the same body on a place sloping upward, no matter how little, above the horizon, does not move up except by force. And so the conclusion remains that on the horizontal plane itself the motion of the body is neither natural nor forced. But if its motion is not forced motion, then it can be made to move by the smallest of all possible forces.
>
> Galileo Galilei, *On Motion and Mechanics*, I. E. Drabkin and
> Stillman Drake (trans. and ann.) (Madison, WI: University of
> Wisconsin Press, 1960), p. 66

It is worth stressing, for readers familiar with modern mechanics, that Galileo *does not* think of this kind of motion as we do of 'inertial motion.' He never suggests that motion could continue without at least "the smallest of all possible forces"; he doesn't think that he has discovered a general law of motion – this is specifically an analysis of motion *on Earth*; and he doesn't think of this motion as rectilinear – it's along the horizon, so will continue in a very big circle *around Earth*. But the change to the analysis of projectile motion that this new concept of 'neutral motion' allowed is dramatic.

The parabola of the projectile's trajectory, Galileo reasons, is a mixture of forced and natural motion – against Aristotle's injunction: rectilinear motion created by force in the direction it's aimed; and rectilinear motion downwards, created by the heaviness of the body. In fact, instead of forced and natural, these motions are better understood as 'neutral' and 'accelerated.'

The neutral motion, like that of the ball rolling, horizontally, proceeds in uniform velocity; the motion down accelerates. But at what rate? The 'neutral,' horizontal ingredient, in its uniform velocity, covers distances proportional to times, and in a parabola the vertical line is proportional to the square of the horizontal line (in modern notation: $y = x^2$). Hence, the distances of the free-fall component of the projectile motion are proportional to the square of times, or: *a freely falling body covers distances proportional to the square of the time of fall.* The first law of modern mechanics has been formulated. And how is this proportion created? After some confusion, Galileo comes up with a proper law of acceleration: *a freely falling body adds equal increments of velocity in equal times.* Weight has nothing to do with the velocity of its fall! The foundations of Aristotelian theory of motion have been heavily battered.

Descartes and the Mechanical Philosophy

A mathematical philosophy of nature required a mathematical world to study, and Galileo provided that as well:

> I do not believe that for exciting in us tastes, odors and sounds there are required in external bodies anything but sizes, shapes, numbers, and slow or fast movements; and I think that if ears, tongues, and noses were taken away, shapes and numbers and motions would remain, but not odors or tastes or sounds. These, I believe, are nothing but names, apart from the living animal, just as tickling and titillation are nothing but names when armpits and the skin around the nose are upset.
>
> Galileo Galilei, *The Assayer*, in Stillman Drake and C. D. O'Malley
> (eds. and trans.), *The Controversy on the Comets of 1618*
> (Philadelphia, PA: University of Pennsylvania Press, 1960), p. 311

There is nothing in Galileo's world but properties fit for the mathematician to study: "sizes, shapes, numbers, and ... motions." All the rest – all those properties that make the world what it is for us – "tastes, odors and sounds" – are just that: the impressions that the objects make *on us*. They are not *real* properties of things, and the challenge for the new natural philosopher is to explain them *away* – to reduce them to those mathematical properties, which are the only real ones.

In the two decades after Galileo's trial, this formulation of the task of the New Science, and this new ontology he devised to make sense of it, came to be called the *mechanical philosophy*, and its main prophet was the Frenchman René Descartes (1596–1650).

Descartes' Life and Times

Descartes' lifetime overlapped with Galileo's and Kepler's, but his biography, almost until his premature death, reads as though belonging to a different era: outside both university and court, it's a biography of an independent, urban intellectual. Born to minor Catholic gentry in La Haye (now named after him – Descartes), Descartes was educated at the Jesuit college of La Flèche, which provided him with a strong footing in both Aristotelianism and the Humanist curriculum. He spent only two years at the University of Poitiers, earning a license in law, and at the age of 20 he moved to Paris. There he wrote his first original tract, on music, for the Minim Friar and Polymath Marin Mersenne (1588–1648), who would become his lifelong intellectual interlocutor. Two years later, in 1618, he traveled to Breda, where he took a practical-math-ematical course in military engineering and joined the Dutch army under Maurice of Nassau, apparently more as a physician than a warrior. For three years, he traveled Europe as a mercenary in the Thirty Years' War until discharging himself and commencing traveling on his own. Descartes then returned to his hometown and, in 1623, he sold his share of the parental estate, bought bonds whose interest would support him, modestly but com-fortably, for the rest of his life, and returned to Paris. Even mighty Paris was apparently too quiet for him and, in 1628, he moved to the most bustling European center of the time – Amsterdam, where in the crowds, he said, he found the solitude and freedom for his ambitious intellectual work. His ear-lier adventures turned out to be of value: he became quite well known for his medical knowledge, mostly by word of mouth, and he formed a working alliance with his erstwhile teacher from military school: Isaack Beeckman (1588–1637), also a physician with interests in the new mechanical philoso-phy. For some ten years (until priority disputes soured the relationship), the two worked together on problems of mechanics and hydraulics, developing the mathematical tools and physical principles for this new science.

Descartes never gave much credit to Galileo, but it's very clear that he was keenly aware that they were both involved in the same project: when the news of the Italian's trial came, he canceled his plans for a big work that was to be named *The World*, and instead, in 1637, published its contents more modestly, as three independent essays collected together and fronted by his first famous philosophical tract – *On Method*. In one of these essays – *La Géométrie* – he laid the foundations for a new kind of mathematics, by teaching his readers how to solve geometrical problems with arithmeti-cal operations. His *Méditations* (1641) and *Principes de philosophie* (1644) made him the most influential natural philosopher in Europe. His *Passions of the Soul* (l650), which presented quite a different approach to knowledge than his youthful ambitions for certainty that he's commonly identified with, had to be published shortly after his death. In the winter of 1649, after years of being self-supporting, he accepted a position as court philosopher to Queen Christina of Sweden. The queen, returning from a horseback hunting expedition just before Christmas, liked to be taught philosophy over breakfast, and Descartes, whose writings are dotted with discussions of sleep and dreams, and whose brilliance bought him a dispensation to rise late even in La Flèche, succumbed to a frigid Swedish winter morning. He caught pneumonia six weeks into his tenure and passed away.

The Mechanical Ontology

In Descartes' confident hands, Galileo's poetically expressed ideas turned into full-blown ontological foundations for a mechanical philosophy of nature. The only elements that this philosophy allowed in nature were those which could be fully captured by mathematics: matter and motion. Matter itself – which for the Aristotelian was such a rich metaphysical category – was reduced to mathematical dimensions – the space it occupied or its 'extension.' Color, smell – all unquantifiable properties, even hardness – were 'secondary'; derivatives of the 'primary,' mathematical properties. And motion was only relative change in the geometrical configuration of parts of matter – it didn't entail places, termini or resistance. It didn't require causal explanations like those of the Aristotelian natural philosopher, but mathematical analysis, like Galileo's analysis of free fall.

The Aristotelian education Descartes received at La Flèche is often apparent in his terminology and modes of argumentation, but the contents of his mechanical philosophy are explicitly and dramatically different. Even though he couldn't allow empty space or true atoms – if matter and extension are one, all space is *ipso facto* full and everything that has dimensions can be divided – his image of the world is much more akin to that of the Atomists. It's a world devoid of any forces, places or forms, and populated only with hard material 'corpuscles' which constantly collide. Collision is the only real form of causation in Descartes' world – no formal or final causes allowed – and this causation is conserved. Namely: both the amount of motion in each collision and the total amount of motion in the world remain the same, even if the directions and speeds of the colliding corpuscles change.

Descartes practiced what he preached, making a concerted effort to formulate the mathematical laws of collision. He wasn't successful with the details – his laws were amended within a generation – but the image of corpuscles exchanging motions, together with Descartes' principles of how to describe these motions mathematically, came to define the New Science for the next generation of natural philosophers. He also provided them with the fundamental laws governing the motions (Figure 9.7):

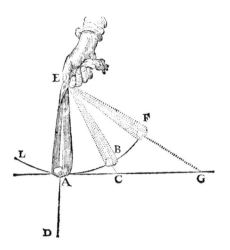

Figure 9.7 Descartes' way of demonstrating his laws of nature. The stone whirled by the hand will only remain in the circular trajectory as long as it is constrained by the sling. Once released (say at A), it will move in a straight line (ACG) and in uniform velocity. This centrifugal pull is also responsible for the sling remaining taut along EAD, although how the tangential tendency becomes radial was only solved only much later by Descartes' disciple Christiaan Huygens.

> *The first law of nature*: that each thing, as far as it in its power, always remains in the same state, and that consequently, when it once moved, it always continues to move …
>
> *The second law of nature*: that all movement is, of itself, along straight lines; and consequently, bodies which are moving in a circle, always tend to move away from the center of the circle which they are describing.
>
> Descartes, *Principles of Philosophy*, pp. 37–40

These laws were Descartes' version of Galileo's conclusion that a body moving along the horizon will continue until stopped by force, but as *Laws of Nature* they were no longer insights about motion on the surface of the Earth. They were universal truths concerning *all* motions of *all* bodies in the world: motion in a straight line and uniform velocity was not a form of change but a state. It didn't need a cause to continue, only to change: to start, stop, or alter direction or velocity.

The New Mechanized Sciences

This was a final divorce from the Aristotelian way of understanding nature which directed scholars throughout Europe, North Africa and much of Asia for two millennia. Historians argue over whether the change should be termed 'revolution,' and that of course depends on how one uses the term. There is no doubt, on the one hand, that Galileo, Kepler, Descartes and others who engaged in the New Science perceived themselves as bringing

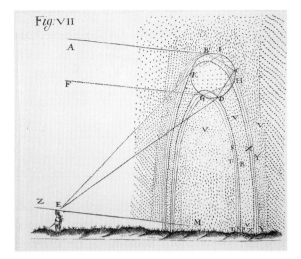

Figure 9.8 Descartes' analysis of the rainbow from his *Meteors* of 1637. The ray coming from A passes through the raindrop, which Descartes mimicked with a water ball. It is refracted at B, reflected at C and refracted again at D, and the angle in which it falls on the eye E determines its color.

about a dramatic and radical change, doing away with old ways of knowledge and creating new ones – almost at any cost (mostly intellectual, but sometimes personal). But neither is there any doubt that the change wasn't instantaneous and wasn't *ex nihilo*, as the cathedral metaphor illustrates: novelties are constructed over time, using available resources. Galileo's work was an essential resource for Descartes, and Galileo himself drew heavily on Archimedes, Buridan and sixteenth-century mechanics. Kepler's work was as essential: his optics (Chapter 7) provided Descartes with a model of purely mathematical-physical science; the unpublished *Le Monde* (*The World*) was to be subtitled *A treatise on Light*.

Light, according to Kepler's new optics, was a physical agent that could be fully analyzed in mathematical terms, and Descartes demonstrated this in his analysis of the rainbow, a phenomenon which baffled traditional natural philosophers. Aristotle taught that transparency was a property of some media, like water and air, while opaqueness was a property of solid objects, and color was a form of opaqueness. So how could colors float in mid-air? Descartes' answer was both mathematical and empirical. The light ray, he explained, is refracted and reflected multiple times within the raindrop, as diagramed in Figure 9.8, and the colors are determined by the angles of these changes in direction – red, he discovered by experimenting with a water ball, is the result of a 42° angle of refraction. It is likely that thinking of light as pure motion, always rectilinear, is what allowed Descartes to turn Galileo's ideas about 'neutral motion' *around* the Earth into the law that straight motion of uniform velocity needs no cause (to continue). Descartes would go on to try to develop, along the same principles, theories of planetary motions, of meteors and comets, of magnetism, physiology and human sensations. Although none was as successful as his version of Keplerian optics, these attempts became the paradigms of the new mechanical philosophy of nature.

Founding the New Science

The Collapse of the Old Order

Europe in the first half of the seventeenth century – the decades that saw the emergence of the New Science in the work of Kepler, Galileo and Descartes – was a troubled place. The tensions building up for a century – since the Reformation – finally burst in 1618 in a war that would last until 1648, earning the moniker 'The Thirty Years' War.' It began as a squabble between Catholic and Protestant countries, but as Descartes' mercenary adventures demonstrate (he was a Catholic; Prince Maurice, his employer and commander – a Protestant), it soon deteriorated into a cruel all-against-all war, driven by greed and opportunism. To the direct wreckage it caused – in some parts of Germany more than 50 percent of the population was lost – a new sort of devastation was added. The plethora of princes and fiefs rushing to enjoy the spoils were making use of new banking techniques (Chapter 5), taking monetary obligations way beyond their means. Promissory notes with princes' signatures and coins with kings' images flooded Europe, losing their value correspondingly. Inflation was a new and strange phenomenon: how could money, the allegedly transparent means of exchange, have a price? What did it represent? And what did the prince's image on the coin stand for, if it didn't guarantee its value?

The crisis extended to all forms of stability and lawfulness: religious, political and economic; and its most rattling expression came, of all places, from England. Physically detached from the continent and institutionally detached from the Church already a century earlier by Henry VIII, it should have been less affected. In fact, the war sent its worst shock through the old relations in England between the divine and political order, which also made sense of the natural order philosophers were searching for (Chapter 1). King Charles, keen to join the seemingly profitable great war, kept raising taxes to finance his (unsuccessful) military ventures until Parliament rebelled. The members of the House of Commons – lesser nobility and representatives of major towns – felt that tradition and common law gave *them* the authority over taxation, and were willing to fight for it. Charles' suspect religious affiliations – he was of Scottish descent and married to a Catholic – and his (failed) attempts to arrest the House leaders exacerbated the rift. In 1642, a proper war erupted between Parliament's army and the king's army, which ended on January 20, 1649 with the king being put on trial:

> Charles Stuart has been and is the occasioner, author, and continuer of the said unnatural, cruel and bloody wars, and therein guilty of all the treasons, murders, rapines, burnings, spoils, desolations, damages and mischiefs to this nation, acted and committed in the said wars or occasioned thereby ... And the said John Cooke [the prosecutor] on behalf of the people of England does for the said reasons and crimes impeach the said Charles Stuart as a tyrant, traitor, murderer and a public and implacable enemy to the Commonwealth of England.
>
> Quoted in Kesselring, *The Trial of Charles I*, p. 34

On January 26, Charles was found guilty and condemned to death; four days later he was beheaded.

This was not the first time in history that a king found his violent death in the hands of his angry subjects, but it was the first time a king was put on trial. But on whose laws? By whose authority? Wasn't the king the "author" of all laws? How could a king be a traitor? Against whom? It was Charles, when he had finally addressed the court – the Parliament whose jurisdiction he never recognized – who put the dilemma most eloquently:

> A king cannot be tried by any superior jurisdiction on earth ... If power without law may make laws – may alter the fundamental laws of the Kingdom – I do not know what subject in England can be sure of his life, or anything that he calls his own.
>
> Quoted in Kesselring, *The Trial of Charles I*, p. 41

But tried he was, and executed.

Europe descended into a deep void absent of religious and political authority, which the New Science, with the new knowledge it was offering and the new practices and instruments by which to gather this knowledge, could have filled. It could have, like other unorthodox movements in centuries past, become a 'heresy' – a populist religion of nature making moral and social demands. Most of these movements were quashed, but the Reformation proved that they could also succeed. Galileo, we saw, gave Bellarmine good reasons to worry that this was exactly the danger that he represented.

Yet the New Science didn't redirect its revolutionary tendencies from knowledge to politics or religion. Instead, it shaped itself as a progressive, reformist project, supportive of and cooperative with the institutions of power. Three persons were particularly influential in formulating this stance – a project of radical epistemology in the service of conservative politics: Francis Bacon (1561–1626), Descartes, in his philosopher's hat, and Robert Boyle (1627–1691).

Bacon's Idols

Bacon belonged to a new erudite English elite. His mother, Lady Anne Cooke, daughter of Edward IV's tutor, the great humanist Anthony Cooke, was famous for her learnedness. His father, Nicholas Bacon, was knighted when he became Lord Keeper of the Seal – the second highest rank in Elizabeth's court. Francis studied in Cambridge, then sought a career in law when his father's death impoverished the family. He became a confidential advisor to the most powerful Earl of Essex (power won and lost through Elizabeth's grace), then a Member of Parliament and, in 1613, Attorney General. When James succeeded Elizabeth, Bacon's stock began to rise, and a series of impressive court intrigues landed him the Lord Keeper position in 1617, and the following year he bettered his father to become Lord Chancellor and Baron Verulam. Similar intrigues brought about his downfall in 1621: he was charged with bribery, found guilty upon his own admission, fined 40,000 pounds, sentenced to the Tower of London and prohibited from holding office or sitting in Parliament. The sentence was reduced, no fine was paid and only four days were spent in the Tower, but his political career was over and he dedicated his last years to philosophy and science. In March 1626, while riding, Bacon decided to experiment with the effect of cold on the decay of meat. He purchased a chicken, stuffed it with snow (Europe was much colder in the seventeenth century), caught a cold that developed into bronchitis, and died on April 9.

As the frozen chicken story macabrely demonstrates, Bacon took himself seriously as an active, empirical philosopher of nature. His *Sylva Sylvarum* (*Forest of Forests*), published only posthumously, is a massive compilation of such work, and as its name attests, very much in the tradition of natural magic (Chapter 6). But Bacon's most influential works were philosophical: *The Advancement of Learning* of 1605 and *Instauratio Magna* (*Great Revival*) and *Novum Organum Scientiarum* (*New Organ for the Sciences* – paraphrasing the title of Aristotle's logical works – the *Organon*) of 1620. Together they lay out a strongly empirical program for the New Science, whose clearly benevolent ambitions he summarized in a 1592 letter to his uncle, Lord Burghley: "I hope I should bring in industrious observations, grounded conclusions, and profitable inventions and discoveries."

As befitting a statesman and a Humanist (see Chapter 5 concerning the Humanists' ideas of knowledge), Bacon's science was to be responsible, concrete and beneficial; it was not aimed at elevating the soul or deciphering God's creation, but at the betterment of life on Earth. He had some practical ideas about how to achieve these goals. He offered a new systematic rearrangement of the disciplines of learning and a method of collecting empirical data in 'tables of presence and of absence' from which one could presumably discover the hidden forms of things by carefully comparing their apparent properties. But all in all, again as the story of his death reveals, he had a limited understanding of the complex, instrumental and mathematical route that empirical research was given by the savants of his generation – notably Galileo and Kepler. His great influence was on the way the budding science began to perceive and present itself – and thus on the way it still does – and he achieved this influence through the critical part of his philosophy even more than through his methodology. His harsh critique of the sciences – both those taught at the universities and those practiced at their fringes – is summarized well in this letter:

> ... if I could [only] purge [philosophy of nature] of two sorts of rovers, whereof the one with frivolous disputations, confutations, and verbosities, the other with blind experiments and auricular traditions and impostures, hath committed so many spoils.
>
> Francis Bacon, in J. Spelding *et al.* (eds. and trans.), *The Works of Francis Bacon* (Cambridge, MA: Riverside Press, 1900), Vol. VIII, p. 109

"Disputations, confutations, and verbosities" is a disparaging reference to Aristotelianism; "blind experiments and auricular traditions" – to traditions of practical know-how, including alchemy and natural magic. Just as Galileo had been doing in Florence and Descartes was beginning to do in Amsterdam, the Englishman was vehemently rejecting tradition. Aristotelianism, on which they were all educated, with its sophisticated terminology and practices (Chapters 4 and 5), was a favorite straw man, but *tradition* was their real target. Knowledge had to start anew.

Knowledge had to start anew because it could get much further than tradition aspired to or allowed. There's an air of complaint to Bacon's words, but they express, in fact, tremendous optimism and confidence in the human capacity for knowledge in general and in the new ways of achieving it in particular. We err because we're deceived, but we can learn to identify and free ourselves from these deceptions. In his *Advancement*, and especially the *Novum Organum*, Bacon classifies these deceptions, poetically, as different *Idols*. The "Idols of the Tribe" arise from our limited nature, deceiving us because "human understanding is like a false mirror, which, receiving rays irregularly, distorts and discolors the nature of things by mingling its own nature with it." "The Idols of the Cave" are the deceptions stemming from one's individual education and habits, "[f] or everyone ... has a cave or den of his own, which refracts and discolors the light of nature" – clearly a reference to Plato's cave. Distortions owing to prejudice, prejudgment, "unfit choice of words," etc. he names "Idols of the Marketplace," and the "Idols of the Theater" are those stemming from "various dogmas of philosophies."

Descartes' Common Sense

Bacon's hopes were mild and responsible; but the means were radical: all we usually take as the basis for gathering knowledge – learnedness, education, tradition, our senses – are obstacles, rather than foundations, and we need to start our inquiry by ridding ourselves of them. Some fifteen years later, Descartes would express a similar sentiment in his *On Method*:

> ... as soon as I was old enough to emerge from the control of my teachers, I entirely abandoned scholarship, resolving to seek no knowledge except what I could find in myself or read in the great book of the world.
>
> Descartes, *Discourse on Method*, Part I

Descartes is much more careful than Bacon in admitting the value of reading books and studying from the learned, if only "in order to know their true value and guard against being deceived by them." He was also much more abreast of the New Science: he conducted sophisticated experiments, built instruments and deeply understood – and worked to shape – the role mathematics would play in it. But most fundamentally, his idea of science was akin to Bacon's: optimistic, unbound, unmediated, personal and open:

> Good sense is the best shared-out thing in the world; ... the power of judging well and of telling the true from the false – which is what we properly call 'good sense' or 'reason' – is naturally equal in all men.
>
> Descartes, *Discourse on Method*, Part I,
> www.earlymoderntexts.com/assets/pdfs/descartes1637.pdf, p. 1

So, like Bacon, he offered a 'method,' which was supposed to ensure that everyone could participate in the marvelous creation of new knowledge:

> As for me, I have never presumed my mind to be in any way better than the minds of people in general. Indeed, I have often had a sense of being less well-endowed than others: I have wished to be as quick-witted as some others, or to match their sharpness and clarity of imagination, or to have a memory that is as capacious ... But ... in one respect I am above the common run of people.

Ever since my youth I have been lucky enough to find myself on certain paths that led me to thoughts and maxims from which I developed a method; and this method seems to enable me to increase my knowledge gradually, raising it a little at a time to the highest point allowed by the averageness of my mind and the brevity of my life ... So I don't aim here to teach the method that everyone must follow if he is to direct his reason correctly, but only to display how I have tried to direct my own.

Descartes, Discourse on Method, Part I, pp. 1–2

Even more than Bacon's, Descartes' idea of method remained vague. Early in his intellectual career, he seems to have truly believed that he could formulate "Rules for the Direction of the Mind" – as he called a treatise he began writing when he moved to Amsterdam. These rules were meant to lead by steps of mathematical certainty, from 'clear and distinct ideas' to indubitable conclusions. But Descartes never completed this project, and not for negligence. As he honed his skills in the new mathematical, experimental natural philosophy, the idea of such a method faded, and his vast scientific work was as eclectic and opportunistic as that of any cathedral builder, real or metaphorical.

Still, the philosophical conviction behind the idea of method was undoubtedly genuine. It embedded the tremendous hope that, done 'the right way,' human knowledge has few boundaries. Galileo had formulated this optimism in the most radical way:

I say that the human intellect does understand some [propositions] perfectly, and thus in these it has as much absolute certainty as Nature itself has. Of such are the mathematical sciences alone; that is, geometry and arithmetic, in which the Divine intellect indeed knows infinitely more propositions, since it knows all.

Galileo Galilei, Dialogue Concerning the Two Chief World Systems,
Stillman Drake (trans.), 2nd edn. (Berkeley, CA: University of California
Press, 1967), p. 103

God knows *everything*, and humans – only very few things. But whatever we *do* know – notably mathematics – we know as well as God. To Church authorities and more conservative-minded scholars claims like this were much more alarming than any argument for the motion of the Earth. Medieval philosophers made the distinction between human and divine knowledge a fundamental part of both their theology and epistemology. God didn't only know everything, He also knew it differently from humans. Our knowledge was *discursive* – it had to be put in into words and sentences, and thus necessitated generalization and categorization, which meant disregarding much of the true essences of particular things. God's knowledge, on the other hand, was *intuitive* – He knew each and every part of His world 'from within,' wholly and individually. God knew the world as *its maker* – to suggest that we could know it in the same way was to suggest that humans could somehow emulate Creation – an inexcusable show of hubris. Yet as we saw in Chapter 6, even the practical Bacon's ambitions were not far removed from this near-heresy, and the priest-scientists of "Salomon's House" in his *New Atlantis* brag that they have, indeed, unlocked the mysteries that would allow them to mimic creation: "we make demonstrations of all lights and radiations; and of all colours ... we practice and demonstrate all sounds and their generation ... We imitate smells, making all smells to breathe out of other mixtures than those that give them ..." (Bacon, *New Atlantis*, pp. 33–35).

The Academies

It is not surprising that Bacon was promoting his idea of the New Science by envisioning a new kind of institution, like Salomon's House. Knowledge needs a place, and the New Science needed a new place. The universities were too entrenched in the traditional conglomerate of beliefs and practices that directed education and scholarship for centuries while giving them a measure of freedom (Chapters 4 and 5). Court sponsorship, as both Tycho and Galileo learned, was too capricious to trust.

New institutions of knowledge indeed started sprouting, informally, already in sixteenth-century Italy. The model may have been the *Accademia Aldina*, which the publisher and humanist Aldo Manuzio (1452–1515) had established in Venice, in 1492. It was supposed to be a "new academy" – a new incarnation of Plato's *Academia* – whose purpose was to help him solve the philological challenges of printing classical texts, and it comprised, as its successors would, gentlemen of independent means and intellectual interests who met regularly to discuss new ideas. Later academies would have more practical, and increasingly experimental, agendas. The *Florentine Camerata*, an academy concentrating on music and music theory, sponsored Vincenzo Galilei, and we've mentioned the *Accademia dei Lincei*, which supported Galileo. To these one can add the *Accademia del Cimento* ('of experiment'), founded by Galileo's students in Florence in 1657.

These academies usually didn't survive the death of their founder, so it was an important milestone when, in 1635, Cardinal Richelieu, the most powerful man in French politics, convinced Louis XIII to grant 'letters patent' to a Parisian group interested in French philology. This created the *Académie Française* – the first state-sponsored academy with an official status – and allowed its almost-continuous existence to our time. It also provided the great French politician of the next generation – Jean-Baptiste

Colbert – with a model he could offer The Sun King, Louis XIV: a state-sponsored scientific academy, whose members and statutes were determined by the state and its representatives. This institution was established officially in 1666 as the *Académie des Sciences*, and together with the *Académie Française* it became the core of the *Institut de France*. Perhaps more than any other institution, the *Académie des Sciences* represents the unofficial pact that the New Science struck with the political elites – quite a different one from that represented by the *universitas*. The state would provide funds and institutional support, while the new scientific institutions would take into consideration the state's priorities – commerce, industry, transportation, arms – and work to enhance them.

Boyle and the Royal Society

It is also not surprising that the Baconian rhetoric appealed to practical men like Richelieu and Colbert: less "verbosities" and more "demonstrations," "perfumes" and "engines" (see full quote in Chapter 6). The activists of the New Science were no less keen about their side of this pact: the flagship of all early modern academies put much of its early energy into acquiring the mandate to call itself *The Royal Society of London for Improving Natural Knowledge.*

The Royal Society evolved from two informal groups. One called itself "The Invisible College," and had been meeting from about 1645 around London, especially in Gresham College. The college was a public education institute of the city, named after its 1597 founder, Thomas Gresham, who endowed a number of resident professors to give regular open lectures. The other group was "The Experimental Philosophical Club," which used to meet in Wadham college, Oxford, where its leader, John Wilkins (1614–1672), was the warden. The two were brought together by a person for whom the realization of Bacon's vision was a life-long undertaking, with deep personal and even religious significance: Robert Boyle (1627–1691).

Robert was the youngest son of Richard Boyle, an 'adventurer' who had made an enormous fortune – becoming perhaps the richest person in England – in the colonization of Ireland as its Lord High Treasurer. Boyle Junior was educated as was appropriate of his father's acquired status as First Earl of Cork; by private tutors at home, then at Eton, then in a long *Grand Journey* through the continent. Returning to England in 1644, he settled at the Dorset estate left to him by his father and began to compose religious-moralistic treatises, but his interests started to shift towards the new experimental sciences. In 1649 he successfully set up a laboratory at the estate and in 1656 he moved to Oxford to join Wilkins' group, spending there twelve years of intensive experimenting and writing. In 1668, he moved to London and to his sister's, Lady Ranelagh, building himself a laboratory there and rarely leaving the city thereafter.

Boyle, then, was well acquainted with both the London and Oxford groups and had the financial and social resources to formalize them. He felt comfortable doing so when England's failed attempt to become a republic ended and the king – Charles II, son of the beheaded Charles I – was reinstated. Relative calm settled among the English ruling classes and intellectual elites – only Cooke, Charles' prosecutor, was made to bear the brunt of the monarchy's fury – and in November 1660, in a meeting following a talk at Gresham College, the *College for the Promoting of Physico-Mathematical Experimental Learning* was established, receiving the Royal charter two years later.

The early membership of the Society tells much about the kind of institute it was and the kind of science it would promote and bequeath. The first president was William Brouncker (1620–1684) – a viscount, commissioner of the Royal Navy and mathematician; Seth Ward (1617–1689) was a bishop, mathematician and astronomer; Sir William Petty (1623–1687) was an economist, geographer and physician-general to Oliver Cromwell (the leader of the Parliament army and England's 'Lord Protector' during its republican decade). We've mentioned John Wilkins, the warden and natural philosopher; in his Oxford group were also John Wallis (1616–1703), a chaplain and a mathematician-cryptographer, and Thomas Willis (1621–1675), a court physician and anatomist. John Evelyn (1620–1706) was the most urbane of the group: diarist and historian, he came from a wealthy gunpowder-producing family. The secretary of the Society was Henry Oldenburg (1619–1677), a German immigrant and theologian by education. Oldenburg developed his ideas about the structure and mode of operation proper for such a society when he visited many of the European academies during his travels through Europe as a diplomat (and a spy). He was one of the two paid members of the Society (at least in theory – the Society was slack in collecting membership fees and in meeting its financial obligations). The other was Robert Hooke (1635–1703), its 'curator of experiments.' Hooke was the one professional among the fellows: he designed and built instruments; planned and performed experiments; suggested theories and reviewed new books. When he was (rarely) unavailable, the weekly meeting would usually be called off. We will return to him at length in Chapter 10.

The official ideology of the Society – the way it understood and presented itself – is well displayed in its emblem (Figure 9.9): Charles II is wreathed by an angel, flanked by Brouncker on the viewer's left and Bacon on the right; books are set to the side; instruments of observation and experiment are ready at hand; and a clear, sunny vista opens at the back. Their motto on the coat of arms above Charles' bust is even clearer: *Nullius in Verba* – "in

Figure 9.9 The Royal Society's emblem, engraved by Wenceslaus Hollar as the frontispiece to the first edition of Thomas Sprat's (1635–1713) *History of the Royal Society of London* (1667). Charles II's bust at the center, Brouncker on its left and Bacon on its right. Note the barometers on the right, the clock on the left, Hooke and Boyle's air pump above Charles' right shoulder, and a large hanging telescope behind it. © The Royal Society.

no-one's word." The emblem served as the frontispiece for *The History of the Royal Society of London*, commissioned in 1662 as part of the effort to secure the royal charter, and the author, Thomas Sprat (1635–1713), expressed this ideology in powerful words:

TO THE KING.

SIR, OF all the Kings of Europe, Your Majesty was the first, who confirm'd this Noble Design of Experiments, by Your own Example, and by a Public Establishment. An Enterprize equal to the most renoun'd Actions of the best Princes. For, to increase the Powers of all Mankind, and to free them from the bondage of Errors, is greater Glory than to enlarge Empire, or to put Chains on the necks of Conquer'd Nations ...

Thus they [the fellows of the Society] have directed, judg'd, conjectur'd upon, and improved Experiments. But lastly ... there is one thing more, about which the Society has been most sollicitous; and that is, the manner of their Discourse: which, unless they had been very watchful to keep in due temper, the whole spirit and vigour of their Design, had been soon eaten out, by the luxury and redundance of speech. The ill effects of this superfluity of talking, have already overwhelm'd most other Arts and Professions ...

They have therefore been most rigorous in putting in execution, the only Remedy, that can be found for this extravagance: and that has been, a constant Resolution, to reject all the amplifications, digressions, and swellings of style: to return back to the primitive purity, and shortness, when men deliver'd so many things, almost in an equal number of words. They have exacted from all their members, a close, naked, natural way of speaking; positive expressions; clear senses; a native easiness: bringing all things as near the Mathematical plainness, as they can: and preferring the language of Artizans, Countrymen, and Merchants, before that, of Wits, or Scholars ...

The Society has reduc'd its principal observations, into one common-stock; and laid them up in publique Registers, to be nakedly transmitted to the next Generation of Men ... as their purpose was, to heap up a mixt Mass of Experiments, without digesting them into any perfect model: so to this end, they confin'd themselves to no order of subjects; and whatever they have recorded, they have done it, not as compleat Schemes of opinions, but as bare unfinish'd Histories.

> Thomas Sprat, *The History of the Royal Society of London*
> (London: J. Martin, 1662), Dedicatory Epistle

One needn't be overwhelmed by seventeenth-century hyperbole nor wonder when it was that Charles gave an example of designing experiments – he didn't (his uncle, Prince Rupert, did have experimental interests). It's the sentiments Sprat is conveying on behalf of the Fellows of the Society that are illuminating, and they are clearly Bacon's, almost as if Sprat is paraphrasing the *New Atlantis*: openness – all are invited; progress – boundless optimism in human capacity for knowledge and responsibility towards "the next Generation"; strict empirical accountability – 'see for yourself'; and an utter distrust in words – in verbosity, in slippery court language, in highfalutin theories.

The Experimental Legacy

Who Was Allowed in?

The Royal Society Fellows must have genuinely believed in these ideas, but reality was much more complex. The Society, as the list of its fellows demonstrates, wasn't and couldn't be open to "Artizans, Countrymen, and Merchants." The Fellows, by nature, shared background and education; they had to be of the class that had the leisure and means to participate. There were also more fundamental, philosophical reasons why it had to be exclusive. Science could only emerge if fellows could "keep in due temper" – if their discourse was civil and their debates were settled by evidence and well-reasoned arguments. And such a civil discourse could only be maintained if people whose views and interests were too divergent had to be excluded. Atheists or religious zealots, political radicals or just quarrelsome characters weren't welcome. Thomas Hobbes (1588–1679), the most prominent English natural philosopher of his generation, was a conspicuous exclusion, which made him wryly aware of this gap between ideology and practice:

> About fifty men of philosophy, most conspicuous in learning and ingenuity, have decided among themselves to meet each week at Gresham College for the promotion of natural philosophy. When one of them has experiments or methods or instruments for this matter, then he contributes them.
>
> Hobbes, *Dialogus Physicus*, cited in Shapin and Schaffer,
> *Leviathan and the Air Pump*, p. 113

Why fifty? "Cannot anyone who wishes come and give his opinion on the experiments which are seen, as well as they?" asks Hobbes, and bitterly answers: "not at all."

Members of this club of experimental philosophers needed to be *trusted*: to dispassionately observe experiments and observations, to calmly deliberate arguments, to civilly express their reasoned opinions to gracefully subject themselves to matters of facts even if they found them surprising or disturbing. This was a manly club – even the great scholar, writer and aristocrat Margaret Cavendish (1617–1673), whose 1667 visit to the Society was awaited with trepidation and carried out with great pomp and circumstance, wasn't considered for Fellowship. Fellows needed to be of independent means – which no woman, even of nobility, was allowed to be – and hence, one would hope, of an independent mind. And they were required to possess the right moral aptitude – to be *gentlemen*. In his writings, Boyle shaped the character of this new person of science after his moral ideal of himself; the *Christian Virtuoso*, as he called his 1690 tract. Hobbes – opinionated, boisterous and suspect of atheism – did not fit the bill. The technicians and servants operating the instruments most certainly didn't, and their names are never mentioned – not even when injured by unyielding experimental apparatus, sometimes fatally.

The Air Pump

Yet it was not only presumptions rooted in class, ethics and religion that shaped the Royal Society's professed epistemology and its – quite different – actual scientific practices. The mode of experimental inquiry that the Society inherited, developed and shaped as modern science was by its very nature exclusive and theoretically laden – quite different from Bacon's dreams and Sprat's rhetoric. This is particularly well demonstrated in the prize project of the Society's early years – Boyle's experiments with the air pump, which were carried out by the *Pneumatical Engine* designed especially for them by Hooke (Figure 9.10).

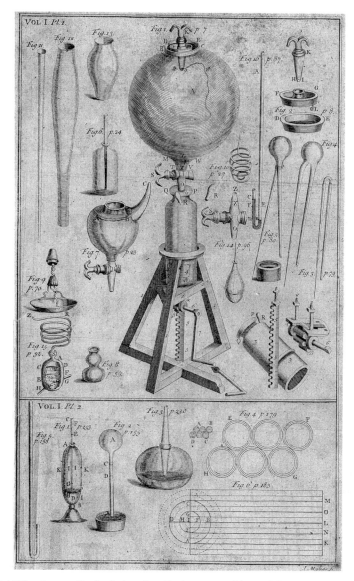

Figure 9.10 The air pump Hooke designed and had constructed for Boyle's experiments. The cylinder (4) contains a piston that is moved by the geared shaft (5) operated by the crank-handle (7). Each time the piston is drawn down, it sucks air from the container (A). The valve (X) is then closed, the piston rolled up and the operation repeated until it becomes too difficult for the persons working the handle. The operation can be reversed to condense air into the container. The lid at the top (BD) with its stopper (K) allows the operator to insert the experimental apparatus into the container.

This was a truly spectacular project. So much of everyday phenomena seemed to be created or affected by air: respiration; combustion; sound; the behavior of fluids; the velocity of fall. What exactly *was* the role of air in all these? Inside the glass vessel, the 'receiver,' questions about air were answered by its removal. This was a bold farther step in the direction set by Galileo and Guidobaldo of interrogating nature by subjecting it to artificial means. Galileo and Guidobaldo were at least mimicking real trajectories by rolling inked balls; Hooke and Boyle were creating a wholly artificial environment; an environment that could not, on principle, exist in nature.

Most excitingly: the gentlemen gathered around the pump could literally *see*, with their own eyes, what the answers were. These were quite definitive: once the receiver was emptied of air, a bell rattled inside would not be heard outside. Exhausting the air would extinguish a candle, as well as the life of a canary – which could be saved if air was quickly pumped back in. An empty bladder would expand as the air was pumped out. Without air, a feather and a metal bead would take the same time to fall from the top of the receiver to its bottom – just as Galileo had predicted.

The spectacle, however, came at a price, both literally and in terms of the Society's avowed values. The air pump was a a set of complex and very expensive piece of equipment, and by this fact alone – exclusive: of the people who might be interested in such experiments, only Boyle could afford one. Most people could not 'see for themselves'; they had to take 'someone else's words,' and Hobbes, again, curtly noticed: "Those Fellows of Gresham ... display new machines, to show their vacuum and trifling wonders, in the way that they behave who deal in exotic animals which are not to be seen without payment." Boyle was also aware of this, and published his experiments in a meticulous account of details – the ancient rhetorical trope of *enumeratio* – which was to mimic, for the reader, the genuine experience of witnessing the experiment:

> First, then, upon the drawing down of the Sucker (the Valve being shut) the Cylindrical space, deserted by the Sucker, is left devoid of Air; and therefore, upon the turning of the Key, the Air contained in the Receiver rusheth into the emptied Cylinder, till the Air in both those Vessels be brought to about an equal measure of dilatation. And therefore, upon shutting the Receiver by re-turning the Key, if you open the Valve, and force up the Sucker again, you will find, that after this first exsuction you will drive out almost a whole Cylinder full of Air ...
>
> Robert Boyle, *New Experiments Physico-Mechanical*
> *Touching the Spring of Air* (Oxford: Robinson, 1662),
> pp. 11–12

And so forth: detailed, accurate, impassive. Bacon's rhetoric pretends to "deliver so many things, almost in an equal number of words," as Sprat promised that same year. But it is still rhetoric: Bacon's readers could enjoy, at best, 'virtual witnessing,' mediated by Boyle's words. Even though Boyle's was a "close, naked," style, it was still a self-conscious style.

Vacuum in Vacuum

Perhaps the most problematic of Sprat's promises was the ideal of a "nakedly transmitted ... mixt Mass of Experiments," free of "compleat Schemes of opinions." One particular experiment demonstrates that this was a pipe dream – the 'vacuum in a vacuum' experiment.

It was perhaps the most spectacular experiment of them all, exporting into the new (and proudly English) apparatus, experimental procedures and ideas developed in Italy, a generation earlier, by Galileo's disciples. The origin of these ideas was in the common knowledge that suction pumps, used to remove water from mines, could draw water only from depths up to 10 meters. The pumps worked by creating a vacuum above the column of water in a pipe, and the Aristotelian principle by which they were understood – that nature didn't allow vacuum and the water rose to prevent it – couldn't explain why they should stop working at any particular depth. For Galileo, the limits of vacuum pumps meant that *Horror Vacuui* was not a metaphysical principle but a physical property, with clear physical parameters – it had certain power and no more. He even concluded that a vacuum can be created, if momentarily. When, for example, two polished marble slabs are separated from each other, the air quickly and forcefully drives in, but in the extremely short time before it refills it, the gap between them constitutes a vacuum.

Galileo's disciples took the argument a step further. The limit of the power of vacuum water pumps, they claimed, was actually a direct consequence of the principle that made them work: these pumps were another example of the ubiquitous principle of the balance. The water column was not *drawn towards* the vacuum, they claimed – there *was* no such thing as *horror vacuui*. Rather, the water was *pushed* by the weight of the atmosphere into the space above it, which, being empty, offered no resistance. Like a balance, they explained, this could work as long as the atmospheric column of air was *heavier* than the column of water on which it rested; apparently, the atmosphere weighed the same as 10 meters of water (or more accurately: an atmospheric column weighed the same as a column of air of the same cross-section).

In 1641, a Roman professor of mathematics by the name of Gasparo Berti (c. 1600–1643) decided to try the hypothesis by direct experimentation (Figure 9.11, left): he sealed the bottom of an 11-meter tube, filled it with water and sealed the top. He then opened the bottom seal into a basin of

Figure 9.11 Two stages of the vacuum experiments. On the left: Berti's original experiment from 1640/41, with the '40 palms' (or so) length of leaden pipe fastened to the wall of his lodging. When the bottom valve *R* was opened, some of the water from the glass flask *CA* on top of the pipe fell into the cask *EF* at the bottom. The bell *M* was apparently an idea of one of the spectators – the great Jesuit scholar Athanasius Kircher (1602–1680): if the space remaining above the water was indeed devoid of air, he reasoned, then striking the bell with the hammer *N* wouldn't produce any sound. This ingenious idea was only carried off in Hooke and Boyle's air pump. On the right: Torricelli and Viviani's 1644 version of the vacuum experiments. The tube is filled with mercury to the top, closed (note the hand at *B*, bottom of the tube) and turned upside down into a vat of the same heavy liquid metal. When the hand at B is removed, the mercury column falls to 76 centimeters (equivalent to the 10 meters for water). Both images are imagined reconstructions by the Jesuit Gaspar Schott from his *Technica Curiosa* (Wurzburg, 1664).

water – and the water column dropped, leaving a space above it. Following Galileo's argument, he claimed that this space was simply empty – a void; a vacuum. In 1644, Galileo's student Evangelista Torricelli (1608–1647), with his own student Vincenzo Viviani (1622–1703), constructed a more efficient version of Berti's experiment (Figure 9.11, right). They replaced the water with mercury, which, being 13.5 times heavier, dropped when the column surpassed 76 centimeters (above line NO).

This version of the experiment became known as the *Torricellian Experiment* and won Torricelli fame as the inventor of the barometer – the device to measure air pressure. Boyle imported it into the air pump: if it were indeed the atmosphere that held the mercury column up, the column should fall as the air was pumped out of the receiver.

And fall it did.

But what was in the space left above the mercury? According to the Galileans and Boyle – nothing. It was simply a vacuum. Hobbes didn't agree – neither with the possibility of a void in general, nor with the idea that the Torricellian tube or the 'Boylean Machine' produced one. Already in May 1648 he wrote to Mersenne that:

> ... all the experiments made by you and by others with mercury do not conclude that there is a void, because the subtle matter in the air being pressed upon will pass through the mercury and through all other fluid bodies, however molten they are. As smoke passes through water.
>
> Quoted in Shapin and Schaffer, *Leviathan and the Air Pump*, p. 86

Boyle wouldn't engage with Hobbes on the metaphysical question of the very possibility of vacuum:

> ... it appears by Mr. Hobbes's Dialogue about the Air, that the Explications he there gave of some of the Phaenomena of the *Machina Boyliana*, were directed partly against the Virtuosi, that have since been honour'd with the Title of the Royal Society, and partly against the Author of that Engine, as if the main thing therein design'd were to prove a Vacuum. And since he now repeats the same explications, I think it necessary to say again, that if he either takes the Society or me for profess'd Vacuists, he mistakes, and shoots beside the mark; for, neither they nor I have ever yet declar'd either for or against a Vacuum.
>
> Robert Boyle, *Animadversions upon Mr. Hobbes's Problemata De Vacuo*
> (London: William Godbid, 1674), pp. 26–27

But was there anything – air or "subtle matter" – in the receiver after pumping or wasn't there? As far as Boyle was concerned, it was enough that

> ... the Instrument ... had been so diligently freed from Air, that the very little that remain'd, and was kept by the Wax from receiving any assistance from without, being unable by its Spring to [affect the experiment].
>
> Boyle, *Animadversions*, p. 94

In the long run, it was Boyle who won. The experimental tradition created by the "Virtuosi ... of the Royal Society" learned to display the same indifference towards metaphysical worries. In their stead, experimenters were happy to admit what came to be called 'operational definitions' like the one Boyle offered here for vacuum: as long as "the very little that remained" didn't affect the experiment, the receiver could be considered empty. But the debate clarifies that theory could always contest experiment. *Pace* Bacon, Boyle and Sprat, nothing could be seen "nakedly."

Conclusion: The Independent Life of the Instrument

Even though openness and language-free observation turned out to be unattainable, progress was not an empty ideal. Yet the "increase [of] the Powers of all Mankind" didn't all come in the way Boyle and the 'gentlemen virtuosi' expected. It had less to do with their virtuosity and more to do with that of their rarely mentioned instrument builders.

Hooke didn't invent the pneumatic engine *ex nihilo* (nor did Galileo the telescope, as we saw in Chapter 7). Glass balls had already been used in optical experiments for centuries, and pistons, cranks and valves were in common use by artisans of different trades (like mining and waterworks). His basic design followed a machine invented in 1650 by the Prussian civil servant and natural philosopher Otto von Guericke (1602–1686), who traveled with it around Germany, demonstrating how marvelously difficult it was to separate two attached hemispheres once the air was exhausted from between them (one could not conduct experiments inside von Guericke's sphere; the option of reaching into the receiver was Hooke's addition). Moreover, Hooke's design was far from the last one. In 1727, another Englishman and a member of both the Royal Society and the Académie des Sciences – Stephen Hales – presented a way to collect gases in Hooke-like glass spheres by heating substances and letting the released fumes bubble through a trap – 'trough' – of water. By the 1760s, a number of experimentalists replaced the water in the trough with mercury, through which gases soluble in water could also pass, and most eighteenth-century chemical research revolved around experimenting on and theorizing about the 'airs' collected in the glass vessels, whose shape was also evolving. And already in 1679, a junior colleague of Hooke and an employee of Boyle – the Frenchman Denis Papin (1647–1713) – used the principles developed by the budding tradition of vacuum machines to invent an apparatus whose future effects would be even more dramatic than the trough's. It was, originally, a 'steam digester' – a pressure cooker for bones – but observing the pressure valve (which distinguished his invention from earlier versions, prone to explosions) bobbing by the power of the steam, he realized that this power could be harnessed. In 1712, Thomas Newcomen (1664–1729), a blacksmith and Baptist lay preacher, received a patent for a working application of this idea, inventing *the* "engine" – literally and metaphorically – of the Industrial Revolution: the steam engine.

In the grand story about erecting the cathedral of science, these mostly (though not always) anonymous instrument-makers were not only the bricklayers, required for their labor and fundamental skills, but also – to press the metaphor a little further – the master masons. They not only carried traditional craft knowledge from generation to generation, but also honed, refined and developed it for new uses. They were obliged to do so: their livelihood depended on selling the artifacts they produced, so those needed to be useful and affordable, evolving to suit changing needs. So when Hooke felt obliged to pay homage to his Royal Society employers, he retreated to the traditional hierarchy between *episteme* and *techne*, writing:

> ... all my ambition is, that I may serve to the great Philosophers of this Age, as the makers and the grinders of my Glasses did to me; that I may prepare and furnish them with some Materials, which they may afterwards order and manage with better skill, and to far greater advantage.
>
> Hooke, *Micrographia*, Preface

But when he came to discuss instruments directly, all Hooke's reticence and humility disappeared. Those "artificial organs," as he called them, brought "prodigious benefit to all sorts of useful knowledge," but even more excitingly,

> By the means of Telescopes, there is nothing so far distant but may be represented to our view; and by the help of Microscopes, there is nothing so small, as to escape our inquiry; hence there is a new visible World discovered to the understanding.
>
> Hooke, *Micrographia*, Preface

The "the makers and the grinders of ... Glasses" didn't only "furnish ... materials"; the mechanical world was very much a world constructed by the mechanic. The innovations of the virtuosi challenged the instrument-makers to come up with new ways to gauge and query nature. At the same time, "the makers," with their own initiative and innovations, created a whole "new visible World" for "the great Philosophers of [the] Age" to study.

Discussion Questions

1. Is the discrepancy between the Royal Society's ideology of openness and its exclusionary practices important? Does it have an ethical significance? An epistemological significance? Does it reflect on science or our times?
2. What kind of relations do you see between early modern science and early modern philosophy? Does science follow a philosophically laid path? Is philosophy recruited to assist science? How?
3. Add Descartes' life to the comparison suggested in question 3 of Chapter 7. What kind of new persona does he present?
4. The inside of the glass vessel of the air pump is the only vacuum on Earth. Does this represent a new stage of experimentation? Does is say something about contemporary empirical science?
5. The 'mechanical philosophy' of Descartes and his disciples is often described as a crucial pillar of modernity, with its mathematical approach to nature and technological approach to the human condition. Do we still live in a 'mechanical era'?

Suggested Readings

Primary Texts

Bacon, Francis, *New Atlantis* (London: Tho. Necomb, 1659), www.gutenberg.org/ebooks/2434.

Descartes, René, *Principles of Philosophy (Principia Philosophiæ, 1644)*, Valentine R. Miller and Reese P. Miller (trans.) (Dordrecht: Reidel, 1983), pp. 59–63.

Elisabeth, Countess Palatine and René Descartes, *The Correspondence between Princess Elisabeth of Bohemia and René Descartes*, Lisa Shapiro (ed. and trans.) (University of Chicago Press, 2007), pp. 61–81.

Hooke, "Preface" to *Micrographia* (London: Jo. Martin and Jo. Allestry, 1665), www.gutenberg.org/ebooks/15491.

Secondary Sources

On the Scientific Revolution:

Floris Cohen, H., *The Scientific Revolution: A Historiographical Inquiry* (University of Chicago Press, 1994).

Koyré, Alexandre, *From the Closed World to the Infinite Universe* (Baltimore, MD: Johns Hopkins University Press, 1957).

On the history of optics:

Lindberg, David C., *Theories of Vision from Al-kindi to Kepler* (University of Chicago Press, 1976).

On Descartes' science and philosophy:

Garber, Daniel, *Descartes' Metaphysical Physics* (University of Chicago, 1992).
Schuster, John, *Descartes-Agonistes: Physico-mathematics, Method & Corpuscular-Mechanism 1618–33* (Dordrecht: Springer, 2013).

On Boyle and the Oxford experimentalists (historically accurate fiction):

Pears, Iain, *An Instance of the Fingerpost* (New York: Berkeley Publishing Group, 1998).
Atkins, Philip and Michael Johnson, *A Dodo at Oxford: The Unreliable Account of a Student and His Pet Dodo* (Oxford: Oxgarth Press, 2010).

On Boyle and the Royal Society:

Shapin, Steven, *A Social History of Truth: Gentility, Civility and Science in Seventeenth-Century England* (University of Chicago Press, 1994).

On Robert Hooke and the Royal Society:

Shapin, Steven, "Who Was Robert Hooke?" in Michael Hunter and Simon Schaffer (eds.), *Robert Hooke: New Studies* (Wolfeboro, NH: Boydell Press, 1989), pp. 253–286.
Pumfrey, Stephen, "Ideas above His Station: A Social Study of Hooke's Curatorship of Experiments" (1991) XXIX *History of Science* 1–44.

On the New Science in its cultural context:

Gal, Ofer and Raz Chen-Morris, *Baroque Science* (Chicago University Press, 2013).
Jacob, Margaret, *The Cultural Meaning of the Scientific Revolution* (Philadelphia, PA: Temple University Press, 1988).

On Galileo's science:

Koyré, Alexandre, *Galileo Studies*, John Mepham (trans.) (Atlantic Highlands, NJ: Humanities Press, 1973).
Renn, Jürgen, Peter Damerow and Simone Riger, "Hunting the White Elephant: When and How Did Galileo Discover the Law of Fall?" (2000) 13 *Science in Context* 299–423.

On the instrumental tradition:

Bertoloni Meli, Domenico, *Thinking with Objects: The Transformation of Mechanics in the Seventeenth Century* (Baltimore, MD: Johns Hopkins University Press, 2006).
Landes, David S., *Revolution in Time: Clocks and the Making of the Modern World* (Cambridge, MA: Belknap Press, 1983).
Lefèvre, Wolfgang (ed.), *Picturing Machines 1400–1700* (Cambridge, MA: MIT Press, 2004).

On the mathematics developed for the New Science:

Mahoney, Michael, "Changing Canons of Mathematical and Physical Intelligibility in the Later Seventeenth Century" (1984) 11 *Historia Mathematica* 417–423.

Mancosu, Paolo, *Philosophy of Mathematics and Mathematical Practice in the Seventeenth Century* (New York: Oxford University Press, 1996).

On Francis Bacon's philosophy:
Rossi, Paolo, *Francis Bacon: From Magic to Science*, S. Rabinovitch (trans.) (London: Routledge and Kegan Paul, 1968).

On Boyle and the Royal Society, 'virtual witnessing,' etc.:
Shapin, Steven, *A Social History of Truth: Gentility, Civility and Science in Seventeenth-Century England* (University of Chicago Press, 1994).

Shapin, Steven and Simon Schaffer, *Leviathan and the Air-Pump* (Princeton University Press, 1985).

On the experimental tradition:
Schaffer, Simon, "Glassworks: Newton's Prisms and the Uses of Experiment" in David Gooding, Trevor Pinch and Simon Schaffer (eds.), *The Uses of Experiment: Studies in the Natural Sciences* (New York: Cambridge University Press, 1989), pp. 67–104.

Shapin, Steven, "The House of Experiment in Seventeenth-Century England" (1988) 79 *Isis* 373–404.

On Charles I's trial and execution:
Kesselring, Krista J. (ed.), *The Trial of Charles I* (Peterborough, ON: Broadview, 2016).

Robertson, Geoffrey, *The Tyrannicide Brief* (London: Chatto & Windus, 2005).

10

Science's Cathedral

The Two Savants

Robert Hooke

Putting the experimental ideology of the Royal Society into practice was the role assigned to Robert Hooke (1635–1703) – who was perhaps the first person whose entire career was defined by the institutions and practices of the New Science he helped shape.

He was born on the Isle of Wight in the south of the English Channel to Cecil Gyles and her husband John Hooke, a Church of England priest. These were origins at once respectable and peripheral: on the Isle, his family was notable enough to move in the circles surrounding the court of the embattled King Charles when it found itself in exile there in 1647. When his father died in 1648, however, the 13-year-old Hooke took his small inheritance and went to London. He became an apprentice to the famous portrait painter Peter Lely (Peter van der Faes), but left a career in painting behind when he moved to study at Westminster School, where he worked for his tuition but was brilliant enough to be spared the famous cane of master Richard Busby. In 1653, he continued to Oxford, again working for his tuition and livelihood, and here his unique combination of practical and theoretical talents found the outlet that would make him such a pivotal figure in fashioning the experimental practices of the New Science.

In 1656, Hooke found employment with Thomas Willis as a "chemical assistant" in the latter's anatomical-physiological experiments. He thus affiliated himself with Wilkins' "experimental philosophical club" (see Chapter 9) and in 1658 was hired by Boyle to be his experimental assistant and instrument designer. Hooke was thus an active participant in the Royal Society's meetings from early on, but, telling of the Society's character, his early contributions were discussed as if presented by Boyle, his gentleman employer. This was the fate, for example, of his discourse and demonstration of capillary action from April 10, 1661. In November 1662, this changed: Hooke was officially appointed the Society's 'curator of experiments,' substituting the more traditional, patronage-like relations with Boyle with a modern salaried position. The salary, however – £60 per annum – was withheld more often than not, and the Society was always eager to count other promised incomes against it. The Gresham Professorship of Geometry (Chapter 9) which he was awarded in 1664 did provide a stable income, but the Cutler Lectureship (after the benefactor John Cutler) that he was awarded in 1665 was very difficult to collect, even though his notes for these lectures contain some of his most innovative work.

When the Society looked for an impressive gift for the king to smooth over their request for support, Hooke was asked to collect his microscopical observations in a book, and the outcome became, unsurprisingly, a bestseller: the microscopic images of the 1665 *Micrographia*, all in Hooke's hand, were spectacular. Although the microscope was about as old as the telescope (its exact origins aren't clear) and a number of books presenting microscopic observations had already been published in the previous decades, none came close to Hooke's in the number, variety and quality of images (see Figure 10.1). In 1666, following the fire that devastated London, Hooke finally found a regular, well-paying position: he became one of the three surveyors nominated by the city. Working alongside his friend and colleague Christopher Wren (1632–1723), who was one of the surveyors on behalf of the king, Hooke and the other surveyors redrew London's map, deciding on allocation of lots and property boundaries. Alongside Wren (the architect of St. Paul's Cathedral), Hooke was the architect of some important edifices, many credited to the former, like the memorial column to the London fire and the Bethlehem psychiatric asylum – immortalized in the colloquial 'Bedlam,' but demolished in the early nineteenth century.

From late 1660 until his death, then, Hooke was a very 'modern,' urban character: moving fast around London, working professionally as a practical mathematician, experimental expert, teacher, architect and inventor, keeping a diary and meeting his colleagues to discuss scientific matters over cocoa in the new coffee houses around town. He supported this feverish lifestyle with an unrestrained regimen of self-medication, passing away in 1703 unhealthy but wealthy (he left some £10,000 in a chest in his rooms).

Isaac Newton

While Hooke was establishing his credentials as an experimenter and instrument builder in Oxford, Isaac Newton (1642–1726) was gaining a name as a mathematical wiz in Cambridge. Like Hooke, he was an orphan of a provincial clergyman, from a little town in Lincolnshire on the east coast of England, and like him he had to work as a servant-student until his talents shone through. Unlike Hooke of the taverns and cafés, Newton was a recluse, yet he seemed to have had an intellectual charisma that

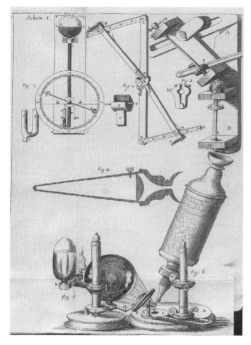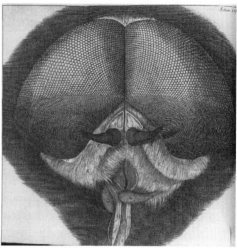

Figure 10.1 Images from Hooke's *Micrographia*: his microscope and accessories on the left, and a fly's eyes, observed through this microscope, on the right. Note how the instruments relate to and support each other. The lenses for the microscope at Figure 6 are ground by the device in Figure 3; Figure 4 is a liquid-filled microscope, and the refrangibility index of the liquids to be used in it is measured by the device in Figure 2, etc. The beautiful drawing of the fly's eye highlights how much of observation (Hooke's and in general) is a work of art – both in the traditional and the modern sense. No one fly can be held still and alive long enough to capture all these segments (Hooke did try to get them drunk) – the image is not a naked reflection of *an* observation, but a construction from many 'sittings.' And note the painterly skills that Hooke invests in making the image look as if it is a direct, 'real' observation: the play of light and shadow and the multiple reflections of the window in the eye's lenses.

Hooke lacked. He became such a prodigy student of the great mathematician Isaac Barrow (1630 – 1677) that in 1669 Barrow resigned in his favor from Cambridge's newly established, prestigious Lucasian Professorship of Mathematics. Newton's notebooks from his student years demonstrate why: they are full of extremely innovative ideas for creating the type of mathematics that could meet Kepler and Galileo's challenge; mathematics that could capture real physical bodies and causes – real motions, real forces (Chapters 7 and 9, and see Figure 10.2 for an example).

With his secure position as a faculty member at Trinity College, Newton settled into a much quieter life than Hooke: the traditional life of a university scholar and teacher. His fame as a mathematician spread, and his work on optics made him a name in the Royal Society (and created a rift with Hooke,

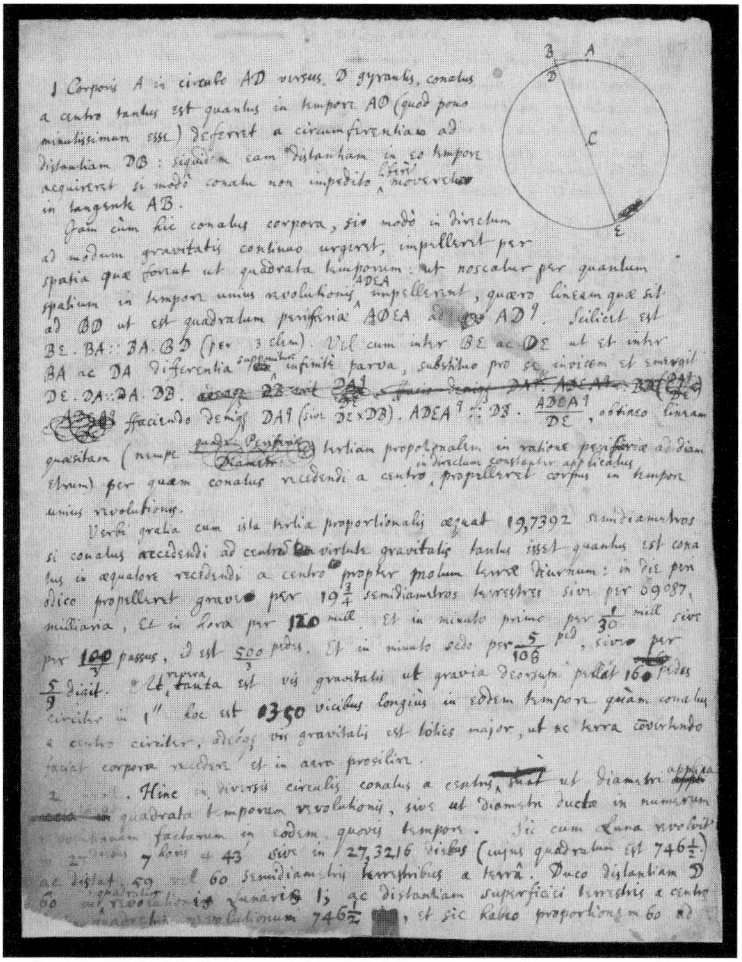

Figure 10.2 An example of Newton's early application of his new mathematical tools to solve a physical problem (from a manuscript written between 1666 and 1669). It is crucial to remember that Newton never had a proof for the mathematical rigor of the bold approximations he is making – he develops this infinitesimal approach as he goes along.

A body moves uniformly in circle ADE around C counterclockwise. In uniform motion, times are proportional to distances, so $t_{AD} \propto AD$. Following Descartes, if the body is released at A to move freely, it will continue uniformly along the tangent AB, and for an infinitesimally small time, Newton assumes AB=AD. The 'little line' BD therefore represents the centrifugal force at point D – it's the distance this force would take the body, if allowed. Newton permits himself to assume that for infinitesimally short time t_{AD}, CDB is a straight line, and this enables him to use the Euclidean theorem $AB^2 = BD \bullet BE$. He then adds the assumption that the centrifugal force works like gravity, so according to Galileo, BD is proportional to the square of the time: $BD \propto t_{AD}^2$ or $BD \propto AD^2$. A series of simple transformations leads Newton to the conclusion that the centrifugal force of the orbiting body is proportional to the radius divided by the square of the period: $f \propto R/T^2$. Merging this with Kepler's Third Law, according to which for the planets $T^2 \propto R^3$ (the square of the period is proportional to the cube of the distance from the Sun) Newton concludes that the centrifugal force on the planets is inversely proportional to their distance from the Sun: $f \propto 1/R^2$.

who found it lacking). But Newton was just as – perhaps more – engaged in his alchemical work and prophetic Biblical exegesis as his mathematics. In alchemy and optics, he demonstrated practical skills that almost paralleled his mathematical ones, and his optical experiments, though contested and debated, gained him almost as much renown as his celestial mechanics. His diagram in Figure 10.3 represents the most celebrated of these experiments: the 'Crucial Experiment,' demonstrating that white light, pervading all creation, is not the homogenous entity it was always assumed to be, but a mixture of colored rays.

The publication of the *Principia Mathematica Philosophiae Naturalis* in 1687 – whose core insight we'll follow below – made Newton a celebrity well beyond the circles of the New Science and earned him the uncompromising hostility of Hooke, who thought that he himself deserved much more credit than he'd received. Newton was involved in a similarly famous priority

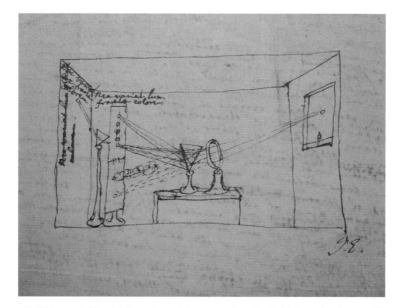

Figure 10.3 Newton's 'Crucial Experiment,' demonstrating that light is not fundamentally white, but a mixture of elementary colored rays. The ray entering the hole in the shutters on the right is focused by the lens and refracted by the prism. Colored rays fall on the screen, arranged vertically. The further away the screen is, the more the colored rays separate. Each separate (colored) ray is allowed through a hole in the screen and through another prism. The important finding is that each ray is refracted at the same angle and keeps its color. It means that the original refraction didn't *change* the light – or the second refraction would have changed it again. Rather, that refraction revealed the essential properties of the primitive elements of light: a color and a refraction index. The colored rays (the colors of the rainbow) are not *modifications* of white light, but its constituents.

debate with the German polymath Gottfried Wilhelm Leibniz (1646–1716) – over the invention of the infinitesimal calculus – but it's very important not to fixate over these debates. Science, as we've discussed from Chapter 1, is not a contest whose goal is set in advance, but a cathedral built by many hands. What we will see in this final chapter is what a complex, collaborative effort the celestial mechanics of the *Principia* was. To the protagonists, however, credit was very important. Newton waited until Hooke's death in 1703 to publish his *Opticks* – the subject of their earlier debate – and become the Secretary of the Royal Society, which he brought back from the disarray into which it had fallen after the death of Oldenburg and most of the early members. He moved to London, became Master of the Mint, was knighted, and devoted most of his energy in his remaining years to consolidating his influence and securing his legacy by having his disciples appointed to influential positions and his detractors kept away from them. Newton's brilliance and innovativeness and his role in shaping modern science are undeniable, but the demigod status he has acquired in the epic of science owes more to those political skills than to his intellectual ones.

The Correspondence: Forging a New Question

The Falling Earth

But in their prime, the two were still on speaking terms, and on November 24, 1679, Hooke wrote a letter to Newton and sent it from London to Cambridge. He had just been elected Secretary of the Royal Society, following the death of its founding Secretary, Henry Oldenburg. Hooke, you'll remember, had been the Society's curator of experiments since 1662, responsible for presenting the gentlemen members of the Society with serious yet entertaining demonstrations at their weekly meetings. It was a thankless role. On the one hand, he was the one professional experimental natural philosopher among amateurs; on the other: a paid employee among leisurely gentlemen, subservient to their whims. Hooke was keenly aware of the discrepancy and his relations with Oldenburg – the only other paid employee of the Society – were tense. Hooke always suspected Oldenburg of making commerce with his ideas (as did Wren) and instigating quarrels between him and other savants. As the newly appointed secretary (an appointment destined to be short and unsuccessful), Hooke was therefore happy to disregard past conflicts and – he wrote to Newton – officially invite the famous mathematician to correspond with the Society.

This wasn't only a show of cordiality. As discussed in Chapter 5, the Royal Society, like other academies, fashioned itself as a 'global institution of knowledge,' and its success was closely tied to its emergence as an international focus of such correspondence. It aspired to be the audience which researchers sought and the institute that donned acceptance and prestige, and indeed letters came from all over Europe: from savants as well as merchants, missionaries and travelers, from Asia and the New World. Hooke was inviting Newton to move under the Society's wings.

So Hooke begins his letter with the appropriate pleasantries, presents some of the Society's recent undertakings and then moves on to the question that truly interests him. "I shall take it as a great favour," he writes,

> if you will let me know your thoughts of that [hypothesis of mine] of compounding the celestiall motions of the planetts of a direct motion by the tangent & an attractive motion towards a central body.
>
> Newton, *The Correspondence*, Vol. II, p. 297

Hooke was proposing to look at the planetary orbit as a *physical effect*, a 'compound' of two causes: rectilinear motion at uniform velocity along the tangent to the orbit, *and* a rectilinear attraction towards the central body – presumably the Sun.

This was a truly incredible idea. Rather than moving in the harmonious circular motion expected of a heavenly body, Hooke was suggesting that the planet is just stumbling along. It's constantly falling towards the Sun, somehow avoiding falling *on* the Sun because of a second motion – an independent motion constantly diverting it. Both motions are in a straight line, and the curved motion *around* the Sun – the orbit itself – has no independent status. It doesn't express any celestial perfection: it's merely a contingent effect of a complex causal process.

In hindsight, we may miss how difficult this suggestion was to understand, let alone accept. First, because we are accustomed to the idea of the planets and the Sun gravitating towards each other. Secondly, it may seem that Hooke's suggestion flows directly from or is an extension of the works of his predecessors, from Galileo and Kepler to Descartes. But in fact, by the time he wrote to Newton, Hooke had been attempting to explain the idea of "compounding the celestiall motions" to his colleagues and employees at the Royal Society for some fifteen years, with little to show for it. Many questions were asked about the properties of planetary orbits, but even after Kepler it didn't occur to anyone but Hooke to ask why the planets *orbit* in the first place; and why they go around the Sun (within the Royal Society milieu, there was no longer a question that they orbit the Sun and not the Earth). The unquestioned assumption, since Antiquity, has been that they just do – it is in their nature. As we shall soon see, even Newton didn't immediately understand Hooke's suggestion nor grasp its significance.

Hooke was presenting a great innovation, but as we have said about every great innovator we've discussed, and as is always the case, he was drawing on existing resources – which takes nothing away from the novelty of his ideas.

Indeed, we should by now be able to recognize these resources. The idea of 'compounding' – the idea that a curved trajectory can result from two rectilinear motions – had been Galileo's great contribution (Chapter 9 and Figure 10.4). Galileo had in turn been developing the insights of Renaissance mechanics, represented in Chapters 5 and 9 by the work of Niccolo Tartaglia. Galileo was interested in terrestrial motion – the motion of projectiles and especially of cannon balls. For this reason, his curves were open – the famous parabolas. But as Tartaglia's struggles testify (remember the discrepancy between the free drawing on his frontispiece and the formal diagrams in the book), even these open curves had been difficult to imagine; it was very difficult to conceive of a continuous 'curving' of one straight-line motion (the forced motion caused by the gunpowder) by another straight-line motion (the natural motion downward caused by heaviness).

Galileo thus offered Hooke one set of resources to draw on. Kepler provided another. In Kepler's work – which he knew well – Hooke found *physica coelestis*, the term Kepler coined for his new astronomy, which was to be the study of the real motions of the heavenly bodies and the real forces operating between them. Hooke's decision to follow Kepler's idea was far from trivial: expanding the dominion of astronomy into physics was an ambitious and controversial novelty, as Kepler himself had been keenly aware. "Physicists, prick up your ears!" he wrote humorously (but completely seriously), "for here is raised a deliberation involving an inroad to be made into your province" (Johannes Kepler, *New Astronomy* (*Astronomia Nova*, 1607), William H. Donahue (trans.) (Cambridge University Press, 1992), p. 89).

When Hooke wrote to Newton about "compounding the celestiall motions of the planetts," he was proposing that Kepler's celestial physics could and should be the same as Galileo's terrestrial physics. This insight was original to Hooke; he couldn't have found it in Kepler or Galileo. As innovative as both had been, neither Kepler nor Galileo had suggested that the innovations each brought to the study of his respective realm – Galileo to terrestrial motions and Kepler to celestial ones – could be applied to the other.

For the idea that *all* motions – terrestrial or celestial – should be explained by the same *mechanical* laws, Hooke had to turn to Descartes. In Descartes, he found that rectilinear motion in uniform velocity is not a type of change (as was assumed since antiquity, we saw in Chapters 2 and 3), but *a state*. This meant that "straight motion" didn't require any force to continue – only to stop, start, or change speed or direction: rectilinear, uniform motion

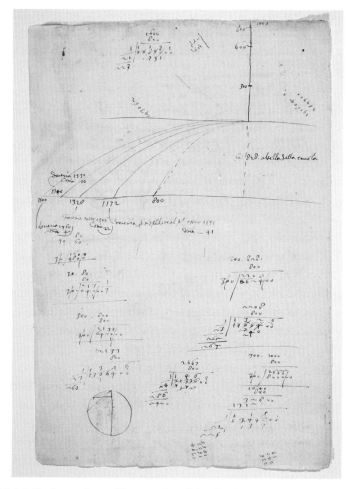

Figure 10.4 Galileo's experimental investigation of the idea that the trajectory of a projectile is composed of two independent motions – a horizontal 'violent' thrust and vertical 'natural' fall – and that together they combine to create a parabolic path. He rolls balls down an inclined plain and lets them shoot freely off the end of it, marking the point of their fall. In this manuscript, he seems to have reconstructed the trajectories, but in others he dips the balls in ink and lets them do the drawing. He then measures the distance they traveled (800, 1,172; 1,328; etc.) and looks for a square proportion with the height from which they were rolled (300; 600; 800) – the proportion constituting a parabola. He's very aware that the material conditions – friction, inaccuracies – prevent the figures from being exact, and the calculations around are attempts to reduce the figures he obtained to the proportions he is seeking. Historians debate whether he found the outcomes of these experiments satisfactory.

required no explanation. And that implied – against Kepler's firm belief – that any *curved* motion is derivative, that it is always a deviation from uniform rectilinear motion, and that this 'curving' is what requires explanation. This cluster of ideas was becoming common wisdom for the mechanical philosophers of the generation after Galileo, but it was Descartes who formulated them in the way that Hooke was to adopt: as a *law of nature*, true of *all* bodies in the world – whether celestial or terrestrial.

The Falling Stone

These were the main resources Hooke was drawing upon. We should stress again: that he had such resources does *not* mean that his suggestion to Newton (which came to be called 'Hooke's Programme') was not profoundly original. We'll discuss this originality later; in the meantime, let's return to their correspondence.

Late in 1679, when Newton received Hooke's letter, he was still, albeit known and respected, the junior of the two, and he answered respectfully and very promptly – on November 28, within four days – although with some reluctance. He had not heard of Hooke's hypothesis, he writes, and has been away from any similar studies for all kinds of personal reasons. But so as not to come across as completely lacking in manners, he offered a thought experiment for Hooke's consideration (Figure 10.5).

Whereas Hooke's hypothesis dealt with the *annual motion* of the planets (including the Earth), Newton's idea concerned the Earth's *daily motion* – its rotation around its own axis. Consider Newton's diagram: imagine a very

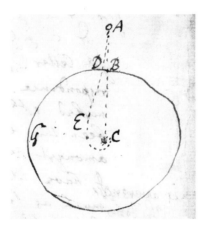

Figure 10.5 Newton's diagram from his November 28, 1679 letter to Hooke. The view is from above the North Pole, as it is, so east is on the left and the Earth rotates from right to left. It represents two contradictory lines of thought. In the first, a stone falls from the top of the tower *A* to the east, towards *D*, because, being further away from the center of rotation than *B* at the bottom of the tower, the eastward component of its motion is faster. This is a demonstration both of the motion of the Earth and of Galileo's insight – that the 'natural' and 'violent' motions of a projectile continue independently, and its trajectory is the combination of both. In the second line of thought, the stone is allowed to fall through the Earth, and it describes a spiral towards *E* until coming to rest in the center *C*. As Hooke would explain in his reply (Figure 10.6), this contradicts the very insight the first argument illustrated.

tall tower *BA*, he suggests, and let a stone fall from its top, *A*. The Earth is moving to the east towards *G* on the left of the diagram – we are looking down from the North Pole. Where would the stone fall? "The vulgar," says Newton, namely those who either follow Aristotle or simply let themselves be guided by common sense, would answer 'backwards' – to the west of the tower or the right of *B* in the diagram. The reasoning here *is* common-sensical (and indeed, experience shows, this is the answer most current *science* students tend to give): the tower seems to be what connects the stone to the moving Earth, so once the connection is severed and the stone starts falling, the Earth – the bottom of the tower – should rotate away from it to the east. This was the traditional argument against Copernicanism: since we *don't* see clouds and birds drifting to the west, the Earth can't be moving. Copernicus, as we saw in Chapter 7, gave a different answer than that of the "vulgar": 'the Earth' is not just the solid part which lies under our feet, he insisted (following Aristotle). It is the sphere of the elements, extending to the Moon, moving together as a whole. The stone, according to the Copernicans, participates in the rotation of the Earth whether or not it touches it solidly (or through the tower), and would therefore fall just under the tower.

The correct and surprising answer, Newton points out to Hooke, is that the stone will fall to the *east* – *with* the motion of the Earth, and further towards *D* in the diagram. This is because the top of the tower is further removed from the center of rotation – which is the center of the Earth – than its bottom, which means it is traveling faster: the further away a point is from the center of rotation, the larger is the distance it travels when covering a given angle. And even though the stone falls down in a straight line, it does not lose the motion it received at the top of the tower. It keeps moving in a straight line east faster than the bottom of the tower. This was Galileo's great insight: that motion in one direction (down) doesn't cancel motion in another direction (east). The final, curved trajectory of the stone would be a combination of these two straight-line motions.

Newton makes the obvious point that the height of the tallest tower is negligible in comparison to the size of the Earth, so this *is* mostly a thought experiment. But if conducted very carefully and repeatedly, he suggested (alongside a clever idea of how it could be done) that one could expect the stone to fall significantly more often to the east of the tower's base than to its west.

Newton's Mistake

At this point, Newton added a casual remark: if the stone could somehow fall *through* the Earth, it would spiral along *DEC* (Figure 10.5), until it would finally come to rest at the center of the Earth, *C*. This is a strange mistake. Not by our criteria – as we have stressed, judging the historical agents by what we believe to be true is not just useless. It truly interrupts our work as historians, because it obscures *their* motivations, dilemmas and resources, blocking our view of how and why *they* shaped the beliefs we have inherited from them. Newton's addition was a mistake according to *his* knowledge.

To think that the falling stone would come to a rest, Newton had to forget the very Galilean insight on which his original thought experiment was based. Namely: that a body in motion would only stop if there is a force to stop it, and that two (or more) motions in one body don't cancel each other. The stone would fall to the east of the bottom of the tower because it continues to move in the tangential motion given to it at the top of the tower. In order to come to a stop at the center, this motion has to cease – but why should it?

Hooke, replying almost as promptly as Newton did (on December 9), seized exactly upon this mistake with the diagram in Figure 10.6. The stone would *not* fall to the center, he corrected Newton (undoubtedly with some glee). Rather, it would orbit *around* it, along the "Elleptueid" *AFGHA*. Only an additional cause – something like resistance of air – would cause it to slow down along *AIKLMNOP* and come to a rest at *C*.

Newton's mistake and Hooke's correction are extremely interesting because they allow a glimpse into science at a moment of great change, when ideas, categories and assumptions were being shaped and reshaped, and when new questions and challenges were intermingling with old habits of thought. Newton's slip-up was not a result of misunderstanding or

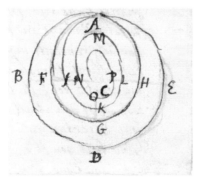

Figure 10.6 Hooke's diagram from his December 9, 1679 reply to Newton. The circle *ABDE* is the Earth, and the stone, like in Newton's thought experiment, is allowed to fall through it. But it will only describe the spiral *AJKLMNOP* and come to rest at *C* if some third force – say friction or resistance – will hinder the original tangential motion. Otherwise, if only that tangential motion and the fall are at work, the stone will orbit the center along *AFGHA* – just as a planet orbits the Sun by a combination of a continuous 'fall' and an independent tangential motion.

disagreement. Quite the opposite: the first part of his thought experiment – the fall of the stone to the east of the tower – demonstrates that he understood the new ideas better, perhaps, than Galileo and Descartes themselves. Certainly, neither of these masters of the previous generations offered such a virtuoso elaboration of what these concepts of motion implied. The closest had been Galileo's famous thought experiment in support of Copernicanism (in a letter from 1624), in which a stone is dropped from the top of the mast of a moving ship. But the case and his analysis were much simpler than Newton's: the ship moves horizontally, and the stone falls directly at "the foot of the mast precisely at that point which is perpendicularly below the place from which the rock was dropped" (Finnocchiaro, *The Galileo Affair* (see Chapter 7), p. 183). The stone falling 'forward' is a truly brilliant explication of the idea that motions 'compound,' which is the very reason why the stone *can't* stop at the center.

So how could Newton be so confused about an idea he had such a good understanding of? How could he forget what he had just preached? Clearly, the instinctive assumption that a falling object ends up at some bottom-most point simply took over once he was no longer thinking carefully about it. Hooke did not make this mistake because it was exactly the idea he was trying to get across: *that the fall of the stone and the motion of the planet are fundamentally the same mechanical process.* This notion follows directly from Galileo and Descartes, yet Hooke was the only one to concentrate on it: given the opportunity, the stone's fall would turn into an orbit. The beauty of this idea is the very reason why it was so hard to come to terms with: the planet's orbit is an outcome of its continuous fall towards the Sun.

To Hooke, his idea didn't seem much of an innovation at all. He'd already introduced it to the Royal Society thirteen years earlier, in 1666, in an almost casual way. "I have often wondered," he said then,

> why the planets should move about the sun according to Copernicus's supposition, being not included in any solid orbs ... nor tied to it, as their center, by any visible strings; and neither depart from it by such a degree, nor yet move in a straight line, as all bodies, that have but one single impulse, ought to do.
>
> Birch, *The History of the Royal Society*, Vol. II, p. 91: May 23, 1666

The planets, Hooke was assuming without much ado, are "regular solid bodies," and as such should follow the same laws that apply to all bodies. This is one of the most fundamental insights of the New Science: there is no cosmos divided into Heaven and Earth, only one world full of "bodies." As bodies, they should follow the laws of mechanics, which means, first and foremost, that they are supposed to "move in a straight line," unless there is "some other cause ... that must bend their motion into that curve." What can this cause be? How does it operate? How does it *create* the planetary orbits the way the astronomers describe them?

Setting the Question Right

This is where Hooke's originality lies. Not in providing a solution – he did not really have one, although he did have some important clues he provided Newton with – but in *setting the question*. The question did follow from the teachings of Galileo, Kepler and Descartes, and in hindsight, it may seem as if one of them should have already asked it. Hooke did not possess any new knowledge, but the fact is that he asked, and they did not. There is a big difference, so we learn, between having all the ingredients for a crucial new idea and actually formulating it. If 'Hooke's Programme' appears in hindsight as a straightforward, even trivial, development of those early ideas, it is because he succeeded (with a lot of effort and after much frustration) to convince his peers to adopt it. The only celestial mechanics *we* know are developed as an answer to his query, and we cannot but read Hooke's immediate predecessors through this lens.

It's therefore telling, historically and philosophically, to consider the three on whose work Hooke drew – Kepler, Galileo and Descartes – and see where he left that work behind.

Kepler had tried to apply physics to the heavens, but his physics had not been mechanics. He neither thought that straight-line motion should be the basis of the explanation of all motions, nor that the planets were "regular solid bodies." For him, it was still the case that "circular motions (and hence revolutions) belong to the everlasting bodies ... ; [and] rectilinear motions certainly [belong] to evanescent bodies" (Kepler, *De Cometis*, translated and cited by J. A. Ruffner, "The Curved and the Straight: Cometary Theory from Kepler to Hevelius" (1971) 2 *Journal for the History of Astronomy* 181). So Kepler didn't ask 'why do the planets move in a curve?' but 'why do the planets *move*?' To Kepler, matter was *inert* – it resisted all motion – so for him the business of celestial physics was to explain how the planets move in the first place. His tentative answer, you'll remember, was that the Sun is the cause of their motion. It rotates and drags the planets along by some "motive force," which may or may not be magnetism; perhaps it was light. So unlike Hooke and his 'compounding' hypothesis, for Kepler the rotation of the planets was explained by the rotation of the Sun; one closed-curve motion could only be created by another closed-curve motion.

Galileo had been closer to thinking about the heavens and the Earth as following the same physical laws. His arguments for Copernicanism had been based on analyzing motions on Earth – of ships and tides – and in the *Dialogues Concerning Two New Sciences* he gestured towards "the beautiful agreement between this thought [about mechanics] and the views of Plato concerning the origins of the various speeds with which the heavenly bodies revolve" (Galileo Galilei, *Dialogues Concerning Two New Sciences*, Henry Crew and Alfonso de Silvio (trans.) (New York: Dover Publications, 1954), p. 261). But this was as far as he went in the direction that Hooke would take. First, although his analysis of projectile motion presented one motion curving another motion, he did not assume that rectilinear motion had some unique metaphysical status. His 'neutral' continuous motion, you may recall from Chapter 9, was actually circular – along the surface of the Earth. Moreover, despite the alleged "beautiful agreement," he never attempted to actually apply his considerations of terrestrial mechanics to celestial motions.

Most surprising, and telling, is that Descartes never asked Hooke's question. After all, "every moving body tends to continue its motion in a straight line" was one of *his* "Laws of Nature." But, even for Descartes, the idea that the motion of the planets was not truly theirs was too far removed from common sense. Even he assumed that they revolve because it is their nature – because such motion, as Kepler said, "belongs to the everlasting bodies" – and didn't see that his own laws of nature implied otherwise. When he came to speculate about the motion of the heavenly bodies, Descartes seemingly forgot that bodies were supposed to move in straight lines, unless their motion is curved, and wrote instead that "the matter of heaven, in which the planets are situated, unceasingly *revolves*" (René Descartes, *Principles of Philosophy* (*Principia Philosophiæ*, 1664), Valentine R. Miller and Reese P. Miller (trans.) (Dordrecht: Reidel, 1983), Part III, Article 30, p. 96 – emphasis added). Look back at Figure 9.7: it shows a body in circular motion – a stone in a sling – created by another circular motion – the whirling hand. It claims that such a body has *a tendency* to continue in a straight line, which for Descartes meant that *if* released from the sling, the stone would continue along a straight line, tangential to its circular trajectory at the point of release. The diagram is crucially different from Hooke's idea: it doesn't treat the circular motion as *created* by this straight-line motion. Descartes never asked the question that Hooke would ask: how do two straight-line motions 'compound' into a curved, orbital trajectory?

We, in the post-Newtonian era, are taught as a matter of course that the planetary orbit is an outcome of inertial motion along the tangent and an attraction that constantly curves it towards the Sun. And in hindsight, it may seem that it was a trivial elaboration of the ideas that

Galileo, Kepler and Descartes had already formulated. But, so it turns out, no one before Hooke had drawn this picture in their mind: that the planets – and we on Earth with them – constantly tumble along, falling towards the Sun but never reaching it. Until it did occur to Hooke, and then it seemed obvious to him, and the obvious question followed: "I have often wondered why ..."

Let's return to the correspondence. We can now see why Newton had missed Hooke's point in the initial letter, and why Hooke's reply to his suggested experiment piqued Newton's attention. Not only did Newton positively dislike being caught in error, he now recognized the importance of Hooke's hypothesis and again replied within four days (December 13, 1679). Newton understood immediately where he got confused – and denied he ever did – and cunningly tried to work out how much of the solution to his own question Hooke already had. Why did Hooke think that he could get a closed-curve *orbit* (presumably an ellipse, but any neatly closed orbit) from two rectilinear motions? Why not something like the complex trajectory in Figure 10.7, where the planet keeps changing its path? Newton presented the diagram as if it was the outcome of a calculation, but there is no reason to take his word for it. What is crucial is that Newton, having now grasped the significance of Hooke's proposal, immediately put his finger on the extreme challenge it represented, which was both cosmological and

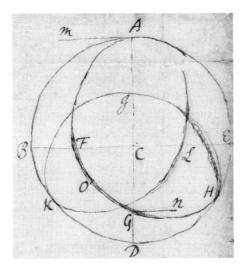

Figure 10.7 Newton's diagram from his letter to Hooke of December 13, 1679. Newton accepts Hooke's correction, but questions his conclusion: why should the body moving by a combination of free fall and tangential motion describe a neat orbit? Why not this complex trajectory? Put the other way: why should we assume that the neat orbits of the planets are produced by such a combination? Does Hooke have an independent argument to support his hypothesis?

mathematical-physical: how can we think of the orbit as an *effect*? The heavenly orbit had always been a symbol of perfect order and harmony. Its circular shape was a direct consequence of this perfection, and required no further explanation. This assumption was so deeply entrenched in cosmology and astronomy that most working astronomers still didn't understand Kepler's idea of physics of the heaven (an idea which by the time of the correspondence was seventy years old). For them, Kepler's ellipse was still a combination of two circular motions – a deferent and an epicycle – as in traditional astronomy from Ptolemy through Copernicus.

How, Newton was asking Hooke, could such orderliness be created by mere mechanical causes? How could the curve be *closed*, so it would repeat itself in an orderly way?

Had Hooke been able to answer the question on his own, he never would have requested Newton's assistance. What he could offer, three weeks later (January 6, 1680 – perhaps Christmas caused a delay), were three principles along which a solution to his answer may be constructed. He had already published these principles some ten years earlier as a short text called "System of the World," which provides another opportunity to peek at science in the making; in it we can recognize both the immediate origins of these principles and the marvelous originality of putting them together.

First, writes Hooke, one must assume that there is a relation between the planet's distance from the Sun and its velocity. Kepler captured this relation with his Area Law, which is clearly what Hooke has in mind, although he formulates it wrongly.

Secondly, the attraction from the Sun that curves the rectilinear motion along the tangent into an orbit must diminish with distance. Hooke again has Kepler in mind; this attraction should be something like light – the other long-distance force connecting the Sun to the planets. Light, Kepler had reasoned, expands like a sphere around the source of illumination. The surface of a sphere grows as the square of the distance from its center (we would say: $\{A = 4\pi r^2\}$; seventeenth-century mathematicians would drop the constants and say: $\{A \propto r^2\}$). Hence, the amount of light falling on each point of this surface *diminishes as the square of the distance from its source*: the famous inverse square law. When he'd written the *System of the World*, Hooke hadn't been sure what the law governing the decline of the attraction from the Sun was. But by the time of his letter to Newton, the similarity between the spherical expansion of light and the spherical expansion of this attraction convinced Hooke that they must follow the same law. This attraction – that both he and Newton would now call 'gravity' – diminishes by the square of distance, and he suggested that Newton should include this in his calculations.

Finally, Hooke suggests a law governing the relation between force – here the attraction from the Sun – and velocity: $\{f \propto V^2\}$. That is: force is proportional to the square of velocity: if a certain force is required to bring a given body from rest to a certain velocity, doubling this velocity requires four times the force. This presumably arises from his own work on springs (although it is difficult to reconstruct his reasons).

Newton must have decided that this was as much as he could learn from his collaborator/competitor Hooke, for he no longer bothered to respond. A couple of weeks later, on January 17, Hooke tried one last time to initiate an exchange, with a summary of what would now become *the* challenge of the new celestial mechanics:

> It now remains to know the proprietys of a curve Line (not circular nor concentricall) made by a centrall attractive power which makes the velocitys of Descent from the tangent Line or equall straight motion at all Distances in a Duplicate proportion to the Distances Reciprocally taken. I doubt not but that by your excellent method you will easily find out what that Curve must be, and its proprietys, and suggest physicall Reason of this proportion.
>
> Newton, *Correspondance*, Vol. II, p. 313

Newton never answered this letter either.

Conclusion: The New Celestial Mechanics

When thinking about the correspondence between Hooke and Newton as a crucial and representative moment in the emergence of celestial and terrestrial mechanics as we know it, two related issues must be noted. One is how complex mechanical thought and celestial physics have become over a couple of generations: forces governed by complex laws; complex ratios between forces and motions; complex relations between distances, velocities and the shape of trajectories. Galileo, Kepler, Descartes and their immediate correspondents and disciples may have hoped that the application of mathematics to natural philosophy would reveal a simple and certain, mathematical-like structure to nature; instead, it made it look more and more complex.

The other issue is Hooke's suggestion that the planetary orbit is not only an effect, but indeed an effect of a 'compound' of causes. This is very far from the harmonious, perfect motion which the heavenly bodies had always been assumed to have. The orbit, according to Hooke, is an outcome of a contingent balance between competing forces and motions. Had either the original velocity of the tangential motion, or the rate of decline of the attraction between the Sun and the Earth (or both), been different, the orbit would have had a different shape. A slight change to this velocity or this force law and there would have been no orbit at all: the Earth would fall on the Sun or shoot into space. The balance of forces and motions is complex and precarious, and Hooke's answer to Newton's question of 'how does one get a neat orbit from all of this' is, in a nutshell: 'one does not.' We should be content if we find that the orbit is more-or-less stable – that we are not going to fall onto the Sun soon.

Four and a half years after the correspondence, in the summer of 1684, Edmund Halley (1656–1742, he of the Halley's Comet fame) undertook the journey from London to Cambridge. Halley, one of the most active members of the Royal Society (and later the Astronomer Royal), had a story and a question to relate to Newton. Had Newton heard about the following speculation? If one assumes that the Sun attracts the planets by a force decreasing by the square of the distance, so it went, one can construct elliptical planetary orbits, as decreed by Kepler. This idea had become a favorite topic of discussion in the coffee houses where the Royal Society circles meet, Halley said. Recently, he told Newton, he had been dining with Hooke and Wren – Hooke's friend, collaborator and business associate – and when they came to this topic, Wren said that he had tried the calculation and failed. Hooke (whose tenure as secretary was short and disappointing and by now in the distant past) responded that he knew how to do it, but did not want to reveal the solution yet. He wanted others to try and fail first, so they would appreciate his achievement. Wren proceeded to offer a wager: a "40 Shillings book" to the first person who would perform the feat. Since Hooke did not claim the book, and knowing "the philosophically ambitious temper that he is of," Halley could not believe he had the solution. Would Newton care to join the discussion?

To Halley's surprise and delight, Newton replied that he had already produced the demonstration a few years earlier, referring presumably to work done immediately following, and instigated by, the correspondence with Hooke. Yet he could not retrieve the alleged paper, promising instead to recover the calculations and send them shortly. Halley returned to London, perhaps skeptical, but by November he could report to the Society that he had received the promised paper (Figure 10.8): *De Motu Corporum in Gyrum* (*On the Motion of Bodies in Orbit*).

De Motu, nine tightly written and drawn (and scratched and marked) pages, did not exactly answer Hooke's challenge or Halley's question, but it offered much in exchange. It comprised three definitions, four hypotheses, four theorems, seven problems and a number of corollaries and scholia. Into those, Newton packed the mechanical knowledge of the time, 'Hooke's

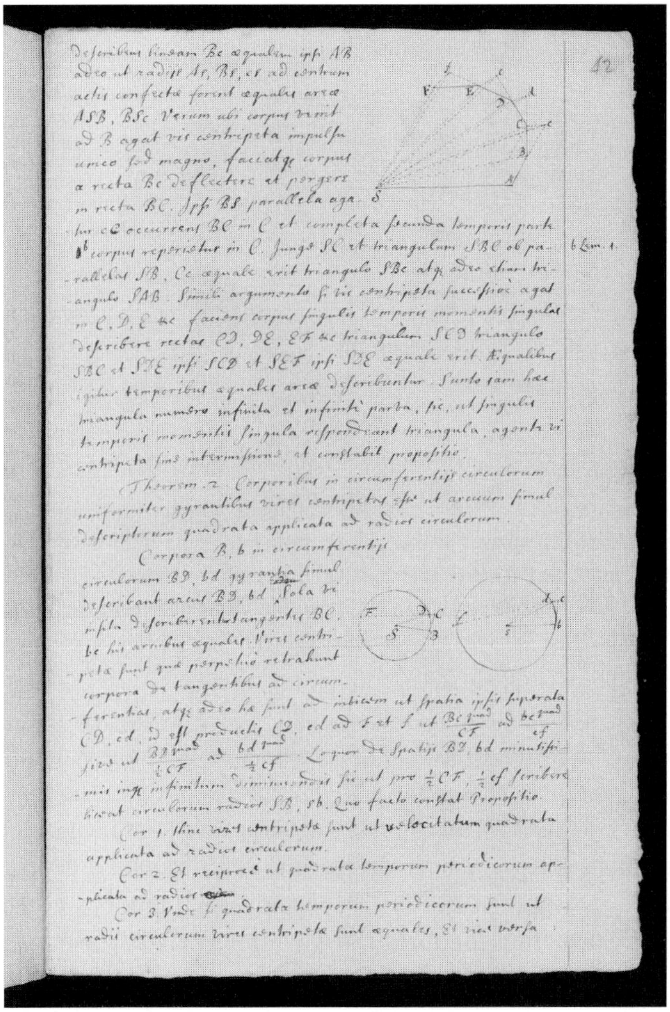

Figure 10.8 Newton's first manuscript version of *De Motu Corporum in Gyrum* (*On the Motion of Orbiting Bodies*) – the series of drafts of the *Principia*. The diagram on the upper right is explained in Figure 10.9. The top-right diagram is a proof that Kepler's Area Law holds for all orbiting bodies: the radius vector of a body moving around a center of force traverses areas proportional to time. The fairly simple proof is explained below. The bottom-right diagram is an advanced version of the construction of the law of force from Figure 10.2.

Programme' for physics of the planetary motions, and his own highly original, highly idiosyncratic mathematical wherewithal. The definitions cover centripetal force – his newly coined term, designating the type of attraction towards a central body that Hooke had in mind; inherent force; and resistance. The second hypothesis is Descartes' law: that bodies in rectilinear motion continue to move in uniform velocity unless forced otherwise. The third and fourth are taken from Galileo: that motions are combined in a

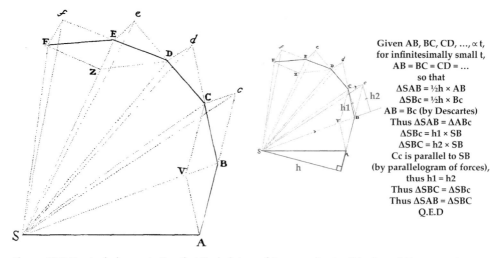

Given AB, BC, CD, ..., ∝ t,
for infinitesimally small t,
AB = BC = CD = ...
so that
ΔSAB = ½h × AB
ΔSBc = ½h × Bc
AB = Bc (by Descartes)
Thus ΔSAB = ΔABc
ΔSBc = h1 × SB
ΔSBC = h2 × SB
Cc is parallel to SB
(by parallelogram of forces),
thus h1 = h2
Thus ΔSBC = ΔSBc
Thus ΔSAB = ΔSBC
Q.E.D

Figure 10.9 Newton's demonstration that Kepler's Law of Areas applies to all bodies orbiting around a center of attraction. The diagram first appears in Newton's *De Motu* (see Figure 10.8) and this version is from the *Principia*. The diagram on the right has lines added to explain the demonstration. Note how Kepler's so-called '2nd law,' which for him was but an empirical approximation, receives in Newton's hands mathematical status. It allows Newton to use areas as representations of time in his analysis of the planetary motions.

moving body (according to the parallelogram rule) and that motion under force covers distances proportional to the squares of time. The first theorem proves Kepler's Area Law in the most general way (and using the simplest of mathematics) for every body forced into orbit around a center of force (the proof is in Figure 10.9). The second theorem develops a geometrical formula for calculating the force law governing any particular orbit from the proportion between periods and distances (a development of the one sketched in Figure 10.2). The fifth corollary to this theorem plugs in Kepler's third law – the proportion between the square of the times and the cube of distances – to deduce the inverse square law for our Sun and planets. Problem Three demonstrates that for a body moving according to Kepler's first law – on an elliptical orbit with its center of power in one of the foci – the 'centripetal force' declines with the square of distance. This is the reverse of Hooke's challenge, which was to construct an elliptical orbit by assuming an inverse square law of attraction, and as close as Newton ever managed to provide an answer to that original question. But this diminishes nothing from his achievement. Kepler's laws, enigmatic and empirical, turned out to follow directly from the mathematical-physical structure of nature. The inverse square law of force, offered by Hooke but without support and inexplicably dropped in idle London conversations, had been turned, in Newton's hands, into the key that tied them together.

Halley was so impressed with *De Motu* that he did not relent until he convinced the Cambridge recluse to turn the little tract into a book. It was to be published under the auspices of the Royal Society, of which Halley was soon to become the 'clerk,' so he could see it to print himself. Newton completed the manuscript within eighteen months, and it was published the following year (1687): *Principia Mathematica Philosophiae Naturalis* (*The Mathematical Principles of the Philosophy of Nature*). A second edition appeared in 1713 and the final, expanded edition in 1726.

Coda: The Principia

Experience tells that a curious twenty-first-century student introduced to the works of Galileo, Kepler, Descartes or Hooke also requires some historical tools to recognize that they are relevant to what she studies in her science classes. The first impression the same student has when faced with Newton's *Principia Mathematica* is usually that of familiarity. Indeed, she requires historical sensitivity to recognize that it is, in fact, quite different from what she is taught as Newtonian Physics.

This, more than historiographic tradition, is what makes the *Principia* a proper ending for this book. But a few words about this cathedral of science are in order, primarily because for Newton's contemporaries, the *Principia* was a cathedral in quite a literal sense, and Newton – its high priest: "Nature and Nature's laws lay hid in night, God said, 'Let Newton be!' and all was light" was the epitaph that Alexander Pope composed for Newton in 1730. Not long thereafter, Voltaire suspended his usual cynicism to echo: "the catechism reveals God to children, but Newton has revealed him to the sages." These words expressed more than just admiration – to Halley, directly involved in the laborious production of the *Principia*, it still seemed closer to revelation than a human creation, and he prefaced it with an "Ode to Newton":

> Lo, for your gaze, the pattern of the skies!
> What balance of the mass, what reckonings
> Divine! Here ponder too the Laws which God
> Framing the universe, set not aside
> But made the fixed foundations of His work
> ... Then ye who now on heavenly nectar fare,
> Come celebrate with me in song the name
> Of Newton, to the Muses dear;
> for he Unlocked the hidden treasuries of Truth:

So richly through his mind had Phoebus cast

The radiance of his own divinity.

Nearer the gods no mortal may approach.[1]

The eulogies to Newton may read as quite over the top, but the three books of the *Principia* are indeed composed as if deciphering the very "Laws which God, framing the universe ... made the fixed foundations of His work," and the approach sketched in *De Motu* is magisterially laid out. Newton begins with a set of "Definitions" and "Axioms, or Laws of Motions." Then comes the First Book – "The Motion of Bodies" – in which Newton lays out his new mathematical tools and demonstrates how to apply them to his definitions and axioms to solve and prove increasingly complex problems and theorems concerning the motions of bodies orbiting according to 'Hooke's Programme.' The Second Book deals with "The Motions of Bodies in Resisting Mediums." These are the motions we know around us – within the atmosphere and in water, so if the First Book prepares the way to deal with what in traditional cosmology used to be the heavenly realm, the Second Book deals with the terrestrial realm. But there are no longer distinct realms: the definitions and axioms laid out at the beginning are true of *all* bodies and *all* motions. This deep insight of the New Science is celebrated in the Third Book: "The System of the World," in which Newton shows how the "phenomena" – namely Kepler's laws, as expressed by the moons of Jupiter and Saturn – follow the "mathematical principles" developed in the First Book.

Calling his book *The Mathematical Principles of the Philosophy of Nature*, Newton points to the work he is attempting to emulate and surpass: Descartes' *Mathematical Principles*. And in one fundamental way, Newton follows Descartes' lead better than Descartes himself. Descartes demanded that only mathematizable elements be allowed into natural philosophy, but with him this demand remained mostly a philosophical aspiration. Newton put it into practice. His very elements are quantities. Instead of 'matter' and 'motion' he defines "Quantity of Matter" ("density and bulk conjoint") and "Quantity of Motion" ("velocity and quantity of matter conjoint"). But by the third Definition Newton veers sharply from Descartes, breaks the latter's decree that there is nothing in the world but matter and motion, and introduces the third element at work in Hooke's Programme and in his own *De Motu*: force. Newton had been studying the mathematics of force since his student years, and the correspondence with Hooke had apparently

---

[1] In this and the following quotes from Newton's *Principia*, I used the 1729 translation by Andrew Motte with slight changes.

convinced him that Descartes' philosophical requirements were too restrictive: a *physica coelestis*, as imagined by Kepler, can't be achieved without giving force an independent ontological status, as Hooke did in his
Programme. So Newton defines "innate force of matter," which is the force
by which bodies resist change of motion (or rest); "impressed force" – the
force which causes such change; and "centripetal force" – like the attraction he had discussed with Hooke, which during the correspondence they
began to simply call 'gravity.' He then returns to quantities, and defines "the
quantity of force" from three different perspectives: absolutely, measured
by acceleration, and by acceleration over time.

Before turning to apply these quantities to the analysis of nature, Newton
does feel the need to provide them with a metaphysical framework. In
Descartes' spartan matter-and-motion universe, all motion was strictly a
relative change of place: whether I walk through the door or the doorway
slides around me (or whether the doorway and I were moving simultaneously in opposite directions) is merely a matter of perspective. But adding
real forces to the fundamental elements of the world, Newton had to establish a real distinction: it is *I* who moved, not the doorway, because it is I
who starts and stops and turns – it is I who accelerates and decelerates. So
it is in me, not in the doorway, that the force resides.

But how can this distinction be made? What is the difference between
a body which accelerates because it experiences real force and the body
whose change of place is only apparent, an artifact of the motion of bodies
surrounding it? The difference between real and relative motion requires a
frame of reference independent of the moving things, a frame of reference
against which I was moving and the doorway – not. Newton provides that
frame by postulating an "Absolute Time" and "Absolute Space": a vessel
in which the objects of the world *are* and a duration in which the objects
move, time and space existing prior to and independently of the objects.
This is not an easy idea – Newton was later sternly criticized for it – and
he illustrates it with a little thought experiment. Imagine a bucket of water
hanging on a wound-up string. When the string is allowed to unwind, the
bucket spins around the water, which first remains still. When the water
gradually starts spinning with the bucket, it rises against the bucket's side.
But when the bucket spins and the water rests, the water's *relative* motion –
its motion relative to its immediate environment, the bucket – is high.
When the bucket and water spin together, the water's relative motion is low.
Why does the water change – why does its surface curve – when it spins
with the bucket? It has to be because it moves *absolutely*, argues Newton –
not relative to the bucket, but with respect to absolute space!

With the elements of his mathematical natural philosophy defined – quantity of matter and motion; force; absolute time and space – Newton continues the difficult but extremely fruitful synthesis of Descartes' and Hooke's insights with his own "Axioms, or Laws of Motion." The first law is a neat combination of Descartes' two Laws of Nature: "Every body continues in its state of rest or of uniform motion in a straight line, unless compelled to change that state by force impressed on it." The second law is a strictly mathematical version of the idea of force entailed in Hooke's Programme: "Change of motion is proportional to the motive force impressed and is made in the direction of the straight line in which this force is impressed." The third law adds symmetry to the developing structure: "to every action there is always opposed an equal reaction." A series of "Corollaries" then teaches the reader how to conceive these laws of motion and force. The first, simplest corollary is a good example: "a body, acted by two forces simultaneously, will describe a diagonal of a parallelogram in the same time as it would describe the sides of those forces separately." This is the 'parallelogram of forces' rule from *De Motu*, and the attentive reader might recognize a strategy in its *Principia* formulation. The parallelogram of forces entails Galileo's insight that different motions can be combined in a body without eliminating one another. It is the idea that played such a crucial role in his correspondence with Hooke, articulated as both a physical law of nature and a workable mathematical tool. It can then be used for constructing (and proving) much more complicated and less obvious claims about nature.

This is indeed a powerful strategy. Newton's use of it to generate increasingly surprising propositions and endow them with seemingly mathematical certainty may explain the enchantment expressed by Halley, Pope and Voltaire – an enchantment shared by many of their contemporaries. The treatment of Kepler's Area Law – taken quite directly from *De Motu* – provides another early and simple, yet powerful, example. Newton begins "Book One" with laying down his innovative "method of first and last ratios," which enables him to deal mathematically with singular points along a trajectory of motion. He then takes the Area Law, which for Kepler had been nothing more than an empirical generalization, and with a combination of simple Euclidean geometry, physical assumptions and mathematical approximations of questionable rigor, proves that this law applies to any body orbiting a center of force (Figure 10.9).

With the Area Law now a theorem, Newton can express time geometrically. A few propositions later he uses this capacity to construct a simple geometrical expression, deducing from the dimensions and period of an orbit the law by which the attraction between the central body and the one orbiting it diminishes with distance. A few propositions later, this geometrical formula for the calculation of force laws is used to reveal the law governing the orbit for which the exercise is carried out: an ellipse with the attracting body in one of its two foci; the orbit Kepler assigned to the planets around the Sun. The force that bends planetary motions into Keplerian orbits, it turns out, declines by the square of the distance between the two bodies. The attraction between the Sun and the planets obeys the inverse square law – the law that Kepler established for light; that Hooke applied to gravity on Earth and later applied to his Programme; and that Newton himself, earlier in life, had calculated for (the centrifugal force of) the planets. As we already saw in *De Motu*, with the inverse square law, Newton ties Kepler's laws into a coherent mathematical-physical whole, and in the Third Book it acquires another crucial significance to which we'll return below. But in the First Book Newton is careful to show that his commitment to the inverse square law is strictly limited by empirical considerations. He teaches his readers to calculate force laws for orbits of a variety of curves and also adds a series of corollaries that show how different force laws would have followed had Kepler's 'Harmonic Law' (the proportion between the square of a planet's period and the cube of its distance) been different. Deep into the book, in Propositions 53–55 and 66–68, he teaches the reader how to handle complex orbits, which are interrupted by attraction from a third body, as with the Earth orbiting the Sun together with the Moon: by treating the orbit itself as if it's rotating.

Originally, Newton had planned to add to the highly technical Book One a second book, explaining this new way of conceiving Heaven and Earth "in a popular method, that it might be read by many" (as he writes in his introduction to the final Book Three). This early "Treatise of the System of the World" was eventually published, but only posthumously (Figure 10.10; the book is still readily available: https://books.google.com.au/books?id=DX-E9AAAAcAAJ). In a decision that clearly marked the exclusive, elitist route that modern science would take, Newton replaced the proposed "popular method" with the deliberately technical and defiantly difficult "Motion of Bodies in Resisting Mediums." He had realized, he wrote to Halley, that he didn't want to be drawn into debates by people whose philosophical agendas outweighed their mathematical skills. Science was now for the professionals.

Book Two, as it came to print, was thus a late addition. Historians often read it as a long argument against Descartes' vortices theory: motion in resisting mediums is too complex to allow explaining planetary orbit. Its Section III, for example, deals with "the motion of bodies that are resisted

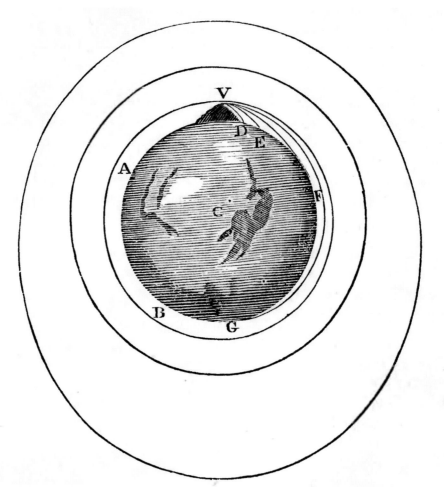

Figure 10.10 Newton's *System of the World*. Note how the diagram turns Hooke's abstract diagram from the correspondence with Newton (Figure 10.6) into a didactic tool. A cannonball shot from the top of a very high mountain V will fall to Earth at D, E, F or G according to the force with which it was shot. If this force is strong enough, the cannonball will never complete its fall, and will continue to orbit around the Earth along BAVAB: a cannonball on Earth and a planet 'falling' on the Sun follow fundamentally the same process.

partly in the ration of their velocity and partly as the square of this ratio," and its concluding scholium summarizes the resistance of spherical bodies in fluids arising "partly from the tenacity, partly from the attrition, and partly from the density of the medium." Section IV analyzes mathematically the point Hooke had made to Newton in his December 9, 1679 letter (Figure 10.6): resistance would turn an orbit into a spiral towards the center of attraction.

Replacing the original "System of the World" with the final, more technical version of Book Three did more than limit it to "readers with good mathematical skills." It turned the book from an explication of the principles of the new celestial mechanics into an argument for their power. Newton presents careful tables of the periods of the planets as well as those of the moons of Jupiter and Mercury; the best measures of planetary distances; outcomes of pendulum comparisons at different heights and in different places around globe; the results of his own experiments on falling bodies (many of those available only or corrected for the second edition); and more. All, Newton shows, fit well within the structure erected with the mathematical-physical theorems of Book One:

> Proposition I: That the forces by which the circumjovial planets (Jupiter's moons) are drawn off from rectilinear motion and retained in their proper orbits tend to Jupiter's center and are inversely as the squares of the distances of the places of the planets as the squares of the distances of the places of those planets from that center.
>
> Proposition III: That the force by which the Moon is retained in its orbit tends to the Earth and is inversely as the square of the distance of its place from the Earth's center.
>
> Proposition XX: To find and compare the weights of bodies in different regions of our Earth.
>
> Proposition XXIV: That the flux and reflux of the sea arise from the action of the Sun and the Moon.
>
> Proposition XL: That the comets move in some conic sections, having the foci in the center of the Sun, and by radii drawn to the Sun describe areas proportional to the times.

This overwhelming array of phenomena establishes more than the empirical efficacy of the mathematical-physical tools developed in Book One: it demonstrates their universality. Newton's "System of the World" succeeds in realizing the plan Hooke had sketched in his own text of the same name: to show that the attraction that curves the motion of the planets into an elliptical orbit is not just a force unique to the Sun (as Kepler had thought). Rather, it is the same force that ties all moons to the planets they orbit and with whom they orbit the Sun; the same force that makes stones fall towards Earth; and the same force that ties every particle of matter to every other particle. Perhaps the most spectacular argument for this universality is the famous "Moon Test": from the astronomical measures of the Moon's orbit, Newton calculates that the Moon's acceleration towards the Earth is $1/60^2$ of the rate of acceleration on Earth. Since the common estimation of the Moon's distance from Earth is 60 Earth radii, it follows that the force which holds the Moon to the Earth also obeys the inverse square law – just like the force between the Sun and the planets. Obeying the same law implies that this is the same force. Emanating from Earth suggests that this force is simply gravity, even if it reaches all the way to the Moon. The conclusion is the core of the *Principia*'s achievement: all celestial and terrestrial attractions are one and the same – universal gravitation, following the inverse square law.

This "catechism [which] revealed him to the sages," however, came at a steep price. Book Three is riddled with approximations, compromises and bold generalizations. In the case of the Moon Test itself, Newton admitted to Roger Cotes, the editor of the second edition, that he had to 'help' the numbers in order to produce the neat calculation. The choice of Jupiter's moons as a prime example was also far from innocent: they provide a much more orderly example than any of the motion with the solar system. Most importantly, Kepler's Laws were an idealization: a planet could be expected to orbit the Sun in an orderly ellipse, covering areas proportional to time, only if the Sun and the planet were the only two bodies attracting each other. Newton was acutely aware that the very idea of universal gravitation prescribed a fundamentally different scenario than that prescribed by Kepler's fascination with divine orderliness and by the Ancient Greek ideals of the perfection of the Heavens. In the final version of *De Motu*, he admitted in a pensive scholium:

> If the common center of gravity is calculated for any position of the planets it either falls in the body of the Sun or will always be very close to it. By reason of this deviation of the Sun from the center of gravity the centripetal force does not always tend to that immobile center, and hence the planets neither move exactly in ellipse nor revolve twice in the same orbit. So that there are as many orbits to a planet as it has revolutions ... and the orbit of any one planet depends on the combined motion of all the planets, not to mention the action of all these on each other. But to consider simultaneously all these causes of motion and to define these motions by exact laws allowing of convenient calculation exceeds ... the force of the entire human intellect. Ignoring those minutiae, the simple orbit and the mean among all errors will be the ellipse.
>
> Newton, *De Motu* III, in Herivel, *The Background to Newton's Principia*, p. 297

Nature is irreducibly complex and human knowledge is limited. The planets don't really revolve in neat orbits, and no science can offer "exact law allowing of convenient calculation."

This scholium, however, is nowhere to be found in the *Principia*. In this *opus magnum*, Newton had no place for a modest admission of the limits of science nor for the idea that "the ellipse" is nothing more than "the mean among all errors." In their stead, the General Scholium concluding his great book offers a brash declaration of unfailing knowledge of orderly nature:

... the whole space of the planetary heavens either rests (as is commonly believed), or moves uniformly in a straight line, and hence the communal centre of gravity of the planets ... either rests or moves along with it. In both cases ... the relative motions of the planets are the same, and their common centre of gravity rests in relation to the whole of space, and so can certainly be taken for the still centre of the whole planetary system. Hence truly the Copernican system is proved *a priori*. For if the common centre of gravity is calculated for any position of the planets it either falls in the body of the Sun or will always be very close to it.

Indeed, with Newton's *Principia*, modern science had its cathedral: incomplete, asymmetric, its divine aspirations never fully hiding the traces of its human makers – an object to marvel at, emulate, enhance and reshape.

Discussion Questions

1. Is there something interesting to be learnt from the similarities and differences between Hooke's and Newton's biographies on the one hand, and the stark difference between their posthumous fame on the other?
2. Can you suggest a way to think about the correspondence between Hooke and Newton and the way knowledge seems to be created in the exchange? Does it make you say something about knowledge as social?
3. Look again at Figure 10.5. How could someone as skilled as Newton make such a basic mistake in material he understood so well? What does it say about the ideas discussed?
4. How can we think of Hooke's innovation? Does it make sense to say that he 'invented a question'? Why didn't this question come to the mind of Kepler, Galileo or Descartes? Is this a reasonable query?
5. Does the idea that the Earth, and the other planets, constantly 'falls' towards the Sun but never makes it still seem strange? If it is, what does it say about science that it assigns such an important role to such a strange idea? If not, what does it say about us as 'consumers' of science that we have gotten used to an idea that was so strange?

Suggested Readings

Primary Texts

Hooke, Robert, "Address to the Royal Society" in Thomas Birch (ed.), *The History of the Royal Society of London* (London, 1756–7). Facsimile reprint in *The Sources of Science*, Vol. 44 (New York: Johnson Reprint Corporation, 1968), Vol. II, pp. 91–92.

Hooke, Robert and Isaac Newton, "Correspondence" in Isaac Newton, *The Correspondence of Isaac Newton*, H. W. Turnbull (ed.) (Cambridge University Press, 1960), Vol. II, pp. 297–313 (esp. pp. 300, 301–302, 305, 307–308, 312–313).

Newton, Isaac, "De Motu III-B" in J. W. Herivel (ed. and trans.), *The Background to Newton's Principia* (Oxford: Clarendon Press, 1965), p. 301.

Newton, Isaac, "Definitions, Scholium, Axioms" in Isaac Newton, *The Principia: Mathematical Principle of Natural Philosophy*, I. B. Cohen and A. Whitman (trans. and ann.) (Berkeley, CA: University of California Press, 1999), pp. 403–418.

Secondary Texts

Brackenridge, J. Bruce, *The Key to Newton's Dynamics: The Kepler Problem and the Principia* (Berkeley, CA: University of California Press, 1996), https://publishing.cdlib.org/ucpressebooks/view?docId=ft4489n8zn;brand=ucpress.

De Gandt, François, *Force and Geometry in Newton's Principia* (Princeton University Press, 1995).

Densmore, Dana, *Newton's Principia: The Central Argument*, William Donahue (trans.), 3rd edn. (Santa Fe, NM: Green Lion Press, 2003).

Dobbs, Betty Jo Teeter, *The Janus Faces of Genius: The Role of Alchemy in Newton's Thought* (Cambridge University Press, 1991).

Gal, Ofer, *Meanest Foundations and Nobler Superstructure: Hooke, Newton and the Compounding of the Celestiall motions of the Planetts* (Dordrecht: Kluwer, 2002).

McGuire, James E. and Piyo M. Rattansi, "Newton and the 'Pipes of Pan'" (1966) 21 *Notes and Records of the Royal Society of London* 108–143.

Schaffer, Simon, "Glass Works: Newton's Prisms and the Uses of Experiment" in David Gooding, Trevor Pinch and Simon Schaffer (eds.), *The Uses of Experiment* (Cambridge University Press, 1989), pp. 67–104.

Westfall, Richard S., *Force in Newton's Physics* (London: Macdonald and Co., 1971).

Index